The Design of
High-Efficiency Turbomachinery
and Gas Turbines

The Design of High-Efficiency Turbomachinery and Gas Turbines

David Gordon Wilson

The MIT Press
Cambridge, Massachusetts
London, England

This book was set in Times New Roman by Asco Trade Typesetting Ltd.,
Hong Kong and printed and bound by Halliday Lithograph
in the United States of America.

Library of Congress Cataloging in Publication Data

Wilson, David Gordon, 1928–
  The design of high-efficiency turbomachinery and gas turbines.

  Bibliography: p.
  Includes index.
  1. Turbomachines—Design and construction.
  2. Gas-turbines—Design and construction.   I. Title.
  TJ267.W47   1984   621.406   84-18741
  ISBN 0-262-23114-X

This book is dedicated to Edward Story Taylor, director of the Gas-Turbine Laboratory at the Massachusetts Institute of Technology from its creation in 1946 to his retirement in 1968, who has had an extraordinary influence on the design, and the designers, of turbomachinery and gas turbines.

Contents

Contents

13
Mechanical-design considerations

Appendixes

Preface

Most texts are created from course notes, and this is no exception. The courses were originally taught by Edward S. Taylor at the MIT Gas-Turbine Laboratory; I took one when I first came to the United States on a postdoctoral fellowship in 1955–56. Philip G. Hill and Carl R. Peterson took over at least one of the courses when E. S. Taylor retired, and they produced their own excellent text, still in active use in many schools. (I have included it, along with other currently available texts on turbomachinery, in the reference list at the end of chapter 1.) When Hill and Peterson moved on to other callings, the courses lapsed, there being little interest at MIT at that time either in energy or its conversion. I had recently joined the faculty after having spent most of the rest of my professional life in industrial gas-turbine and turbomachinery design, and I took over the courses to save them from extinction. The emphasis was changed from analysis to design, and the notes grew and changed in the universal way of all course notes. They eventually become unique, as did these. While there are texts that cover parts of the subject material, there is none as yet, despite some formidable and excellent competition, that treats the subject of turbomachinery design and gas-turbine engine design, from what are, I believe, the vital fundamentals. I persuaded myself that the world needs this book. Whether I have been deluding myself and others will be a matter for you (many of you, I hope) to judge.

The book is intended to serve two sets of readers. One set includes those college instructors and their senior undergraduate and graduate students in courses on turbomachinery design. Most current courses are, I believe, taught from the viewpoint of analysis, particularly the analysis of the fluid-mechanical complexities of turbomachinery, and this book will not suit them because I have tried to avoid these areas insofar as possible without being condemned for undue escapism.

My emphasis has been on thermodynamics, and design approaches, and on the use of very simple rules for choosing cycles, vector diagrams, blade forms, heat-exchanger configurations, and so forth. There are sufficient data on, for instance, diffuser performance and gas-turbine-cycle characteristics to go somewhat beyond the needs of most students and to serve the second set of readers: the practising engineers. This book is intended to give them all the fundamentals to carry out a thorough preliminary design, which in many cases can be refined by the same methods into a final design of the flow path. (The book treats the mechanical design in a very cursory manner.) For critical machine designs, such as aircraft engines and other high-cost or extreme-duty applications, or those units to be produced in quantity, it would usually be appropriate to take the blade forms, for instance, produced by the simple methods described here as a starting point for refinement by one of the computer methods now available.

Such computational refinements have been known to increase the blade-row efficiency of successful, established designs by half a percentage point, an amount that can have a major effect on the energy use of a utility or an airline. I have incorporated the words "high-efficiency" into the title of the book even though these complex computations are not explicitly treated. My reason is that I have encouraged throughout a fresh approach to turbomachinery and gas-turbine design which in many cases can lead to more substantial improvements in efficiency through better basic design choices. For instance, a precise definition of compressor and turbine efficiency leads to an economic optimum choice of multistage components in many cases. I have therewith campaigned against the "macho" influence in design, which seeks to prove, for instance, that a high pressure ratio can be attained in a single stage, while ignoring, or being unaware of, the usually concomitant lower design-point efficiency and much poorer off-design performance of single-stage designs in most cases. In gas-turbine cycles it is demonstrated that low-pressure-ratio cycles incorporating heat exchangers of high thermal effectiveness can give design-point efficiencies higher than those of any other heat engine, together with good off-design efficiencies.

Two necessary conditions to high efficiencies are the reduction of kinetic-energy losses, which can be attained through suitable choices of stage vector diagrams and of the number of stages, and the achievement of efficient diffusion. More attention and data are devoted to diffusion and diffusers here than in any preceding texts on turbomachinery.

Although the overall treatment is applicable to any class of turbomachinery, including steam turbines and centrifugal- and axial-flow pumps, the emphasis is on shaft-power gas-turbine engines and components.

My aim has been to produce a general turbomachinery text that is useful, relevant, accurate, thermodynamically honest, and yet is easily used. It has not been a simple task. To anyone thinking of embarking on a similar venture, I would recommend long, hard thought before making a firm commitment. I enjoy writing, and I have always been able to dash off a speech, article, or even a paper without great pain and with reasonably happy results. But writing a technical book incorporates a degree of difficulty proportional to at least the third power of the number of words and to a higher power of the number of symbols and equations. One therefore not only imposes a very heavy load on oneself, but one spills over the toil, and its effects, on to many others. I would like to try to acknowledge some of these.

Anne Sears Wilson, John Michael Boulton Wilson, and Erica Sears Wilson have long put up with lost family weekends, missed summer opportunities, and my late nights and 4:30 A.M. arisings, and I hope that they will welcome me back to combined operations.

Anna M. Piccolo, who looks after the secretarial needs of several of us at MIT, has typed and retyped the notes and tables and manuscripts in between her many other duties over several years. She has also made many useful suggestions that have improved the presentation.

Noralie Barnett has produced almost all the line drawings not taken directly from other publications, and has done so with a superb quality that will undoubtedly greatly improve the readability and usefulness of the charts. Several of the excellent line drawings of turbomachinery types were made by Matt Russell.

An early version of the text was read in part by several friends, among them Donald M. Dix, now on the scientific staff of the Department of Defense; Willem Jansen, technical director of Northern Research & Engineering Corporation; James P. Johnston, and A. L. London, professors of mechanical engineering at Stanford University; and Edward S. Taylor, professor emeritus of aeronautical and mechanical engineering at MIT. All are busy people who spent more time reading the manuscript than I had the right to ask, and none should be assigned any of the blame for deficiencies in the final book.

Since 1969, graduate students in two courses ("turbomachinery design" and "gas-turbine design") have puzzled over successive versions of the course notes and have made many suggestions, some of them polite, on improvements. Several have made the improvements themselves by supplying me with computer-generated data, charts, and curves. Some who deserve especial mention are Duncan Apthorpe, Paul Ballentine, Andre By, Robert Bjorge, Jeff Gertz, Bert Vermeulen, and Thomas Wolf. In some cases I have been able to acknowledge their work on the data or charts themselves, as I have in many other cases where I am indebted to the work of others.

A solutions manual is available to instructors: a generous grant to help in the production of this manual has been made from the Bernard M. Gordon (1948) Engineering Development Fund, MIT.

To all who have given me such support and help in this undertaking, I wish to record my most sincere appreciation. I hope that you, readers of this text, will judge that their effort and sacrifice have been worthwhile. I hope also that you will write to me with comments, suggestions for improvement, and with notes of those errors that get through our combined screen.

The Design of
High-Efficiency Turbomachinery
and Gas Turbines

A brief history

Somewhat mischievously, I like to ask this question of people who proudly describe how they have arrived at successful designs by going through successive iterations of refinement: "Why didn't you design your final machine right away, saving all the waste motion of the preliminary phases?"

One can similarly ask: "Why didn't the Romans make water turbines?" and the answers can be equally instructive. In general, there are two types of responses to these questions. In one, the answer may be that the best course was revealed by trial and error, or by structured experiments, or by analysis. These steps may take hours or centuries, but when the knowledge is acquired, it can be applied to the original design problem without other essential influences. Thus the Romans had the technology to cast and machine and mount a high-efficiency water turbine. They also had the need for water power, and they made water mills. But they lacked an understanding of hydrodynamics and mechanics, an understanding mankind acquired over many centuries.

The second type of response is more complex and would answer what seems a sillier question: "Why didn't the Romans make gas-turbine engines?" The first part of the response is obvious: they had no need for them. But if we repeat the question and apply it to the engineers and entrepreneurs of the nineteenth century, when there was a need for engines of increasing power and efficiency, when there were repeated attempts to design and make gas-turbine engines, and when appropriate gas and liquid fuels were available, we find that a sufficient knowledge of fluid mechanics was still lacking, particularly of how to accomplish deceleration efficiently. But also needed was a large set of developments in other fields, such as metallurgy, high-speed gearing and bearings, and combustion. In part developments in related fields may allow developments to take place in turbomachinery and gas-turbine engines, and in part these external developments can be, and have been, stimulated by the perceived need for better high-temperature turbine materials, for instance. Technological history is therefore not only a study of development but of the background and incentives for development.

It is not true, of course, that anyone who is ignorant of history is condemned to repeat it. A reading of this summary is not an essential part of a program of instruction. But we should learn from mistakes. For instance, there is much current (early 1980s) enthusiasm for radial-inflow gas-turbine expanders. None of the several design teams involved in this movement has been aware, when questioned by this author, of Parsons's multistage radial-inflow steam turbine which lasted less than an hour, or of the many similar cases of erosion in radial-inflow gas turbines since then. (High-speed radial-inflow turbines will not pass high-density solid or liquid particles, which become trapped and erode the nozzles and rotor blades.) The designers' skills may enable them to avoid these problems without a knowledge of history. But design skills are sharpened by experience, and vicarious experience—history—is still valid experience.

The material for this section has been culled from several sources (listed at the end of the section), including historical review articles in the *Chartered Mechanical Engineer* and *Power* and from one on radial-inflow turbines by John D. Stanitz; but the major source has been an excellent book: *The Origins of the Turbojet Revolution* by Edward W. Constant II (Johns Hopkins University Press, 1980). Obviously, I have had to omit many interesting details to compress a long period of varied developments into a single short piece.

Early turbomachinery

The Romans introduced paddle-type water wheels—pure "impulse" wheels—in around 70 B.C. for grinding grain. It seems likely that the Romans were the true initiators, because Chinese writings set the first use of water wheels there at several decades later:

The Greek geometrician and inventor Hero (or Heron) of Alexandria devised the first steam-powered engine, the aeolipile, a pure-reaction machine, in A.D. 62. A simple closed spherical vessel mounted on bearings carrying steam from a cauldron or boiler with one or more pipes discharging tangentially at the vessel's periphery is driven around by the reaction of the steam jets.

Through succeeding centuries, water wheels of the impulse or gravity-fed types can be seen in texts about grinding mills, water supply, or mining. For instance, Georgius Agricola has many illustrations of water wheels in *De Re Metallica* (1556) (tr. by Hoover, Dover, 1950) often driving various positive-displacement pumps for mine drainage. Taqi-al-Din, writing in 1551 about Islamic engineering, described an impulse steam turbine that

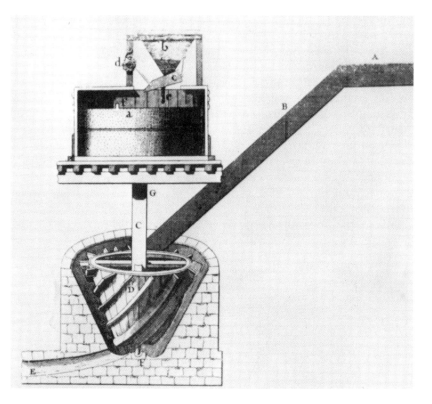

Garonne or tub wheel, 1730. From Stanitz 1966.

drove a spit (Ludlow and Bahrani 1978). He may have inspired Giovanni Branca who proposed similar steam turbines in 1629. Branca was formerly considered to be the originator of the impulse steam turbine.

The movement toward modern turbomachinery really started in the eighteenth century. In 1705 Denis Papin published full descriptions of the centrifugal blowers and pumps he had developed. Four volumes on "Architecture hydraulique" were published in France in the period from 1737 to 1753 by Bernard Forest de Belidor, describing waterwheels with curved blades. These were also called "tub wheels" and were precursors of what are now called radial-inflow or Francis turbines.

The great Swiss mathematician Leonhard Euler (1707–1783), then at the Berlin Academy of Sciences, analyzed Hero's turbine and carried out experiments with his son Albert in the period around 1750. He published his application of Newton's law to turbomachinery, now universally known as "Euler's equation," in 1754 and thereby immediately permitted a more scientific approach to design than the previous cut-and-try methods.

Waterwheels

The study of turbomachinery by models was introduced by the British experimenter John Smeaton (1724–1792) in 1752 and subsequently. He showed that the efficiency of "overshot" waterwheels could exceed 60 percent, versus the usual maximum of 30 percent for the more-usual "undershot" type. He also defined power as equivalent to the rate of lifting of a weight, a concept that is still fundamental in thermodynamics.

Stream-tube analysis and the study of ideal waterwheels were introduced in 1767 by Jean Charles Borda.

A pure-reaction waterwheel, similar in principle to Hero's turbine and to some modern-day lawn sprinklers, was produced in 1807 by the Marquis d'Ectot (1777–1822). However, up to this point there were no turbines—the word was coined in 1822 by another Frenchman, Claude

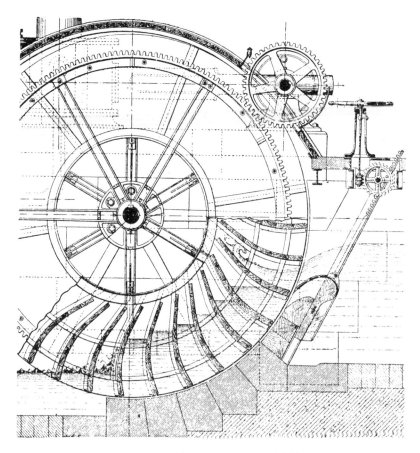

Poncelet type of water wheel, 1824. From Buchetti 1892.

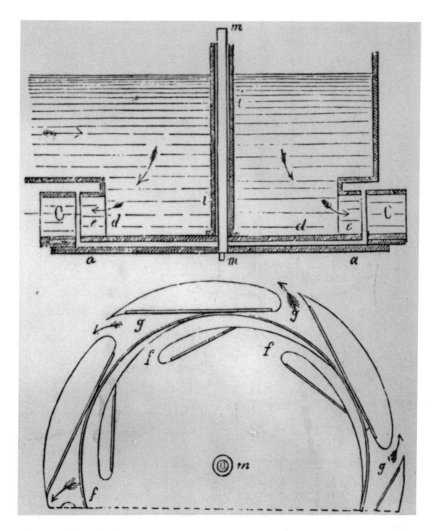

Burdin radial-outflow hydraulic turbine, 1824. From Stanitz 1966.

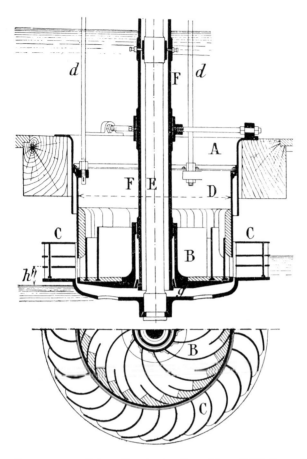

Fourneyron radial-outflow hydraulic turbine, 1827. From Buchetti 1892.

Burdin, from the Latin for "that which spins, as a spinning top—*turbo, turbinis.*"

It is always interesting to look for incentives behind any series of developments. Smeaton's excellent work virtually destroyed incentives for further improvements in waterwheels in Britain, where overshot waterwheels gave way directly to steam power. France had more rivers and less coal, and at that time there was an artificial but powerful incentive—the offer of a prize of 6,000 francs by the French Academy of Sciences for improvements in waterwheels. General Lazare Carnot had already identified two vital principles for efficient turbomachines: the fluid should enter the wheel without "shock," and it should leave with as little energy as possible. Jean-Victor Poncelot (1788–1867), a French mathematician and engineer, followed these principles in 1824–27 in designing an undershot waterwheel with blades curved so that the incoming water flowed smooth-

A Brief History

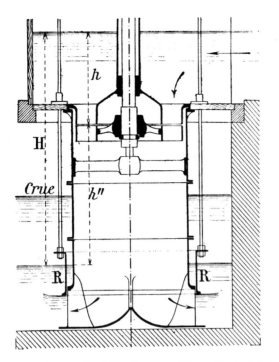

Jonval type of axial hydraulic turbine with draft tube. From Buchetti 1892.

ly upward, came relatively to rest, and dropped gently into the tail race, producing a continuous torque on the blades. He doubled the efficiency from 30 to 60 percent, in this case by dynamic action rather than Smeaton's use of gravity in a relatively static manner.

Burdin tried to win the prize with a radial-outflow turbine with poor blade angles, but it was his student Benoit Fourneyron (1802–1867) who constructed the first true high-efficiency water turbine in 1824–27. This was also of the radial-outflow type, but it used efficient blade angles and ran "full," rather than having "partial admission" in the form of a single incoming jet. In the development of his turbine Fourneyron used the brake produced in 1822 by the Baron Riche de Prony. The Prony brake permitted much greater accuracy in measuring efficiencies, and with the greater candor in reporting which appeared to take over in that period, the efficiencies of engines and motors such as turbines began to increase rapidly. Water turbines in general had efficiencies in the range of 20 to 30 percent before 1800, but Fourneyron's turbine eventually reached a maximum efficiency of 85 percent. The output per unit had been increased to 50 hp by 1832, and in total he made about 100 machines, all of the radial-outflow type.

A significant but apparently isolated development of an axial-flow

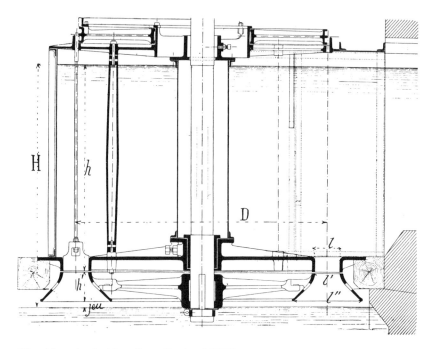

Girard impulse turbine. From Buchetti 1892.

reaction turbine with a draft-tube diffuser, which simultaneously increased the machine efficiency and enabled it to be placed above the tail-race level, was introduced by Jonval of Mühlhausen in 1841. Another foretaste of things to come was a partial-admission impulse turbine giving improved part-load efficiency made by Dominique Girard in 1850, but again this was not followed up. Perhaps Fourneyron's success inhibited other improvements, as had Smeaton's earlier work in Britain.

In any event the water-turbine action moved to the United States, where Elwood Morris had produced a turbine of the Fourneyron type in 1843, followed by Uriah A. Boyden's similar turbines in 1844 and 1846. Boyden added a vaneless radial diffuser, the first known use of this device. This added three points to the efficiency, bringing it up to 88 percent, and his turbines were produced in sizes giving more than 190 hp.

In 1851 James B. Francis began a series of turbine tests with Boyden and A. H. Swaim in the Lowell, Massachusetts, canals, first published in 1855 as the *Lowell Hydraulic Experiments*. The high-efficiency radial-inflow turbines which resulted by 1875 are now almost universally known as Francis turbines.

In Britain James Thomson (1822–1892) started work on an efficient radial-inflow turbine with a spiral inlet casing and adjustable inlet guide vanes in 1846–47, and patented it in 1850 as the Vortex turbine. This

A Brief History

Boyden radial-outflow turbine with diffuser, 1846. From Stanitz 1966.

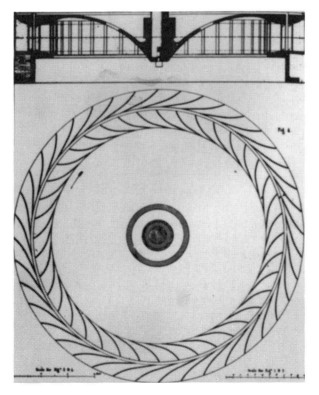

Francis center-vent turbine. From Stanitz 1966.

James Thomson's "Vortex" turbine, 1847–50. From Stanitz 1966.

continued in production for eighty years. In 1883 Osborne Reynolds established the foundations of flow similarity for laminar-turbulent transition in channels, and in 1892 Rayleigh introduced dimensional analysis.

In the California mines prior to 1872, so-called "hurdy-gurdy" wooden waterwheels with teeth like those of a circular saw working in a high-velocity jet were made and installed by local carpenter-smiths. In 1872 S. N. Knight introduced curved, cast-iron buckets, which increased the efficiency and the life of these wheels. A year later Nicholas J. Coleman patented the idea of curved split buckets on an impulse turbine wheel, whereby the incoming water jet was divided evenly to either side of the wheel, virtually eliminating axial thrust loading. An improved version of the split bucket was developed by Lester G. Pelton in 1878 and patented in 1880. His improvements plus his business acumen resulted in a rapid

A Brief History

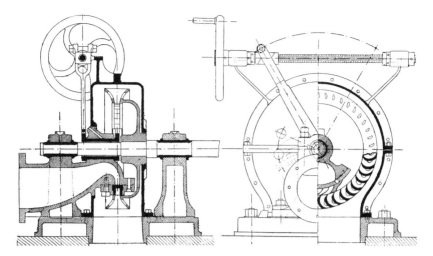

French pre-Pelton impulse hydraulic turbine. From Buchetti 1892.

adoption of the partial-admission impulse water turbine, now generally called the Pelton wheel, for high heads.

In water turbines the axial type, despite Jonval's pioneering work, had not been developed to high efficiency. Forrest Nagler designed an axial turbine in 1910, which was installed in 1916, but it was Victor Kaplan's proposals for a propeller turbine with adjustable-pitch blades and with adjustable "wicket" (nozzle) blades (1910, Czechoslovakia) that resulted in the general use of axial turbines of high efficiency at both full and partial loads, suitable for large flows and low heads. This type now bears his name.

Pumps

After Denis Papin's descriptions of his centrifugal blowers and pumps in 1705, there is little information on further developments until the nineteenth century. Crude centrifugal pumps were used in the United States and probably elsewhere by 1818, and improved versions were introduced in the New York dockyards in 1830. Eli Blake used a single-sided impeller (rather than paddle-wheel-type blading) in 1830, and in 1839 W. D. Andrews added a volute.

The first mention of a vaned diffuser was in a patent by Osborne Reynolds in 1875 for a so-called "turbine pump," which was the combination of a centrifugal impeller with a vaned diffuser. In that year Reynolds (the son of a Belfast minister, made head of department at Owens College, later Manchester University, in 1868 at the age of twenty-six) also made a multistage axial-flow steam turbine which ran at 12,000 rpm. In 1885

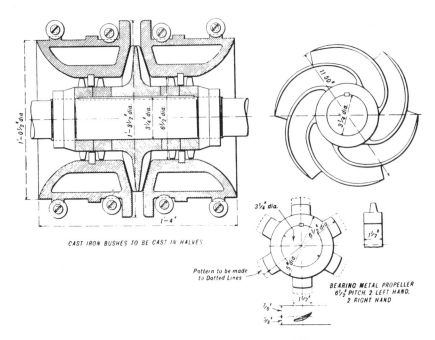

Parsons axial-radial double-flow pump. From Parsons 1936.

he described something of great relevance to future steam turbines, the convergent-divergent nozzle. From 1887 his centrifugal pumps were made by Mather & Platt, a still-flourishing British manufacturer. In the next decade turbine pumps began being produced by Sulzer, Rateau, and Byron Jackson, and in the 1900s by Parsons, de Laval, Allis Chalmers, and Worthington.

Compressors

The major fluid-mechanic problem in turbomachinery is the design of compressors. A turbine, with its flow usually going from a high to a low pressure, will always work, and with reasonable care it will work at high efficiency. A compressor, particularly the axial-flow type, may not produce a pressure rise at all. Until almost the beginning of the present century, isentropic compressor efficiencies were generally less than 50 percent.

We do not know the efficiencies of Denis Papin's centrifugal blowers (1705). There seemed to be little development between that time and 1884, when Charles Parsons patented an axial-flow compressor. Three years later he designed and sold a three-stage centrifugal compressor for ship ventilation. He returned to experiments on axial-flow compressors in 1897,

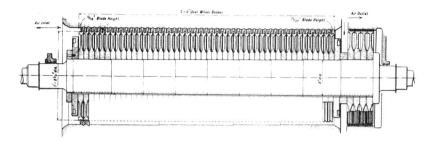

Forty-eight-stage axial-flow compressor (Parsons) for gold mine, 1904. From Parsons 1936.

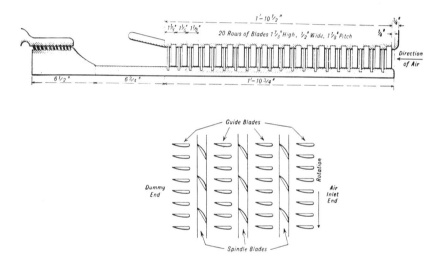

Blade settings used in Parsons' axial-flow compressors, showing potential for stalling. From Parsons 1936.

and made an eighty-one-stage machine in 1899 which attained 70 percent efficiency. The number of stages is probably an all-time record. By 1907 his company had made or had on order forty-one axial-flow compressors, but they were plagued by poor aerodynamics, and he ceased production in 1908. Parsons used far too high a spacing/chord ratio for the blade settings, and all blade rows would be likely to be stalled over much of the operating range. Parsons returned to making radial-flow compressors.

The other major pioneer working on compressors at that time was Auguste Rateau, who published a major paper on turboblowers in 1892. However, a turbocompressor he designed to give a pressure ratio of 1.5 at 12,000 rpm and tested in 1902 gave an isentropic efficiency of only 56 percent. Subsequently Rateau designed and built compressors of in-

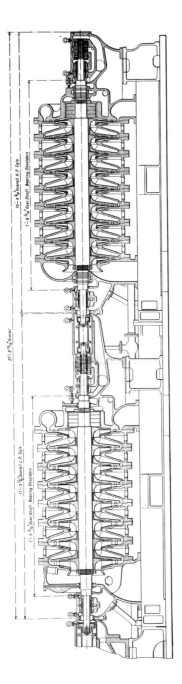

Parsons' sixteen-stage radial-flow tandem intercooled compressor, 1909. From Parsons 1936.

A Brief History

creasing pressure ratio and mass flow, and gradually increasing efficiency. In 1905 he demanded a more rigorous definition of efficiency.

The continued story of compressor development is largely involved with the long struggle to produce a working gas-turbine engine. Many attempts failed simply because of poor compressor efficiency.

Steam turbines

Hero's turbine of A.D. 62 was a toy or experiment having no power output, and Giovanni Branca's proposal for an impulse steam turbine in 1629 was not built, so far as we know. Interest in turbine possibilities must have continued, because in 1784 James Watt wrote to Matthew Boulton that he was glad he did not have to consider a turbine for his locomotives because of the high shaft speeds. The first useful steam turbine we know of was the spinning-arm reaction (Hero's) turbine of William Avery (U.S.) who in 1831 and following years made about fifty as drivers for circular saws and similar duties.

Following the engineering tradition of Scottish clerics (Robert Stirling, for instance), Rev. Robert Wilson of Greenock patented in 1848 some forms of steam turbine, including a radial-inflow turbine. We have mentioned the multistage axial turbine constructed and run by Osborne Reynolds in 1875. We have no evidence that either of these efforts had any influence on other designers. In fact, the much earlier and still astonishingly advanced theory of ideal thermodynamic reversible cycles published by Lazare Carnot's son Sadi in 1824 also had no measurable influence on design in his century: in steam turbines, theory tended to follow practice.

The first steam turbine that had a major impact on the engineering world was that of Charles Parsons (1854–1931), who made a multistage axial-flow reaction turbine giving 10 hp at 18,000 rpm in 1884, his first year as a junior partner at Clarke, Chapman & Co. in Gateshead in northern England. He used brass blades on a steel wheel and an advanced self-centering bearing, presumably to avoid whirling-speed problems. This was to fill a new market: a shipboard electrical generating set, and Clarke, Chapman made over 300 of Parsons's sets in the next five years. But at the end of that period he quarreled with Clarke, Chapman, dissolved the partnership, leaving the ownership of the axial-turbine patents with them, and formed C. A. Parsons & Co. He had in mind what he thought was a better idea, a multistage radial-inflow turbine. He made a thirteen-stage version called the "Jumbo" which ran beautifully on first startup, but the power dropped almost to zero during the first hour. On taking it apart, he found that the blading had almost disappeared, because "foreign matter and water, imprisoned in each wheel case (was) thrown

Parsons' first steam turbine. From Parsons 1936.

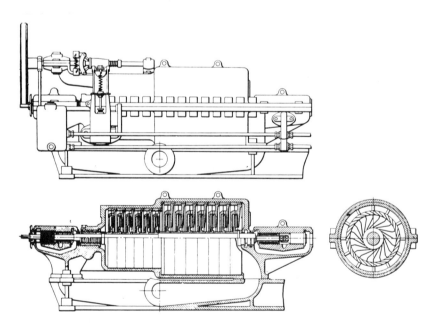

Parsons' "Jumbo" radial-inflow steam turbine, 1889. From Parsons 1936.

A Brief History

alternately inward by the steam and outward by centrifugal force" (from Richardson, quoted by Stanitz).

Parsons then turned to radial-outflow turbines but decided that the future lay with axial-flow machines, and in 1893 he bought back his patents from Clarke, Chapman for £1,500. He had what seems to be an essential combination for a successful innovator: not only superb design skill and judgment but business acumen and a flair for showmanship. Having failed to persuade the Admiralty of the advantages of turbine power, he designed not only the propulsion machinery but the hull (after many model experiments) of a small vessel, which he named the Turbinia, powered by a multistage radial-outflow turbine giving 2,000 hp at 2,000 rev/min. At the great naval review of 1897 celebrating the diamond jubilee of Queen Victoria, the Turbinia dashed around the fleet at the astonishing speed of 34-1/2 knots. The Admiralty was indeed impressed, along with everyone else, and placed an order for a 30-knot turbine-driven destroyer, HMS Viper. By 1907 his marine turbines were powering the 38,000-ton Atlantic liner Mauretania with an output of 70,000 hp. In 1912 Parsons's company supplied a 25-MW station for Chicago, and in 1925 his turbines reached 50 MW in size, with a plant thermal efficiency of 30 percent.

Parsons was not of course the only developer of steam turbines. In 1878 Carl Gustav Patrik de Laval (1845–1913) had made a new cream separator that needed to be driven at 12,000 rev/min. He had tried a Hero-type reaction turbine, which was unsatisfactory. He turned to an impulse wheel and a convergent-divergent (supersonic) nozzle, which ran the turbine up to 30,000 rev/min, with a peripheral speed of 360 m/s (1,200 ft/s). Realizing that the high kinetic energy of the nozzle was not being used efficiently, he constructed in 1897 a velocity-compounded impulse turbine (a two-row axial turbine with a row of turning-vane stators between them, all fed by a single set of high-velocity nozzles). In 1897 de Laval developed double-helical gearing, especially useful for turbine-powered ships, and used very high temperatures and pressures (3,400 psia, 23 MPa) for his turbines.

In France Auguste Rateau (1863–1930) experimented with a de Laval turbine, in 1894, and developed the pressure-staged impulse turbine by 1900. As with Parsons and de Laval, the company bearing his name is still very active.

In America Charles G. Curtis (1860–1953) patented in 1896 the velocity-staged turbine, similar to a two-stage de Laval turbine, but after further development he sold all his rights to General Electric in 1901 for $1.5 million.

Meanwhile Allis Chalmers and Westinghouse had taken licenses on the Parsons designs in 1895, followed by Brown-Boveri (a Swiss-German company founded in 1892 partly by a Briton) in 1899.

Some miscellaneous steam-turbine developments worth noting are

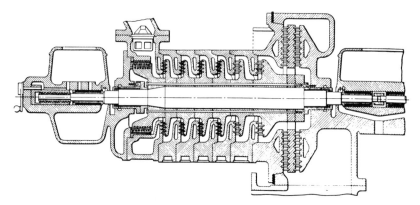

Parsons' multistage radial-outflow turbine, 100 kW, 1891. From Parsons 1936.

The Turbinia at $34\frac{1}{2}$ knots, 1897. From Garnett 1906.

A Brief History

High-pressure rotor and low-pressure cylinder of twenty-five-megawatt Chicago turbine, 1912. From Parsons 1936.

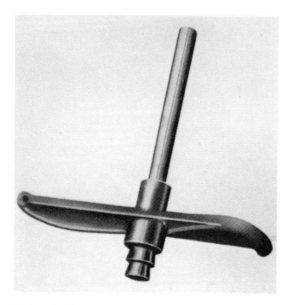

De Laval "Hero"-type reaction turbine, 1883. From Parsons 1936.

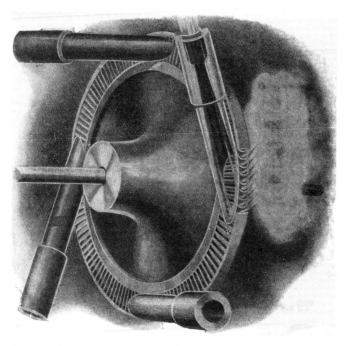

De Laval impulse steam turbine. From Garnett 1906.

A Brief History

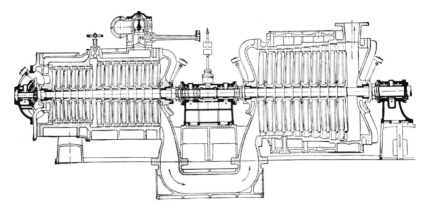

Rateau pressure-staged impulse turbine. From Stodola 1905.

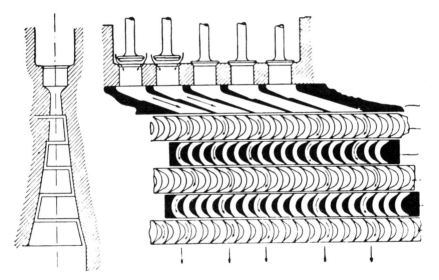

Curtis velocity-compounded impulse turbine. From Stodola 1905.

A Brief History

these. Berger Ljungström (Sweden) introduced his counterrotating radial-outflow turbine in 1912. This author did part of his apprenticeship in Britain on Ljungström turbines, and I remember a 1913 turbine coming back for repair in 1949. W. L. R. Emmot recommended the mercury high-temperature cycle in 1914, and one was given a full-scale trial at Dutch Point in Hartford, Connecticut, in 1923. Several other sets followed, but they never achieved their thermodynamic promise.

Regenerative feed heating was brought in in 1920, and the reheat cycle in 1925. In the same year the Weymouth, Massachusetts, set used 1,200 psig (8.2 MPa) steam. Huntley station had a single-casing 75-MW set in 1929, and in the same year the State Line station installed a triple-cross-compound unit of 208 MW. It was 1958 before the units grew to 500 MW with a single steam generator.

Gas-turbine engines

Just as the name "steam turbine" is used for both the complete plant and the turbine expander itself, so the name "gas turbine" is used for the gas-turbine engine and for the gas expander. We have discussed separately some of the developments in compressors and expanders, and this section concerns the complete engine.

It has had a long and frustrating history. The steam engine was relatively easy to design, build and run because so little work is required to force water into a boiler; little sophistication is required to boil the water; and when steam is formed at high pressure and led to either a piston engine or a turbine, it will produce more power than required by the feed pump. Internal-combustion piston engines also function fairly easily because, although relatively much more work is required to compress air than is needed to get water into a boiler, the combustion temperatures are so high (several thousand degrees C) that the piston expansion work is much greater than the piston compression work, even if the compression and expansion processes themselves involve considerable flow losses. Gas-turbine engines could not use such high temperatures at the start of expansion, and therefore, to produce net positive work, the compression and expansion losses must be low. For many decades the compression efficiencies in particular were just too low for positive work to be given at the temperatures that turbines of the day could withstand. Accordingly many inventors produced machines that never even ran without an external power input.

The earliest patent for a gas-turbine engine was John Barber's of 1791, but nothing resulted from this. In 1872 Dr. J. F. Stolze designed a hot-air turbine which was built in 1900–04, but it failed to produce power. In the same period there were several more patents for gas-turbine engines and

A Brief History

three serious full-scale experiments, none of which was successful in producing power.

The first gas-turbine activity to result in a working engine was started in Norway by Aegidius Elling (1861–1949), whose work has been unaccountably overlooked in most historical reviews of turbomachinery development. He started working on gas-turbine designs in 1882 and constructed the first shaft-power constant-pressure-cycle gas turbine to produce net output (11 hp) in 1903. It had a relatively advanced six-stage centrifugal compressor with variable-angle diffuser vanes and water injection between stages; a combustion chamber; a small heat exchanger producing steam at the same pressure as the air, and which was mixed with the combustion gases before the nozzles; and a turbine described as centripetal but appearing to be an axial impulse type from contemporary photographs.

In 1904 Elling built a regenerative gas turbine: instead of the combustion gases raising steam in a heat exchanger, the turbine-outlet gases transferred heat to the compressor-delivery air. This turbine used a turbine-inlet gas temperature of 500 C, versus the 400 C of his first machine, and produced 44 hp.

The third successful gas turbine was produced in France by Charles Lemale, who was granted a patent for a constant-pressure (Brayton or Joule) cycle in 1901. He formed, with René Armengaud, the Societé Anonyme des Turbomoteurs in Paris in 1903. In 1905–06 the company commissioned Rateau to design a 25-stage centrifugal compressor in three casings on one shaft running at 4,000 rev/min and absorbing 245 kW (328 hp), giving a pressure ratio of three to one. This was made by Brown Boveri, and it achieved an isentropic efficiency of 65 to 70 percent. This would normally not be high enough to allow a gas-turbine engine to produce net power, but an astonishingly high combustion temperature of 1,800 C was reported to be attained in a carborundum-lined combustor, and the two-stage Curtis turbine was water cooled. The steam raised in the cooling circuit was led to nozzles and passed through the same turbine wheel, a concept that is receiving renewed attention at the present time. This ambitious engine produced positive power, albeit at only 3.5 percent thermal efficiency. When Armengaud died in 1909, the experiments were stopped. Brown Boveri continued in the compressor business. The Societé des Turbomoteurs turned to making kerosene-combustion-heated compressed-air-powered torpedoes.

Another early engine to achieve partial success was proposed by Hans Holzwarth in 1906–08 and constructed in 1908–13 by Brown Boveri. This was an explosion or constant-volume cycle, in which the potentially high temperatures of combustion that are obtained in a periodic-firing system, as in an engine cylinder, can compensate for poor compression and expansion efficiencies. Holzwarth's 1,000-hp design produced only 200 hp, and both its size and efficiency were less favorable than then-available reciprocating engines. Both Brown Boveri and M. F. Thyssen of Mulheim continued to persevere with the Holzwarth engine, producing eight prototypes between 1908 and 1938, and a ninth and last was made later. The association of gas turbines with diesel engines in a modified form of the constant-volume cycle has continued to this day, ranging from simple turbocharging in which the turbomachinery is a small engine appendage to cases where the diesel engine, in the form perhaps of a free-piston "gasifier," is simply a replacement for the more conventional constant-pressure combustor.

A potentially significant, but alas fruitless, effort began in the United States when Sanford Moss, influenced by Professor Fred Hesse's thermodynamic classes at the University of California, devised a constant-pressure (Brayton or Joule) cycle in 1895. He wrote his master's thesis in 1900 on gas-turbine-engine design, with a proposal for a locomotive, and did his doctorate on a similar topic at Cornell. He used a steam-driven air compressor, but his engine did not produce positive work. Moss returned to his company, General Electric, in 1903 and carried out experiments on compressor design at Schenectady, New York, before moving to the Lynn, Massachusetts, works in 1904. He consulted with Professor Elihu Thomson and Richard H. Rice, with whom he made the important contribution of rationalizing diffusers into subsonic and supersonic types. GE ended work on his gas-turbine engines in 1907 when it seemed that they could not exceed a thermal efficiency of about 3.5 percent, well below the capability of competing engines. However, GE made and sold turbocompressors to Moss's designs from then until 1925, when the company sold the compressor business. Moss became a research engineer on steam turbines but worked also on aircraft-engine superchargers until the 1940s. Another gas-turbine-engine enthusiast, Glenn B. Warren, who wrote his bachelor's thesis on combined gas-steam cycles at the University of Wisconsin in 1919, joined GE in 1920 and subsequently made a design study of a 10,000 kW engine. Management decided that the company did not have sufficient knowledge in compressor design or heat transfer, however, and Warren was directed to work as well on steam-turbine problems. He made major advances in several areas.

There were many other abortive attempts to produce working gas-turbine engines in the first two decades of this century, including Adolph Vogt (1904–05), Barbezat and Karavodine (making a constant-pressure engine which barely produced net power in 1908), and Hugo Junkers (who made free-piston engines with gas-turbine expanders, unsuccessfully, in 1914). Exhaust-gas turbine-driven engine superchargers were successfully developed by Alfred Buchi and Rudolph Birmann.

The various efforts leading to the modern successful gas-turbine engine

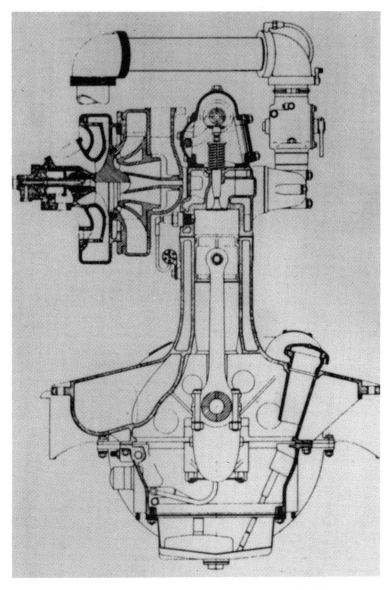

Birmann radial-flow turbosupercharger, 1929. From Stanitz 1966.

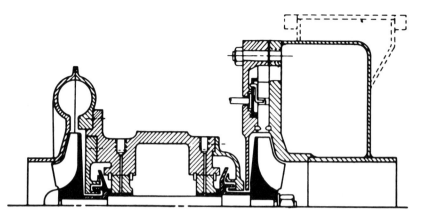

Aircraft turbosupercharger developed in Germany, 1943. From Stanitz 1966.

can be said to have started in the few years from 1927 to 1936 by different people, some of whom were unaware of earlier or contemporary work. The first of these people, however, and one who has had a major impact, was certainly fully cognizant of past attempts: Aurel Stodola, professor at Zurich Polytechnic. He was involved in testing a Holzwarth engine at the Thyssen steel works and noted the high heat transfer in the water jacket around the turbine. Noack of Brown Boveri recommended that the waste heat be used in a steam generator, and from 1933–36 the so-called Velox boiler was developed. This had an axial compressor "supercharging" a boiler in which gas or liquid fuel was burnt, with the hot gases subsequently being expanded through a gas turbine which drove the compressor. For shop tests of the turbomachinery, a high-intensity combustor was substituted for the boiler in 1936, and net power was produced at a reasonable thermal efficiency. Thus the first successful industrial gas-turbine engine was arrived at by chance. Adolf Meyer designed true gas-turbine engines which Brown Boveri sold from 1939, and one was fitted in a locomotive in 1942, but the major gas-turbine sales were of Velox boilers for Houdry catalytic-cracking plants from 1936 on. The twenty-stage axial compressor gave the high efficiency of 86 percent isentropic and was driven by a five-stage axial turbine.

Stodola published the first edition of his classic *Steam Turbines* (called in later editions *Steam and Gas Turbines*) in 1903, which has had a major influence on design.

Much of the other work on shaft-power gas-turbine engines also started in Switzerland. In 1936 Sulzer studied three alternative types of gas turbines and went on to produce axial-flow engines working on the constant-pressure open cycle. Jacob Ackeret and Curt Keller at Escher Wyss designed, in 1939, the closed-cycle gas turbine using, at that time, air. Escher Wyss continues to be the leading proponent of closed-cycle

A Brief History

gas turbines and produces them now using principally helium as the working fluid for coal and nuclear heat input.

Christian Lorensen in Berlin began experiments on axial-flow turbines and hollow air-cooled blades, which were made by Brown Boveri in 1929. His work led directly to German turbojet cooled-blade designs.

In Sweden, A. J. R. Lysholm, chief engineer of the Ljungström Steam Turbine Company, began investigating gas-turbine engines and designed some units. One of his designs was made by Bofors in 1933–35, but success was prevented by severe surging in the centrifugal compressors. Lysholm turned to the Roots straight-lobe-type blower, which cannot surge but has no internal compression and is therefore inefficient for anything above a very small pressure ratio. Lysholm then invented the helical-screw compressor with internal compression, versions of which are now widely used in the niche between piston and centrifugal compressors. He did no further work that we know of on gas turbines. The Elliott Company tested a gas turbine that incorporated a Lysholm compressor, but it had unacceptable noise and vibration.

The turbojet

In the same decade that saw the shaft-power gas-turbine engine reach successful operation, the turbojet was independently developed by four people, Whittle in Britain and von Ohain, Wagner, and Schelp in Germany.

Frank Whittle, a Royal Air Force cadet, was first with his invention, 1929, and patent, 1930, although he was not first at successful running or in the first flight. His patent was for an axial-plus-centrifugal compressor and a two-stage turbine and included the possibility of using silica-ceramic blades. He was rejected by all the major British aero-engine manufacturers (in fact no turbojet development in Britain or Germany originated with an engine builder) and was not given government support, except for an education at Cambridge, until 1936. He formed Power Jets Ltd. in 1935, but his patent lapsed because he could not afford the £5 required to renew it. His design included free-vortex-twist turbine blades, and his incorporation of blade-angle variations which allowed for radial-flow equilibrium gave British engines a higher efficiency than their German opponents.

Whittle had two principal problems: to make a combustor with about ten times the previous maximum combustion intensity for liquid-fuel combustion, and to overcome the mechanical failures that plagued his turbines. A major contribution to the combustion problem was made by Ian Lubbock, of Shell, in 1940. A year later the Henry Wiggin Company produced its first Nimonic series of nickel-chromium-cobalt turbine-blade materials, effectively solving this problem. Meanwhile the Whittle engine was taken to the United States and started production at the General

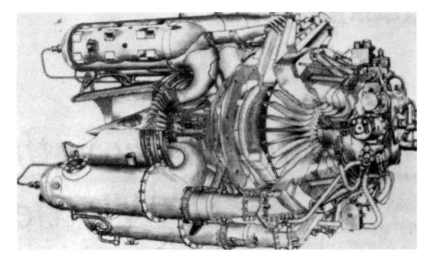

Whittle W2/700 turbojet engine, 1943. From Wailes 1981.

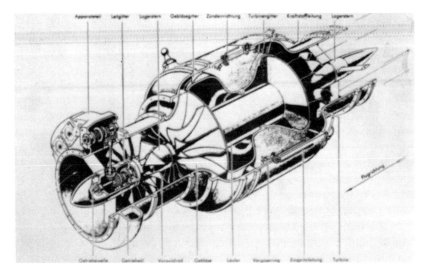

Heinkel HES8A turbojet engine with radial-flow compressor and mixed-flow turbine. From Stanitz 1966.

A Brief History

Electric plant at Lynn, Massachusetts; it had its first U.S. run in 1942 and its first flight later that year in a Bell XP-59. This was then only a year later than the first British flight. The Whittle engine was developed into the Rolls Royce Welland in Britain and the J 33 in the United States.

Whittle's counterpart in Germany was Hans von Ohain. Although he started later—he received his Ph.D. in physics from Göttingen in 1936 and joined Ernst Heinkel that year—he achieved his first engine run and first flight sooner. This was partly due to better backing—Heinkel wanted to build the world's fastest plane; partly to excellent assistance—Heinkel hired Max Hahn to help von Ohain; and partly to wise tactical decisions. Von Ohain knew that, to get continued support from Heinkel, he had to demonstrate quickly something that showed promise. The configuration he chose was a centrifugal compressor plus a radial-inflow turbine made mostly of mild steel, and he avoided the liquid-fuel-combustion problem by using hydrogen for his first run in 1937. With greater support forthcoming subsequently, he was able to solve the combustion problem ahead of Whittle, and the first flight of a jet aircraft, the Heinkel 178, was on August 27, 1939, with an engine weighing 361 kg (795 lbm), giving 4,890 N (1,100 lbf) thrust at 13,000 rev/min.

In 1935 Herbert Wagner, professor of aerodynamics in Danzig and Berlin, was on leave at Junkers and designed a turboprop engine. In 1937 he and Max Adolf-Müller designed an engine with a five-stage axial-flow compressor, a single annular combustion chamber and a two-stage turbine. The compressor had 50-percent-reaction blading and a pressure ratio of three to one—too great for only five subsonic stages—and accordingly had a poor operating range. The engine never ran under its own power. It was later included in the program of the German Air Ministry (RLM).

Helmut Schelp returned to Germany in 1936 having taken a masters degree at Stevens Institute of Technology. He joined the RLM and studied the problems of high-speed aircraft. In 1937 he decided that a jet engine incorporating an axial-flow compressor would be optimum. Unlike the other pioneers, he was aware of preceding work. With his superior, Hans A. Mauch, he visited all the major German engine manufacturers in 1938 and persuaded Junkers and BMW to accept study contracts on reaction propulsion. (Nowadays governments have less difficulty in getting companies to accept funds.) A contract was also given to Messerschmidt for a turbojet fighter, which became the highly successful Me 262. Anselm Franz, head of supercharger development at Junkers, was put in charge of the new engine project. In the same way as von Ohain before him, he decided on a design with the best chance of running even if some sacrifice in performance must be taken. The Junkers Jumo 004 resulted, with an eight-stage axial-flow compressor giving a pressure ratio of three to one and an isentropic efficiency of 78 percent, and with a single-stage

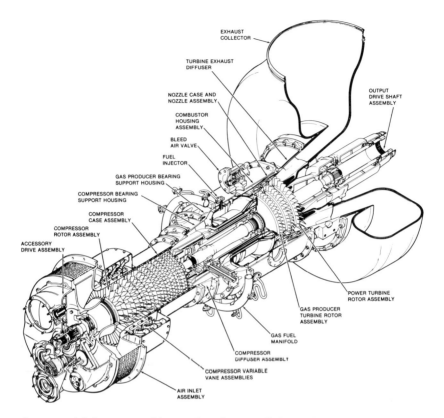

EXHAUST
COLLECTOR

TURBINE EXHAUST
DIFFUSER

NOZZLE CASE AND
NOZZLE ASSEMBLY

OUTPUT
DRIVE SHAFT
ASSEMBLY

COMBUSTOR
HOUSING
ASSEMBLY

BLEED
AIR VALVE

FUEL
INJECTOR

GAS PRODUCER BEARING
SUPPORT HOUSING

COMPRESSOR BEARING
SUPPORT HOUSING

COMPRESSOR
CASE ASSEMBLY

COMPRESSOR
ROTOR ASSEMBLY

ACCESSORY
DRIVE ASSEMBLY

POWER TURBINE
ROTOR ASSEMBLY

GAS PRODUCER
TURBINE ROTOR
ASSEMBLY

GAS FUEL
MANIFOLD

COMPRESSOR
DIFFUSER ASSEMBLY

COMPRESSOR VARIABLE
VANE ASSEMBLIES

AIR INLET
ASSEMBLY

Centaur axial-flow gas-turbine engine. Courtesy Solar Turbines International.

turbine. Initially, the blades were solid, but later they were made from a manganese alloy sheet and were air cooled. There was no nickel and only 2 kg of chromium in the engine. The combustor was aluminized mild steel and lasted twenty-five hours. Six-thousand Jumo 004 engines were made by the end of the war.

A similar diversity was taking place in Britain. Griffith at the Royal Aircraft Establishment had been very influential in axial-compressor fluid mechanics, and he and Hayne Constant designed an eight-stage axial-flow compressor in 1937 which ran in 1938 but failed mechanically. It was rebuilt but was destroyed in an air raid in August 1940. C. A. Parsons Ltd. made a similar test compressor in 1940. Armstrong Siddeley ran an engine in 1939, following Griffith's theories of having several "spools"—concentric shafts each with a turbine stage driving a few compressor stages, now an established procedure for large turbojets. In 1939 Griffith went from the RAE to Rolls Royce to speed up jet-engine work there. In January 1941 Frank Halford, designer of the highly advanced piston engine the Napier Sabre, was asked to design a turbojet at de Havilland, and the

A Brief History

DH Goblin was made by August 1941, ran in April 1942, and flew in a Gloster Meteor in March 1943. It was giving 13,450 N (3,000 lbf) thrust by 1945, with a very good fuel consumption. Bristol Engines Co., which had rejected Whittle in 1930 because of its development problems with its double-sleeve-valved radial piston engines, started work on turbojets in 1944.

In the forty years since the end of World War II, gas-turbine engines, including turbojets, have developed rapidly. But the development has been along the lines that Edward Constant calls "normal technology"— no revolutionary changes have occurred. Much of the development can be attributed to one overriding factor—the increase in turbine-inlet temperatures.

References

A century of power progress—growth marks the years 1907–57. *Power*, June 1981.

A century of power progress—boosting powerplant performance. *Power*, July 1981.

Agricola, Georgius. 1556. *De Re Metallica*. Tr. by Herbert Clark Hoover and Lou Henry Hoover. Dover, New York, 1950.

Andersson, S. B. 1974. Development in air compression. *I. Mech. E.* (September): 66–72.

Bamford, L. P., and S. T. Robinson. 1945. *Turbine engine activity at Ernst Heinkel Aktiengesellschaft Werk Hirth—Motoren*. Combined Intelligence Objectives Sub-Committee, file no. XXIII-14. London.

Bowden, A. T. 1964. Charles Parsons—purveyor of power. *Chartered Mechanical Engineer* (September): 433–438.

Buchetti, J. 1892. *Les moteurs hydrauliques*. Published by the author, Paris.

Constant, E. W., II. 1980. *The Origins of the Turbojet Revolution*. Johns Hopkins University Press, Baltimore, Md.

Garnett, W. H. Stuart. 1906. *Turbines*. Bell, London.

Holmes, R. 1967. The evolution of engineering—the rocket engine. *Chartered Mechanical Engineer* (October): 436–442.

Johnson, Dag, and R. J. Mowill. 1968. *Aegidius Elling—A Norwegian Gas-Turbine Pioneer*. Norwegian Technical Museum, Oslo. March.

Legat, E. S. 1950. The development of air and gas compressing plant. *Institution of Mechanical Engineers* (February): 468–471.

Ludlow, C. G., and A. S. Bahrani. 1978. Mechanical engineering during the early Islamic period. *Chartered Mechanical Engineer* (November): 79–84.

Lysholm, A. J. R. 1964. An interview with Fredrik Ljungström. *Chartered Mechanical Engineer* (May): 261–263.

Parsons, R. H. 1936. *The development of the Parsons steam turbine*. Constable, London.

Schlaifer, Robert. 1950. *Development of aircraft engines.* Harvard University Graduate School of Business Administration, Boston.

Seippel, Claude. 1983. Personal letter to the author, with a translation of an article. Constant volume versus constant-pressure combustion—a chapter in the history of the gas turbine. February 22, Zurich.

Smith, N. 1980. The origins of the water turbine. *Scientific American*, 242 (January): 138–148.

Stanitz, J. D. 1966. History of Radial-Inflow Turbine Development. Unpublished monograph.

Stodola, A. 1905. *Steam turbines with an appendix on gas turbines and the future of heat engines.* Second German edition translated by Louis C. Loewenstein. D. Van Nostrand, New York.

Stodola, A. 1927. *Steam and Gas Turbines.* 6 German ed. Transl. by Louis C. Loewenstein. McGraw-Hill, New York, 1927.

Wailes, Rex. 1961. An interview with Sir Frank Whittle. *Chartered Mechanical Engineer* (June): 353–358.

Whittle, F. 1981. CME interview Sir Frank Whittle. *Chartered Mechanical Engineer* (July): 41–43.

1

Introduction

1.1 Aims

The goals of this text are to make available to nonspecialist engineers and to students a simple but fundamental approach to the design of turbomachinery, including the choice of configuration and the determination of close approximations to optimum dimensions and flow angles; to develop simple methods for finding the performance characteristics of turbomachinery under various operating conditions and to show how the designs may be modified to give favorable "off-design" operation; and to show how the interrelationships among material limitations, fluid-mechanic laws, thermodynamics, heat transfer, and such considerations as mechanical vibration set boundaries to design choices.

The particular combination of turbomachinery and other components (combustors and heat exchangers) that forms the power-producing gas turbine is considered in some detail. The gas turbine's close relative, the exhaust-gas turbocharger, is also treated.

This work has been undertaken because of the apparent lack of a modern text dealing with turbomachinery from the design, or problem-solving, standpoint. The problems arise when satisfactory answers cannot be obtained from handbooks or from manufacturers' offerings. If, for instance, the requirement is to compress a given flow of gas from given inlet conditions to a given discharge pressure, with power available at a given shaft speed, and given trade-offs for first cost, efficiency, size, and so forth, what type of compressor represents an optimum choice—radial, mixed flow, axial flow, in a single stage or in several? What are the outer and inner diameters, blade lengths, flow and blading angles, and the diffuser size? Or if handed a certain configuration of turbomachine, on paper or in hardware, what will be its off-design performance?

It is one of the aims of this work to answer these and similar questions with sufficient accuracy for most purposes. The accuracy will be sufficient for many areas of industrial and experimental design, and for systems-engineering studies. In those applications where the value of fractions of a percentage point in efficiency or in specific power is very large, for instance, in aircraft gas turbines and fans and in large steam and hydraulic turbines, it is necessary to treat the designs obtained by the methods given here as inputs to highly sophisticated analytical and experimental methods (not treated here) of flow analysis, stress analysis, and vibration prediction, in particular. Such advanced methods have been developed by or for government laboratories, for instance, the Lewis Laboratories of the National Aeronautics and Space Administration in the United States and the National Gas-Turbine Establishment in Britain, as well as in private companies.

The savings to be realized by approaching such sophisticated and extremely expensive programs with a near-optimum design are very large. The technical literature and the unwritten histories of virtually all firms involved with turbomachinery manufacture are full of cases where many millions of dollars have been wasted because development work was started on designs that were very poor choices for one or more reasons.

In other cases needless expense was incurred and inappropriate machines produced simply because designers and engineers have become accustomed to thinking only of a certain range of types of turbomachinery and have neglected other types far better suited to their needs. The chief engineer of a successful manufacturer of jet engines was so proud of his ability to design large pressure ratios into single-casing axial-flow compressors that he regarded two-casing intercooled machines as effeminate. In fact, they have very large economic and performance advantages. In so doing, he reduced to zero the chances of his company successfully diversifying in the direction of supplying air compressors for large oxygen plants. Another chief engineer lost his company a great deal of money and, perhaps more important, time, because he refused to change his predisposition for very-high-reaction compressors even when faced with performance requirements that plainly called for lower-reaction machines with variable-setting-angle stator blades. Other examples could be the overlooking of partial-admission turbines by a generation of engineers brought up on full-admission machines and the automatic use of radial-flow machinery even when all the conditions cry out for the use of axial-flow units.

No amount of refinement by advanced analytical methods or by experiment can overcome the severe disadvantage of an initially incorrect choice of configuration or type. A principal aim of this book is to help the reader avoid such errors.

Introduction

1.2 Definitions

Turbomachine

A turbomachine produces a change in enthalpy in a stream of fluid passing through it and transfers work through a rotating shaft; the interaction between the fluid and the machine is primarily fluid-dynamic lift.[1]

A change in enthalpy can also be brought about by heat transfer, and, although there is often considerable heat transfer in a turbomachine (for instance, in a high-temperature cooled gas turbine), its contribution to the enthalpy change is virtually always almost negligible.

It is not possible in the real world to have lift forces without at the same time having drag forces. Much of the skill in achieving good designs is to keep the drag forces and associated energy losses small. But they are always far from negligible. By stating in the definition of a turbomachine that the interaction must be "primarily" lift, we exclude that class of machines in which the fluid and the rotor interact purely through viscous forces—so-called "drag" turbines and pump-compressors. This definition may not be universal: advocates of viscous-action machines may regard them as true turbomachines.

Lift and drag

Lift and drag forces are normally defined with reference to airfoils (figures 1.1a), lift being normal to the direction of relative motion and drag along this direction. Airfoils very similar to airplane-wing sections can in fact be used in a particular class of turbomachines: axial-flow compressors and pumps (figure 1.2). However, most other classes of turbomachinery employ shaped channels rather than airfoils (figure 1.1b). Lift and drag forces must then be defined with reference to the normal and tangential directions of the channel walls.

Turbomachines can be divided into categories in several different ways. The categories shown in figures 1.3 and 1.4 are, first, turbines on the left of each figure and pumps, fans, blowers, and/or compressors on the right. Turbines produce shaft work from a flow of fluid which drops in enthalpy (which, if the concept of enthalpy is unfamiliar, can be thought of as approximating to a drop in pressure) from inlet to outlet. Pumps, fans, blowers, and compressors absorb shaft work from a "driver," and increase the enthalpy (and the pressure) of a flow of fluid.

The word "turbine" is used for all such machines, whether they work on liquid or gas, over a low or a high pressure ratio. There is no such one

---

1. For those unfamiliar with enthalpy, it can be thought of as related to the energy content of a substance. Enthalpy is defined thermodynamically in chapter 2.

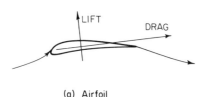

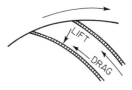

(a) Airfoil

(b) Passage or channel
(in a radial-flow rotor)

Figure 1.1 Lift and drag forces.

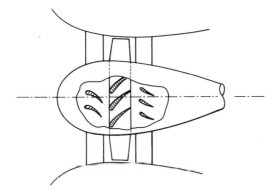

Figure 1.2 Axial-flow pump or compressor.

word for power-absorption machines. Pumps are pressure-increasing devices that operate on liquids. Fans work with gases, usually, but not always, giving so low an increase in pressure that the gas can be considered incompressible. In fact most fans merely consist of a single row of rotating blades which increase the velocity of the throughflow. The name "fan," however, is also given to the high-speed devices sometimes used on jet engines ("fan jets") to produce a combined jet of greatly increased size (and mass flow) and reduced velocity, which gives higher propulsion efficiency and lower noise than a small, high-velocity jet (see figure 13.8). Compressors give pressure ratios that result in a significant increase in density. A machine with a pressure ratio of 1.2:1 or above would be called a compressor. The word "blower" is, like "fan," used sometimes to mean a fan of negligible pressure ratio and sometimes (as in a blast-furnace blower) to mean a compressor of considerable pressure ratio.

Machines in which the fluid flow through the blading is axisymmetric are shown in figure 1.3, while in figure 1.4 there are so-called "partial-admission" turbomachines. There is often a considerable advantage in designing a rotor that, if it "ran full," would be much too large for the flow available, and in ducting the flow to only a portion of the rotor blading. A

Introduction

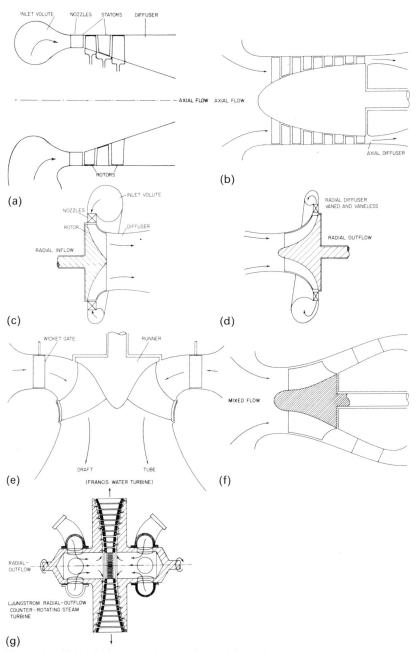

Figure 1.3 Full-admission turbomachinery. (a) axial-flow turbine, (b) axial-flow compressor, (c) radial-inflow turbine, (d) radial-outflow compressor, (e) radial-inflow hydraulic (Francis) turbine, (f) mixed-flow compressor, (g) multistage radial-outflow counter-rotating Ljungstrom steam turbine.

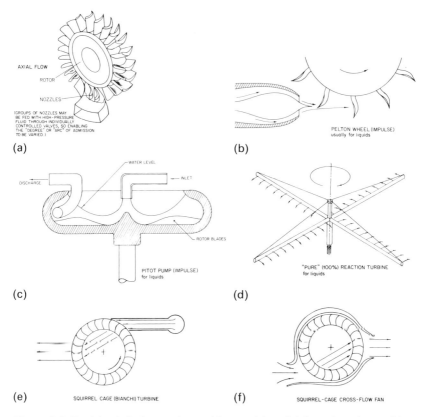

Figure 1.4 Partial-admission turbomachinery: (a) axial-flow impulse turbine, (b) Pelton hydraulic (impulse) turbine, (c) Pitot pump (impulse) for liquids, (d) "pure" (100-percent) reaction turbine for liquids, (e) squirrel-cage (Bianchi) turbine, (f) squirrel-cage crossflow fan.

partial-admission turbine in the final drive of a ship-propulsion power plant, for instance, gives a lower shaft speed than would a full-admission design (because the same rotor-blade speed occurs at a larger radius). Another advantage is that the high-pressure fluid (usually steam for this application) may be ducted to several "nozzle boxes" covering different arcs of admission, and these may be connected, by valves, to the supply separately or in combination, giving a wide range of power outputs.

A special case of partial-admission turbine is the Pelton wheel, used with water under high heads (pressures). The water leaves the nozzle as a jet with a so-called "free" surface in air. In most turbomachinery the fluid stream is confined in ducts.

The Pelton jet can exist with a free surface because the entire pressure drop to atmospheric pressure takes place in the nozzle. The rotor converts

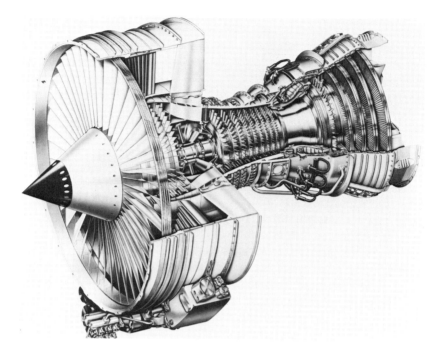

CFM56 high-bypass turbofan. Courtesy General Electric Company.

only kinetic energy to shaft power. This type of arrangement is called an "impulse" turbine. If there is any pressure drop in the rotor, the fluid is accelerated in the rotor passages, and the acceleration results in a reaction force on the rotor blades. Even if the reaction force is small compared with the "impulse" force from turning (from the change of momentum of the fluid), a machine with flow that accelerates relative to the rotor is called a "reaction" turbine. In later chapters we shall introduce the concept of "degree of reaction," which can be used to quantify the extent to which a turbine, or a compressor, or fan, differs from "pure" impulse conditions.

The flow in the Pelton-wheel turbine is tangential to the rotor. The direction of flow forms another category that can be used to characterize turbomachines. In most cases the flow is either axial, that is, in an annulus whose axis coincides with that of the rotor and whose flow direction is approximately parallel to the axis, or radial. In radial-flow machines the flow may travel radially inward or radially outward. Near the axis the flow in a radial-flow machine must turn to or from the axial direction. Radial-flow steam turbines usually have radially outward flow because of the rapidly increasing specific volume of steam during expansion. Radial-flow gas-turbine expanders use radially inward flow because the increase in volume flow is relatively small and can easily be accommodated in the flow path near the axis, and because there are advantages in locating the inter-action of the nozzle exit flow in a region of high rotor-blade velocity.

In some machines the flow at entrance to and/or exit from the rotor blades is close neither to the axial nor the radial direction. These are called "mixed-flow" turbomachines.

Another category concerns the number of separate rows or rings of blades. The "pure-reaction" machines have simply a rotor containing nozzles in the case of the rotating-pipe liquid distributor (frequently used over gravel filters in waste-water-treatment plants as shown in figure 1.4). Most turbomachines have a stator as well as a rotor. The stator may be as simple as the nozzle of the Pelton turbine, or it may have the somewhat greater complexity of the volute, blade ring, casing, and ducting of the radial-flow turbine and compressor. The combination of rotor and stator (in either sequence) is called a "stage." Machines can therefore be cate-gorized as being just rotors, or single stages, or having multiple stages.

The methods developed in this book apply to all these different ma-chines, although we have the space to treat only a few particular classes with any degree of thoroughness.

Gas turbine

The term "gas turbine" is most often used as an abbreviation for a gas-turbine engine—something that accepts and rejects heat and produces work. This work may be given as output torque in a turning shaft or as the velocity and pressure energy in a jet, which would produce thrust on a moving airplane. The term "gas turbine" can also be used more narrowly for just the turbine expander in a gas-turbine engine.

A gas-turbine engine (figure 1.5) consists of a compressor, which con-tinuously compresses gas from a low pressure to a higher pressure; a heater or heaters, in which the temperature of the compressed gas is raised; an expander, which continuously expands the hot gas to a lower pressure; and a cooler or cooling system, in which the temperature of the gas is reduced to that established for the compressor inlet. Two heaters and two coolers are shown in figure 1.5 because sometimes the first heater receives heat from the first cooler. The combination of heater plus cooler is known as a heat exchanger. It is defined more fully in this section, but its design is treated in chapter 10.

An open-cycle gas turbine (figure 1.6) is one in which there is no engine cooling system: the atmosphere performs this function. The gas entering the compressor is atmospheric air, and the hot gas leaving the expander or the heat exchanger is discharged to the atmosphere, the way the earth's flora regenerate the oxygen from the carbon dioxide produced.

An internal-combustion gas turbine (figure 1.7) is one in which at least one of the heaters is dispensed with, and the gas is heated by combustion of fuel in the gas steam. For the fuel to burn, the gas must be either atmo-

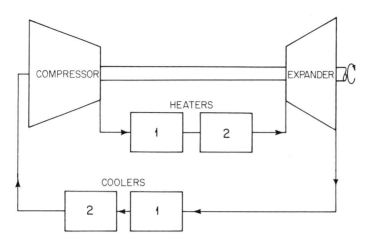

Figure 1.5 General gas-turbine engine.

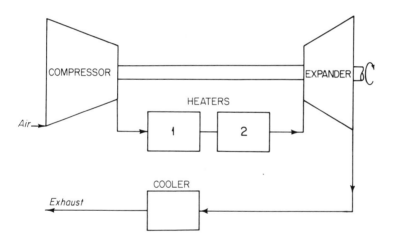

Figure 1.6 Open-cycle gas-turbine engine.

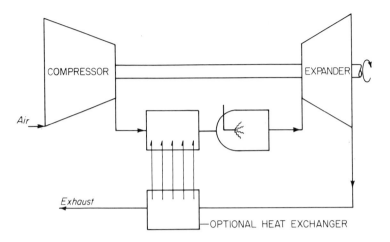

Figure 1.7 Internal-combustion gas-turbine engine.

spheric air or oxygen. Internal-combustion gas turbines are invariably open cycle, working with air.

A regenerative gas turbine supplies heat to the compressed gases in the first heater after the compressor from heat given by the hot expanded gases in the first cooler. The heat exchanger so required is usually called a "regenerator" if it involves submerging a matrix alternately in the hot and then in the cold streams, and a "recuperator" if the heat is transferred through tube or duct walls. (These definitions are accepted in Britain and are gaining ground in the United States, but they are not universal.)

1.3 Comparison of gas turbines with other engines

The gas-turbine engine cycle has some resemblance to both other internal-combustion-engine cycles and steam-turbine engines.

In a spark-ignition (Otto) or in a compression-ignition (Diesel) engine, air goes through a similar sequence of processes to those in an open-cycle internal-combustion gas turbine. Atmospheric air is compressed, heated, expanded, and discharged to the atmosphere for cooling and regeneration. There are two vital differences, however. First, in the Otto and diesel engines the processes occur in the same place (the engine cylinder) at different times (the different strokes). In the gas-turbine engine the processes occur in different places (in the various components) at the same time (figure 1.8). The piston engine uses a "batch" process, while the gas-turbine engine works on a continuous process. The second difference is that much of the combustion in an engine cylinder takes place at constant (or nearly constant) volume. In a gas turbine combustion is carried out at constant pressure. (Constant-volume combustors for gas turbines are,

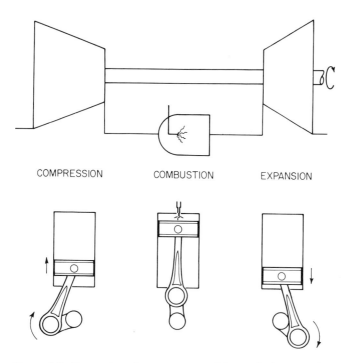

COMPRESSION   COMBUSTION   EXPANSION

Figure 1.8 Comparison between gas-turbine and piston engines.

however, possible. The Holzwarth gas turbine mentioned in the brief history at the beginning of the book had a constant-volume combustion system.)

The result of the continuous nature of the processes in a gas turbine are that it can pass very much larger volume flows of gas than can a piston engine of the same engine volume. The turbine is very compact and has a high power-weight ratio. On the other hand, although it is easier to design the component efficiencies of the gas turbine to be above the efficiencies of the individual processes of a piston engine, it is easier to use a higher peak cycle temperature in a piston engine because it occurs momentarily in a cylinder rather than continuously in a combustor. In the past the thermodynamic benefit of a higher peak temperature in the diesel has out-weighed the benefits of better component efficiencies in the gas turbine, and diesel engines have had higher thermal efficiencies than gas turbines. The use of turbine-blade cooling and higher-temperature materials is enabling the gas turbine to overtake the diesel engine in maximum thermal efficiency in some cases.

The similarity of a gas turbine to a steam-turbine engine is perhaps more obvious. Both employ continuous processes taking place simultaneously in separate components. The components bear a strong resemblance in some cases—both engines have turbine expanders which to the uninitiated eye may seem virtually identical. The steam-turbine engine has a pressurizer in the form of a feed pump paralleling the gas turbine's compressor; feed pumps in large electrical-generating stations may absorb tens of thousands of kilowatts and look a little like multistage centrifugal compressors. And the steam generator or boiler of a steam-turbine engine can look rather similar to a gas heater of a closed-cycle gas-turbine engine.

The principal difference of course is that the steam-turbine engine works with a fluid that changes phase during the cycle, whereas the gas turbine, as its name implies (its name has nothing to do with the fuel it uses) works on fluids that remain in the gaseous phase. The feed pump pressurizes liquid, which means that the work required is relatively small—less than two or three percent of the turbine output. A great deal of heat must then be added per unit mass of fluid in the steam-turbine engine to provide sensible and latent heat. In contrast, the compressor in a gas-turbine engine absorbs a large proportion of the power produced by the turbine expander, and the combustor, being required to add a proportionally small amount of heat and working with high inlet-air density, shrinks to an almost negligible size in comparison with that of a steam generator in a cycle of similar net output.

The gas turbine as a Micawber engine

Mr. Micawber used to say (in very rough translation) "Income twenty shillings, expenditure twenty-one shillings: result, misery. Income twenty shillings, expenditure nineteen shillings: result, happiness."[2] The gas-turbine engine, with its large compressor work, has this kind of energy economics (figure 1.9).

Early gas turbines produced misery because their compressors wanted to absorb more power than their turbines would deliver. Successful, "happy," gas-turbine engines were made only when the compressor efficiencies had been improved to the point where the power required was less than the turbine power.

Such an engine, where the net power produced is the difference between two large quantities, is obviously very much affected by component efficiencies. When the compressor power almost equals the turbine power, a small change in the performance of any component can have a delightful or a disastrous effect on the net output of the machine. In early gas-turbine engines a 1 percent improvement in compressor efficiency could yield a 5 percent increase in engine thermal efficiency. And the turbine output could be increased either by improvement in its efficiency or by increasing the turbine-inlet temperature. Accordingly, there was an enor-

---

2. In Charles Dickens' *David Copperfield*.

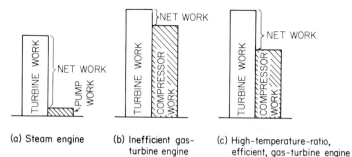

(a) Steam engine   (b) Inefficient gas-   (c) High-temperature-ratio,
                        turbine engine   efficient, gas-turbine engine

Figure 1.9 Energy accounting for different engines.

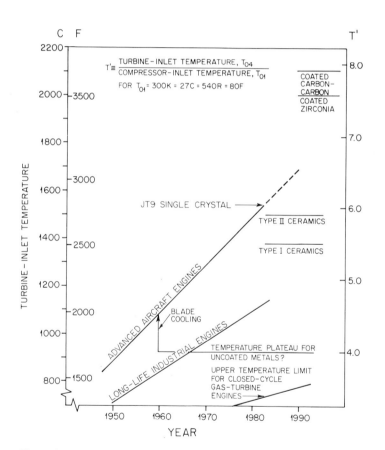

Figure 1.10 Increase of turbine-inlet temperatures.

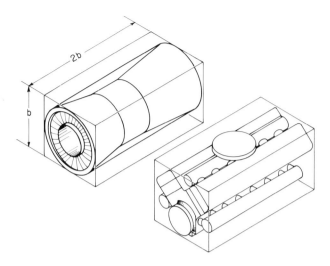

Figure 1.11 Gas-turbine and piston-engine envelopes.

mous effort made in many countries, in many engine companies and governmental and academic laboratories, to understand the fluid mechanics of turbines and compressors, to produce higher-temperature materials, and to improve cooling of surfaces scrubbed by hot gases (figure 1.10). The gas turbine quickly became the most highly researched engine in history. The resulting design methods have become classics of the scientific method. In contrast, piston engines, which have received a much longer period of continuous, worldwide development, are still designed to a considerable extent by rule of thumb.

The "swallowing" capacity of turbomachinery

We mentioned earlier that the gas turbine can process much larger volume flows of gas than can piston engines of similar size because of the steady, rather than discontinuous, nature of the flow. In addition the flow area used by a gas turbine will be a larger proportion of the cross section than will that of a piston engine.

As a simplification, consider two engines that are contained in an envelope formed by a solid of square cross section of side b and of length $2b$, figure 1.11. The gas turbine will draw in air through an annulus of outside diameter a little less than b, say $0.9b$, and inner diameter typically one-half of this, $0.45b$, with a Mach number in the range of 0.3 to 0.5, say 0.4. If the speed of sound is a_{gt}, the volume flow handled is

$$\dot{V}_{gt} = \frac{\pi}{4}(0.9b)^2(1 - 0.5^2)0.4a = 0.191a_{gt}b^2.$$

Introduction

The speed of sound in air is given by

$$a = 20.056\sqrt{T}, \quad (K) \text{ m/s.}$$

If we use a temperature of 300 K, $a_{gt} = 347.4$ m/s, and the inlet velocity is about 139 m/s. Therefore

$$\dot{V}_{gt} = \frac{\pi}{4}(0.9b)^2(1 - 0.5^2)139 = 65.3b^2 \text{ m}^3/\text{s.}$$

In the same envelope the piston engine might be a V-8 arrangement. The cylinder bores, in a typical engine, are about $0.2b$ (and the stroke is similar).

The carburetor venturi throat has an area which, for standard automobiles, is about one seventy-fifth of the piston cross-sectional area. At maximum power the flow through the venturi throat will be a fraction, say, seven-eighths, of sonic velocity.

Then the air volume flow, $\dot{V}_{pe}$, will be given by

$$\dot{V}_{pe} = 8 \times \frac{\pi}{4}(0.2b)^2 \times \frac{1}{75} \times \frac{7}{8}a_{pe} = 0.003a_{pe}b^2,$$

where a_{pe} is the sonic velocity. Then the ratio $\dot{V}_{gt}/\dot{V}_{pe} = 0.191a_{gt}b^2/0.003a_{pe}b^2 = 64\,a_{gt}/a_{pe}$.

The speed of sound at the gas-turbine inlet will be higher than that at the piston-engine carburetor venturi because the temperature will not have been lowered to the same extent. From equation (2.39) in the next chapter,

$$\frac{a_{gt}}{a_{pe}} = \sqrt{\frac{T_{gt}}{T_0} \times \frac{T_0}{T_{pe}}} = \sqrt{\frac{1 + [M_{pe}^2(\gamma - 1)/2]}{1 + [M_{gt}^2(\gamma - 1)/2]}} = 1.0783.$$

Then $\dot{V}_{gt}/\dot{V}_{pe} \simeq 68.7$. The power output of an engine is a direct function of mass flow, rather than volume flow, but the densities of the intake air flows to the two engines are identical within the accuracy of this estimate.

A comparison of the cost (in 1981 dollars) of diesel and gas-turbine engines is shown in figure 1.12 by Woodhouse (1981). He demonstrates the correlation of both engines' specific cost with unit size and production rate. The automobile industry shows repeatedly that the manufacturing cost of any hardware assembly which is produced at a rate of a million a year is about 25 percent greater than the material cost. This is a principal reason why future nonmetallic gas-turbine engines should have a considerable cost advantage over diesel engines.

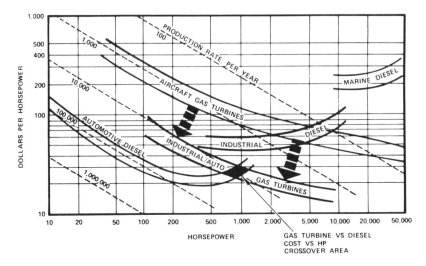

Figure 1.12 Specific cost of gas-turbine engines and competitors. From Woodhouse 1981.

The power output for a unit mass flow of air, termed the specific
power $\dot{W}'$ and given nondimensionally as

$$\dot{W}' \equiv \frac{\dot{W}}{(\dot{m} C p T_{01})},$$

where $\dot{W}$ is the power output, $\dot{m}$ the air mass-flow rate, Cp the air specific
heat, and T_{01} the absolute temperature of the inlet air, can vary widely.
In Brayton cycles it is generally in the range of 0.4 to 1.5 (figures 3.5 to
3.11), with 1.0 being a representative figure for current (1983) high-
performance gas turbines. A typical value for diesel engines is 3.0 and for
spark-ignition engines, 2.5.

If we use mean values of 1.0 for the Brayton-cycle engines and 2.8
for the piston engines, we arrive at a ratio of 66.7/2.8 (or about 25),
the power output of a gas turbine relative to the power output of a piston
engine of similar size. This ratio would apply reasonably well to the larger,
axial-flow aircraft-engine-derived gas turbines but not to simple, small
radial-flow gas turbines, whose volume tends to become dominated by
ducts, and so, to ease manufacturing difficulties, they are purposely
designed to be larger than the minimum possible.

References

(This list includes some general references on turbomachinery and on comparative
performance of engines.)

Anderson, Richard H. 1979. *Material properties and their relationship to critical jet-engine components*. In Proc. Workshop on High-Temperature Materials for Advanced Military Engines. Institute for Defense Analyses, Arlington, Va.

Balje, O. E. 1981. *Turbomachinery*. Wiley, New York.

Cohen, H., G. F. C. Rogers, and H. I. H. Saravanamuttoo. 1973. *Gas-Turbine Theory*. Wiley, New York.

Csanady, G. T. 1964. *Theory of Turbomachines*. McGraw-Hill, New York.

Dixon, S. L. 1975. *Fluid Mechanics, Thermodynamics of Turbomachinery*. Pergamon, New York.

Harman, Richard T. C. 1981. *Gas-Turbine Engineering*. Wiley, New York.

Hill, Philip G., and Carl R. Peterson. 1965. *Mechanics and Thermodynamics of Propulsion*. Addison-Wesley, Reading, Mass.

Jennings, Burgess H., and Willard L. Rogers. 1969. *Gas-Turbine Analysis and Practice*. Dover, New York.

Karassik, Igor J., William C. Krutzsch, Warren H. Fraser, and Joseph P. Messina. 1976. *Pump Handbook*. McGraw-Hill, New York.

Kerrebrock, Jack L. 1977. *Aircraft Engines and Gas Turbines*. The MIT Press, Cambridge, Mass.

Logan, Earl, Jr. 1981. *Turbomachinery*. Marcel Dekker, New York.

Pfleiderer, C., and H. Petermann. 1964. *Strömungsmaschinen*. 4th ed. Springer-Verlag, Berlin.

Shepherd, D. G. 1969. *Introduction to the Gas Turbine*. Van Nostrand, New York.

Shepherd, D. G. 1956. *Principles of Turbomachinery*. Macmillan, New York.

Shepherd, Dennis G. 1972. *Aerospace Propulsion*. American Elsevier, New York.

Stepanoff, A. J. 1957. *Centrifugal- and Axial-Flow Pumps*. Wiley, New York.

Vavra, M. H. 1960. *Aero-Thermodynamics and Flow in Turbomachines*. Wiley, New York.

Wilson, David Gordon 1978. Alternative automobile engines. *Scientific American* 239 (July): 39–49.

Woodhouse, G. D. 1981. *A new approach to vehicular-gas-turbine power-unit design*. Paper 81-GT-152. ASME, New York.

Problems

This first chapter has dealt principally with definitions, and with some discussion of the characteristics and capabilities of turbomachinery and gas turbines. Answering questions on definitions does not engender a love for the subject matter.

Accordingly, the following questions probe background knowledge. Few readers will have more than a few of the data asked for in the first question. Its purpose is to stimulate a survey of the engineering-society papers in the library, and perhaps the advertisements in some of the engineering magazines. Some of the other questions ask for opinions. Again, the purpose is to stimulate thought and perhaps some reading. There are not necessarily "correct" answers to these questions. August committees of eminent scientists and engineers have been frequently totally wrong when they have tried to forecast the future.

1.1
Complete as much of table P1.1 as possible, from your own knowledge or from library study. In general, every entry will be for a different machine, although in some cases there will be relationships between two figures. For instance, the highest-power steam turbine may not operate at the highest pressure used for a steam turbine but may well have the largest low-pressure flow volume. Use numbered references to footnotes to identify your sources.

1.2
Gas-turbine engines have not reached the power-output levels of the largest steam turbines. Why?

1.3
Estimate the design power output of the smallest gas-turbine engine produced in the last decade. Why aren't smaller engines made?

1.4
Give your opinion of the two most promising new applications for gas-turbine engines in the next twenty years, and give reasons for your opinions.

1.5
Why is the maximum temperature of steam turbines so much lower than the current turbine-inlet temperature of gas-turbine engines?

1.6
What do you think are the two principal problems preventing gas-turbine engines from having a much wider application?

1.7
Do you think that gas-turbine engines will be used in outboard-motor boats by, say, 1990?

1.8
For the applications in table P1.8, which do you think that the gas-turbine engine, as a prime mover, would be? (a) Suitable now, and, if so, how would it be used, or in what form? (b) Suitable after certain developments have been successfully completed, and, if so, which? (c) Unsuitable, and if so, why?

1.9
Discuss any present applications of, and future prospects for, vapor-cycle engines using fluids other than water.

Table P1.1

Give the maximum known value of	Steam turbine in generating station	Boiler-feed pump in generating station	Gas expander in gas-turbine engine	Compressor in gas-turbine engine	Axial compressor, any duty, per single casing	Centrifugal compressor for any duty, per casing	Water pump for pumped-storage system	Pelton water turbine (high-head impulse)	Francis water turbine (medium-head inflow)	Kaplan water turbine (low-head axial-flow)
Power (MW)										
Pressure (MPa)										
Temperature (deg C)										
Pressure ratio (max/min)										
Temperature ratio (max/min)										
Number of stages										
Low-pressure flow volume (m³/s)										
Mass flow (kg/s)										
Efficiency (percent, which?)										

Table P1.8

As engines for	To exploit the heat output in
automobiles	geothermal energy
highway trucks	solar energy
city buses	nuclear energy
long-distance interurban buses	seawater thermal gradient with depth
motor cycles	
snowmobiles	
lawn mowers	

2

Review of thermodynamics

Turbomachinery and gas turbines may be fully analyzed, to the degree reached in this book, by rather simple thermodynamic relations. All are based on the first law, expressed for application to a so-called "flow" system, and the second law, in the form of Gibbs' equation. From these, powerful and widely applicable tools, such as the energy-enthalpy relations for various components and the one-dimensional compressible-flow functions, can be quickly and easily derived.

The very simplicity of these tools can lead to their misuse. Serious errors can result when a relation derived for incompressible flow is used for a high-Mach-number gas stream, or when an isentropic-flow function is employed to relate upstream to downstream conditions in a frictional flow. Rigor is vital. We shall derive the significant relations from first principles so that the conditions under which they may individually be used may be fully understood. This review will also serve to establish a language and a system of symbols which will be used in the remainder of the book.

2.1 The first law of thermodynamics

The law of conservation of energy is a postulate. Common experience and careful experiments have shown that it generally holds, but it cannot be formally proved.

In words, it can be stated as follows.

The energy passing into a given mass of material in a given time will be equal to the energy passing out of the material plus the added energy stored within it.

The "mass of material" is any defined collection of molecules, whether solid, liquid, gas, plasma, or a mixture of more than one of these phases.

$$\delta q = dE + \delta w$$

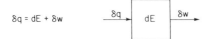

Figure 2.1 Energy flow and storage.

Thermodynamic textbooks often refer to this collection of molecules as a "system." There is no restriction on the form of the energy transfers (as work or heat transfers): they may follow some ideal patterns, or they may be far from ideal.

In symbols, the first law for a fixed mass of material is written as follows (illustrated in figure 2.1):

$$\delta q = dE + \delta w, \tag{2.1}$$

where $\delta q \equiv$ the heat transferred in the process (positive for heat transfer into the material),

$dE \equiv$ the increase of energy level of the material, termed the "internal" energy,

$\delta w \equiv$ the work transferred from the material to the surroundings (positive for energy transfer from the material).

The change in internal energy, dE, includes all forms of energy storage, such as kinetic, gravitational, elastic, electric, magnetic, and thermal which, because of its relative significance, is given the special symbol du.

Of these quantities, only E, the internal energy, is a property of the material. That is, its value depends only on the state of the material, and not on how that state was reached. The energy transfers, δq and δw, are not properties, and their values depend absolutely on the processes used to attain the final state. Later the apparently similar Gibbs' equation will be compared with this statement of the first law.

In many engineering examples there may be several identifiably different heat flows and work transfers, and the material may be composed of parts having different internal energies. The first law can be applied to the whole mass of material with the various quantities summed algebraically, or to individual components.

The first law derived for a flow process
Whereas the statement of the first law is useful for the analysis of, for instance, positive-displacement machinery, because such devices (e.g., piston compressors) confine an identifiable mass of material within boundaries before undergoing various processes, it is less applicable to turbomachinery, in which material crosses boundaries.

Review of Thermodynamics

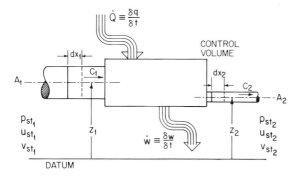

Figure 2.2 Energy flows for a steady stream of fluid passing through a control volume.

We can easily adapt the first law to the material-crossing-boundary case, usually known as a "flow" system, by modeling a device or a space in which a process takes place as a box or "control volume." We can then look at the fixed boundaries of the control volume and the moving boundaries around a defined mass of material (usually a fluid) as it moves into and out of the control volume in some increment of time dt.

For simplicity we shall specify steady conditions, so that we do not have to be concerned with changes of energy storage in the control volume. It is an easy, though somewhat complex, task to extend the treatment to transient conditions. In fact the characteristic response time of the flow in turbomachinery is usually so short that virtually all transients encountered in practice can be defined as slow. That is, they may be analyzed by examining a series of steady-state conditions.

Also for simplicity we shall consider only a single inlet flow and a single outlet flow, each using a duct of constant diameter, figure 2.2. This is not a necessary assumption, and using it results in no loss of generality. Likewise we shall consider only one (constant) heat transfer and one (constant) work transfer.

For the mass of material confined within the control volume and the inlet and outlet ducts, the first law as previously stated holds: $\delta q = dE + \delta w$. We applied this to quasi-stationary systems.

For the control volume alone, we have to take into account not only the various categories of energy that the material takes with it as it passes into and out of the control volume but also the effect of high velocity on the measurement of property values. We shall henceforth use the subscript st to denote property values measured with instruments stationary ("static") with respect to the fluids in systems involving relative motions. The energy categories are internal thermal energy, dmu_{st},

kinetic energy, $\dfrac{1}{2g_c}dmC^2/2g_c$,

potential energy, $dmzg/g_c$,

flow energy (the work of the fluid pushing an incremental volume of material into the control volume against a static pressure),

$$p_{st}A\,dx = dmv_{st}p_{st};$$

where $dm \equiv$ mass of material passing into and out of control volume in time dt,

$\quad C \equiv$ velocity of material,

$\quad g \equiv$ gravitational acceleration,

$\quad g_c \equiv$ constant in Newton's law, force = (mass × acceleration)/g_c, ($g_c = 1$ for SI and other "consistent" systems of units),

$\quad z \equiv$ height above datum,

$\quad v_{st} \equiv$ specific volume (static conditions), where v is the reciprocal of density, ρ,

$\quad p_{st} \equiv$ static pressure (pressure measured at the speed of the flowing material).

If we exclude other energy forms, for instance, from surface tension and electrostatic forces, we refer to the more restricted internal energy by the symbol u.

The energy flows into and out of the control volume can now be equated:

$$\delta q + dmu_{st1} + \frac{dmC_1^2}{2g_c} + \frac{dmz_1g}{g_c} + dmv_{st1}p_{st1}$$

$$= \delta w + dmu_{st2} + \frac{dmC_2^2}{2g_c} + \frac{dmz_2g}{g_c} + dmv_{st2}p_{st2}.$$

We divide by dt to give energy and mass flow rates ($\dot{m} \equiv dm/dt$, $\dot{Q} \equiv \delta q/dt$, and $\dot{W} \equiv \delta w/dt$) and rearrange to obtain

$$\frac{\dot{Q} - \dot{W}}{\dot{m}} = (p_{st2}v_{st2} + u_{st2}) - (p_{st1}v_{st1} + u_{st1}) + \frac{(C_2^2 - C_1^2)}{2g_c} + (z_2 - z_1)\frac{g}{g_c}.$$
$$(2.2)$$

This is known as the "steady-flow energy equation." It has very wide usefulness for the analysis of turbomachinery and of associated steady-flow processes such as combustion and heat transfer. Some examples of its use are given here. However, it is convenient to simplify the equation by defining new properties.

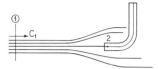

Figure 2.3 Isentropic flow into a pitot tube.

Enthalpy

The group of terms $(pv + u)$ are all properties of the material at a given plane or point. Properties, by definition, are functions only of the thermodynamic state of the material. This uniqueness means that new, unique, properties can be formed from any combination of other properties. It is obviously convenient to define $(pv + u)$ as one property. It is given the name "enthalpy" and the symbol h:

$$h \equiv pv + u. \tag{2.3}$$

While pressure and volume can be defined in terms of measurable length and force quantities, the internal thermal energy, and therefore the enthalpy, can be measured only from a defined datum state. Different tables often use different datum states. When using two sets of data to cover widely differing states, care must be taken that they have the same datum or base state.

The steady-flow energy equation incorporating this "new" property, enthalpy, is written

$$\frac{\dot{Q} - \dot{W}}{\dot{m}} = \Delta_1^2 \left(h_{st} + \frac{C^2}{2g_c} + \frac{zg}{g_c} \right), \tag{2.4}$$

the delta symbol implying that the difference between the quantity in parentheses is to be taken from outlet to inlet.

The enthalpy has the subscript st because, as mentioned previously, it is the value for so-called "static" or "stream" conditions. These are the conditions measured with instruments that are "static" or stationary with respect to the fluid. It is difficult to measure static properties— pressure and temperature—in high-speed flows. The stagnation or total properties are more easily measured, by means of upstream-facing "pitot" or "stagnation" or "total" probes (alternative names for the same instrument), shielded to reduce heat transfer by radiation and conduction, figure 2.3. The steady-flow energy equation can be written for conditions upstream and within a stagnation probe, assuming that the velocity of flow within the probe is vanishingly small, that is, $C_2 \rightarrow 0$:

$$\frac{\dot{Q} - \dot{W}}{\dot{m}} = 0 = \left(h_{st2} + 0 + \frac{z_2 g}{g_c} \right) - \left(h_{st1} + \frac{C_1^2}{2g_c} + \frac{z_1 g}{g_c} \right). \tag{2.5}$$

If the streamline is horizontal $z_2 = z_1$. Let us give the stagnation or total enthalpy the suffix 0, h_0:

$$h_{st2} \equiv h_{02} \equiv h_{st1} + \frac{C_1^2}{2g_c}, \tag{2.6}$$

or in general

$$h_0 \equiv h_{st} + \frac{C^2}{2g_c}. \tag{2.7}$$

Using this definition, we can write the steady-flow energy equation as

$$\frac{\dot{Q} - \dot{W}}{\dot{m}} = \Delta_1^2(h_0) + \Delta_1^2 \left(\frac{zg}{g_c} \right). \tag{2.8}$$

In air and gas turbomachinery changes of height have a negligible effect on the enthalpy change, and this general form of the steady-flow energy equation can be approximated to

$$\frac{\dot{Q}}{\dot{m}} \frac{\dot{W}}{\dot{m}} = \Delta_1^2(h_0) \equiv h_{02} - h_{01}. \tag{2.9}$$

The steady-flow energy equation (SFEE) has wide applicability simply because most of the components of plants incorporating turbomachinery are virtually thermally isolated, so that the heat transfer is negligible and the process undergone is adiabatic ($\dot{Q} = 0$), for instance, compressors, pumps, and turbines; or there is considerable heat transfer but no work transfer ($\dot{W} = 0$), for instance, boilers, combustion chambers, and heat exchangers; or there is neither heat transfer nor work transfer, as in nozzles, throttles, diffusers, and ducts. Table 2.1 summarizes how the restricted SFEE (equation 2.9) is used in components with gas, vapor, or with liquids in horizontal flow.

Use of the SFEE is especially convenient for steady-flow devices that have gases as their working fluids because over a range of commonly encountered pressures the enthalpy of gases at ranges well above their critical temperatures is a function of temperature only. Tabulated data, such as those in *Gas Tables* by Keenan and Kaye (1948) give enthalpies for air and other common gases as functions only of temperature for "low pressures." Comparison with the only high-pressure data available when the *Gas Tables* was published (shown in that work as table A) confirms that the tabulated values are fully valid for pressures up to

Table 2.1
Applicability of the steady-flow energy equation

	$\dot{W}$	$\dot{Q}$	$\Delta_1^2 h_0$
Nozzles	0	0	0
Valves	0	0	0
Throttles	0	0	0
Ducts, pipes, diffusers	0	0	0
Boilers	0		$\dot{Q}/\dot{m}$
Heat exchangers	0		$\dot{Q}/\dot{m}$
Combustion chambers	0		$\dot{Q}/\dot{m}$
Compressors, pumps, etc.		0	$\dot{W}/\dot{m}$
Turbines		0	$\dot{W}/\dot{m}$

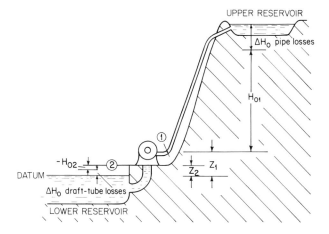

Figure 2.4 Hydraulic-turbine heads.

1·36 MPa (200 psia) and that "the characteristics of the isentropic as given by table 1 (of the *Gas Tables*) are suitable for precise work even to pressures in the hundreds of pounds per square inch" (quoted from the appendix).

Energy equation for liquid flow

For machinery working on liquids, such as hydraulic turbines and pumps, the SFEE is less useful because the enthalpy cannot easily be measured. Consider a hydraulic turbine working between two reservoirs, with the head measurements shown in figure 2.4. The form of the SFEE given in equation (2.2) is suitable for liquids, with the incorporation of the total pressure which can be found from equation (2.6) for incompressible (constant v) fluids only and for lossless deceleration of the flow to total conditions to be

$$p_0 = p_{st} + \frac{C^2}{v2g_c}.$$ (2.10)

(Note that this form cannot be used for gases except at low Mach numbers. For high Mach numbers, the incompressible-flow functions, developed next, must be used.)

The incompressible form for the SFEE is then

$$\frac{\dot{Q} - \dot{W}}{\dot{m}} = \Delta_1^2\left(p_0 v + u + \frac{zg}{g_c}\right).$$ (2.11)

It is usually more convenient to express the pressures as heads of liquid (from $p = \rho g H/g_c = (H/v)(g/g_c)$):

$$\frac{\dot{Q} - \dot{W}}{\dot{m}} = \Delta_1^2\left(\frac{H_0 g}{g_c} + u + \frac{zg}{g_c}\right).$$ (2.12)

In applying this form of the SFEE to the hydraulic-turbine example, we can assume that the heat transfer will be negligible. But we cannot assume that the change of internal energy will be negligible. In fact the change of temperature of the water will be extremely small—usually measured in hundredths of a degree Celsius—but this change of internal energy is a consequence of the frictional losses in the feedpipe and the turbine itself. In large hydraulic turbines, where the measurement of the mass flow and power output is difficult, efficiency measurements are often based on the precise measurement of these very small increases in water temperature.

Therefore the change of total head across a hydraulic turbine is not analogous to the change of total enthalpy across it or a gas turbine or some other turbomachine. The change of total enthalpy (together with the change of potential energy, if applicable) represents the actual work output of the turbine rotor in all cases, however poor the efficiency of the machine. In the absence of friction the change of total head gives the ideal work output per unit mass.

This concept may be better understood by reducing the case to the limiting one of a turbine so lossy that it gives no work at all. It would in this case be precisely similar to a throttle valve. There would be no change of enthalpy across it, but there would be a large change in total head.

In the general case the work output of the hydraulic turbine is then

Review of Thermodynamics

$$\frac{\dot{W}}{\dot{m}} = [(\Delta H_{01} + z_1) - (\Delta H_{02} + z_2)]\frac{g}{g_c} - [u_{st2} - u_{st1}], \qquad (2.13)$$

in words, actual specific work = [ideal specific work] − [losses]. The units of specific work are in SI units (and British-U.S. units) Joules per kilogram (and British thermal unit per pound-mass) or kilowatts per kilogram per second (and horsepower per pound-mass per second): J/kg (and Btu/lbm) or kW/kg-s (and hp/lbm-s).

2.2 Examples of the use of the SFEE

Gas-turbine expander
A gas-turbine expander is measured on test to be passing 84 kg/s of combustion products of known mean specific heat 1,130 J/kg-°K. What is the shaft power output if the mean inlet temperature is 1,250 C and the mean exhaust temperature is 550 C, both measured with stagnation probes?

For gases, the change in enthalpy between two temperatures can be calculated from the change in temperature multiplied by the mean specific heat at constant pressure between those two temperatures. Therefore

$$-\dot{W} = \dot{m}\,\Delta h_0 = \dot{m}\overline{C}_p\Delta T_0$$

$$\dot{W} = 84\frac{\text{kg}}{\text{s}} \times 1{,}130\frac{\text{J}}{\text{kg-°K}} \times (1{,}250 - 550)\,°\text{K}$$

$$= 66.44\;10^6\frac{\text{J}}{\text{s}} = 66.44\;\text{MW}$$

(°K = °C for temperature differences).

Air compressor
This and the following examples of the use of enthalpy tables are given British-U.S. units because some readers feel more comfortable using them. Most other examples, and the property data in appendixes A1 and A2, are given in SI units.

What will be the outlet total temperature of an air compressor taking in 25 lbm/s of atmospheric air at 72.3 F and absorbing 1,950 hp?

From the *Gas Tables*, table 1, the inlet total enthalpy is 127.14 Btu/lbm. The work input per pound is

$$1{,}950\;\text{hp} \times \frac{550\;\text{ft-lbf}}{\text{s-hp}} \times \frac{\text{Btu}}{778\;\text{ft-lbf}} \times \frac{\text{s}}{25\;\text{lbm}} = 55.14\frac{\text{Btu}}{\text{lbm}}.$$

Then the outlet total enthalpy is $(127.14 + 55.14) = 182.28$ Btu/lbm. This corresponds in table 1 of *Gas Tables*, to an outlet total temperature of 760.8 R $= 301.1$ F.

In these two examples of the application of the SFEE, we are unconcerned with the efficiency of the internal processes, except that we should know the destination of the bearing, seal, and windage (disk friction, etc.) losses. If the power is measured at the machine shaft, and if the machine is sufficiently well insulated so that the friction losses go to increasing the enthalpy of the working fluid, the use of the SFEE is correct. It is also correct if the power is that at the rotor blading and if the friction losses are ducted away in oil drains and ventilation ducts, external to the working fluid. When none of these situations applies, a correction must be made for the additional energy transfer across the control-volume boundaries.

Intercooler

Find the energy losses in the air passing into an air-compressor intercooler at 50 psia, 650 R, and leaving at 550 R. There is a total-pressure loss in the air of 10 psi. The air mass flow is 10 lbm/s.

From the *Gas Tables*,

$$h_{01} = \frac{155.50 \text{ Btu}}{\text{lbm}},$$

$$h_{02} = \frac{131.46 \text{ Btu}}{\text{lbm}}.$$

Therefore

$$\Delta_1^2 h_0 = \frac{-24.04 \text{ Btu}}{\text{lbm}}.$$

The energy losses, or simply the heat transfer $\dot{Q}$, is -240.4 Btu/s.

Note that the pressure losses do not enter the calculation. The pressure drop does not constitute a loss of energy. It produces a drop in thermodynamic availability—meaning that the air would be able to give less work in expansion to ambient pressure through a thermodynamically ideal engine than if there were no pressure drop. But available work is a second-law concern, and the SFEE is strictly a first-law, conservation-of-energy, expression. Pressure losses, even to the extent of throttling of the flow, do not produce energy losses.

2.3 The second law of thermodynamics and Gibbs' equation

The second law of thermodynamics is the second principal postulate upon which the science is based. It is known in many forms, which are corollaries of one another. The most popular, perhaps, is, "Heat cannot pass from a cooler to a warmer body without the expenditure of work."

Although this seems a straightforward statement, it requires an understanding of temperature. In the early days of modern science there was much confusion between temperature and heat (a confusion that persists today among many people). As we have not yet defined temperature, we shall start with what might seem a more abstruse statement, yet which in fact requires a smaller degree of intellectual faith: that interacting bodies can approach a state of equilibrium. We shall present a simplified and compressed treatment of the statistical development given in such texts as Reynolds's *Thermodynamics* (McGraw-Hill, 1968).

Matter consists of molecules and atoms, which are themselves composed of many elementary particles. The particles have various forms of energy, and continually exchange energy in quanta with interacting particles.

Consider an assembly of particles, each of which can exist in any of several energy states. Suppose that somehow (Maxwell's demon?) one could force them all into their mean energy states and then release them to interact. Rapidly the assembly would progress from a state of complete order (uniformity) to a state of maximum disorder. The second law states simply that this process from order to disorder, from disequilibrium to equilibrium, occurs. (As a simple analogy, consider an assembly of dice, each with faces of six different colors. At the start, all show the same color uppermost. Then the dice are thrown, one at a time, in a random sequence. The state of the assembly—the "macrostate"—would proceed from order to disorder, from maximum disequilibrium to maximum equilibrium.)

We could conceptually quantify the approach to equilibrium by calculating the number of ways the particles could be arranged among the "microstates" to give the "macrostate" of the assembly. Let us call this number Ω. When the particles are all at the identical (mean-energy) microstate, there is only one way the particles can be arranged, since they are all identical, and the number is 1. As equilibrium is approached, the number becomes extremely large, even for a very small collection of particles (figure 2.5).

If two collections or assemblies of particles, A and B, are allowed to interact, the new number of ways the combined macrostate C could be arranged is the product of the number of ways A and B could individually be arranged:

$$\Omega_C = \Omega_B \times \Omega_B. \tag{2.14}$$

It is more appropriate that this measure of disorder of the combined system should be the sum, rather than the product, of the "disorders" of the constituents, and that the numbers be smaller than they would otherwise be for realistic collections of molecules. We therefore define an

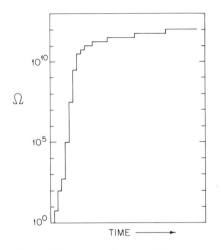

Figure 2.5 Approach to equilibrium.

"extensive" property (one that is proportional to the mass of the system) S, which we call entropy, for reasons which will be more obvious later, and which is proportional to the logarithm of Ω:

$$S \equiv K \ln \Omega; \tag{2.15}$$

$$\therefore K \ln \Omega_C = K \ln \Omega_A + K \ln \Omega_B, \tag{2.16}$$

$$\therefore S_C = S_A + S_B. \tag{2.17}$$

It is more rigorous to define entropy as the time average of $K \ln \Omega$. This overcomes to some extent the major objection of this model, which is that it allows the entropy of an isolated assembly to decrease occasionally even though the general direction of the approach to equilibrium is for the entropy to increase. But to decrease through even one interaction would contravene the second law, which can be restated:

The entropy of an isolated assembly must increase or in the limit remain constant.

Derivation of Gibbs' equation

We use the model just described to consider two bodies, or assemblies of particles, brought into thermal contact and allowed to reach equilibrium by exchange of heat, figure 2.6. They are isolated from the surroundings, and they are not allowed to perform work upon each other. We define (or choose) these assemblies to be "simple" substances, which by defini-

Review of Thermodynamics

Figure 2.6 A thermal interaction to equilibrium.

tion means that the state is fixed by any two independent, intensive properties. (Thus air at normal temperatures and pressures is a simple substance, while a ionized gas or a mixture of reacting substances is not. Gibbs' equation requires additional terms for increased degrees of freedom.)

The entropy of each assembly A and B and of the combined assembly C can therefore (because entropy is obviously a property) be stated as a function of any two other properties. It is convenient to use the internal energy, u, and the specific volume, v. We shall deal with quasi-stationary states and omit subscripts differentiating "static" from "total" because they will be identical:

$$S = S(u, v);$$ (2.18)

$$\therefore dS = \left(\frac{\partial S}{\partial u}\right)_v du + \left(\frac{\partial S}{\partial v}\right)_u dv.$$ (2.19)

Also

$$S_C = S_A + S_B,$$ (2.20)

$$dS_C = dS_A + dS_B$$ (2.21)

$$= \left(\frac{\partial S_A}{\partial u_A}\right)_v du_A + \left(\frac{\partial S_A}{\partial v_A}\right)_u dv_A + \left(\frac{\partial S_B}{\partial u_B}\right)_v du_B + \left(\frac{\partial S_B}{\partial v_B}\right)_u dv_B.$$ (2.22)

At equilibrium, $dS_C = 0$. Because no work interaction has been allowed, $dv_A = dv_B = 0$. A and B will have increased and decreased their internal energies by the amount of heat transferred δq, so that

$$du_A = -du_B = \delta q.$$

Equation (2.22) then simplifies to

$$\left(\frac{\partial s_A}{\partial u_A}\right)_v = \left(\frac{\partial s_B}{\partial u_B}\right)_v,$$ (2.23)

where we have used lower-case s for "specific entropy."

This, then, is the condition for thermal equilibrium. We can define this condition as one in which the temperatures of the two assemblies are equal. We could in fact define the quantities in equation (2.23) as thermodynamic temperature. The result, however, would be the inverse of what we commonly call temperature. Therefore we make the following definition:

$$T \equiv \frac{1}{(\partial s/\partial u)_v}. \tag{2.24}$$

We now incorporate this definition into equation (2.23) and apply it to a second interaction between the assemblies A and B in which we allow the common boundary to move by an increase in the volume of one and an equal decrease in the volume of the other, while maintaining thermal equilibrium, until pressure equilibrium has been attained:

$$dS_C = \frac{1}{T_A} du_A + \left(\frac{\partial S_A}{\partial v_A}\right)_u dv_A + \frac{1}{T_B} du_B + \left(\frac{\partial S_B}{\partial v_B}\right)_u dv_B. \tag{2.25}$$

At equilibrium, by arguments similar to those already made,

$$T_A = T_B,$$

$$du_A = -du_B,$$

$$dv_A = -dv_B,$$

$$dS_C = 0.$$

Then equation (2.25) reduces to

$$\left(\frac{\partial S_A}{\partial v_A}\right)_u = \left(\frac{\partial S_B}{\partial v_B}\right)_u. \tag{2.26}$$

This condition applies for equality of pressure, and of temperature, for the two assemblies. The quantities in equation (2.26) have the dimensions of pressure divided by temperature and are given that definition:

$$\frac{p}{T} \equiv \left(\frac{\partial S}{\partial v}\right)_u. \tag{2.27}$$

Incorporating this definition, equation (2.25) becomes the Gibbs' equation for a simple substance in the absence of energy storage due to motion, gravity, electricity, magnetism, and capillarity:

$$Tds = du + pdv. \tag{2.28}$$

We shall refer to these restrictions as "simple circumstances."

Comparison of Gibbs' equation with the first law

Gibbs' equation has some similarities with the first law written for the same circumstances as equation (2.28), and is often confused for it:

Gibbs' equation $\equiv Tds = du + pdv$.

First law $\equiv \delta q = du + \delta w$.

Gibbs' equation applies to any change, ideal or not; its only restriction is that it must apply to a simple substance. It is composed entirely of properties or changes in property values.

The first-law equation applies to a defined collection of material but is otherwise generally applicable. It includes only one property, du. The quantities δq and δw are dependent on the type of process undergone.

If one considers a thermodynamically reversible process in simple circumstances, one can write for δw the work of a reversible process, equal to pdv. Then, by comparison with Gibbs' equation, δq for a reversible process is Tds, and the equations are identical. This discussion underscores the essential difference in application of the first and second laws. The first law makes no condition about the type of process undergone, only about the end states. The second law is concerned about the process, and allows the end state to be predicted, given the initial state, or vice versa.

We examine process descriptions in the following derivations from Gibbs' equation.

2.4 The perfect-gas model

The definition of a perfect gas includes the equation of state:

$$pv = RT, \tag{2.29}$$

where R is the "gas constant," and the requirement that the specific heats at constant volume and constant pressure, Cv and Cp, be constant.
The perfect-gas model also includes the following relationships:

$$Cp = Cv + R, \tag{2.30}$$

$$dh = CpdT, \tag{2.31}$$

$$du = CvdT. \tag{2.32}$$

It may be easily shown (see, for instance, J. H. Keenan's *Thermodynamics*, The MIT Press, 1970, pp. 96–105) that the following relations may be derived from equations (2.28) and (2.29) and from the relation for the velocity, a, of propagation of a small pressure wave (a sound wave) through a fluid.

$$a^2 = -g_c v^2 \left(\frac{\partial p}{\partial v}\right)_s ,$$ (2.33)

so that for a perfect gas

$$pdv + vdp = RdT,$$

$$vdp = RdT + CvdT - Tds;$$

$$\therefore v\left(\frac{\partial p}{\partial v}\right)_s = Cp\left(\frac{\partial T}{\partial v}\right)_s = -\frac{pCp}{Cv} = -\frac{\gamma RT}{v};$$

$$\therefore a_{st} = \sqrt{g_c \gamma R T_{st}},$$ (2.34)

where $\gamma \equiv Cp/Cv$, and where we have noted that the appropriate temperature is T_{st}, for the actual speed of sound, a_{st}. Later we will introduce hypothetical values of a.

Perfect-gas T-s and h-s diagrams
The enthalpy-entropy diagram is the most used vehicle for illustrating and analyzing turbomachinery, because the high-efficiency streamline flow and the short fluid residence times reduce heat transfer to the level at which it can usually be neglected. Then the work transfers are given by the change of enthalpy, from the steady-flow energy equation (equation 2.9).

The ideal adiabatic process is isentropic, as can be seen from Gibbs' equation (2.28) and the first law written for an ideal, reversible, process, as described earlier.

Then $Q_{rev} = Tds = 0$ for a reversible adiabatic, by definition;

$$\therefore ds = 0.$$

In a perfect gas enthalpy changes are proportional to temperature changes, so that a temperature-entropy diagram may be used interchangeably with an enthalpy-entropy (Mollier) diagram with only a scale change for the temperature axis.

Constant-pressure lines on the T-s diagram
The shape of the constant-pressure lines on a temperature-entropy diagram can be found from the perfect-gas law (equation 2.29):

$$pv = RT,$$

$$pdv + vdp = RdT,$$

and from Gibbs' equation (2.28):

$$Tds = du + pdv.$$

In a perfect gas $du = CvdT$ (and $dh = CpdT$).
Also

$$Cp = Cv + R,$$

$$Tds = CvdT + Rdt - vdp,$$

$$= CpdT - vdp,$$

$$\left(\frac{\partial T}{\partial s}\right)_p = \frac{T}{Cp}. \tag{2.35}$$

This result is that the slope of any constant-pressure line is a function only of temperature. (Cp for a perfect gas is constant. For a real gas it is a function only of temperature for all normal pressures.) The constant-pressure lines on a T-s diagram are therefore identical, have an increasing slope with temperature, and are displaced horizontally from one another (figure 2.7). They do not "diverge," as is often stated; the frequently seen T-s diagram (figure 2.8) in textbooks in wrong and misleading.

The apparent virtue of assuming diverging rather than parallel constant-pressure lines is that the temperature drop in a high-temperature expansion process between two pressures can be larger than the temperature (and therefore the enthalpy) rise in the compression process between the same pressures.

That this inexactitude is not necessary can be seen from the following.

Isentropic work for a small compression or expansion process
We wish to find the temperature rise for an isentropic process over a small pressure rise. We use the same relations as before:

$$Tds = CpdT - vdp,$$

$$\left(\frac{\partial T}{\partial p}\right)_s = \frac{v}{Cp} = \frac{RT}{pCp}. \tag{2.36}$$

The work done in steady-flow compression or expansion process is dh, which for a perfect gas is $CpdT$. The change in pressure δp can be expressed as a normalized pressure change $\delta p/p$.
Then

$$\left(\frac{Cp\partial T}{\partial p/p}\right)_s = \left(\frac{\gamma - 1}{\gamma}\right)CpT = RT. \tag{2.37}$$

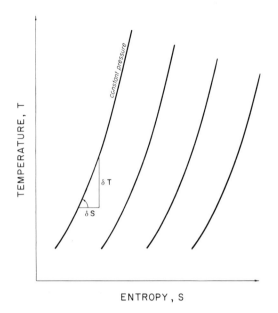

Figure 2.7 Slope of constant-pressure lines on the *T-s* diagram for a perfect gas.

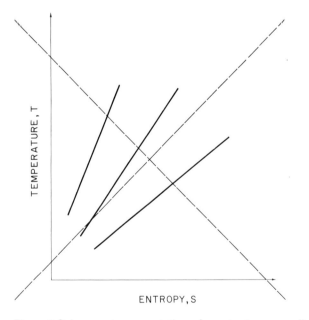

Figure 2.8 Incorrect representation of constant-pressure lines.

Review of Thermodynamics

The work required for an isentropic process between two pressure levels is therefore proportional to the absolute temperature. An isentropic turbine expanding from 1,500 K (2,700 R, 2,240 F) would produce five times the work required by an isentropic compressor working at 300 K (540 R, 80 F) between the same pressure levels.

2.5 One-dimensional compressible-flow functions for a perfect gas

We can apply equation (2.7), relating total and static conditions at a point or along an adiabatic stream tube in a moving fluid, to a perfect gas:

$$h_0 = h_{st} + \frac{C^2}{2g_c};$$

$$\therefore Cp(T_0 - T_{st}) = \frac{C^2}{2g_c},$$

$$\frac{T_0}{T_{st}} = 1 + \frac{C^2}{2g_c Cp T_{st}} = 1 + \frac{C^2}{2(Cp/R)g_c R T_{st}};$$

$$\therefore \frac{T_0}{T_{st}} = 1 + \frac{\gamma - 1}{2}\frac{C^2}{g_c \gamma R T_{st}} = 1 + \frac{\gamma - 1}{2}\frac{C^2}{a^2}, \tag{2.38}$$

$$\frac{T_0}{T_{st}} = 1 + \frac{\gamma - 1}{2}M^2, \tag{2.39}$$

where $M \equiv$ Mach number $\equiv C/a$. This is the fundamental equation of one-dimensional compressible flow. It is based on the first law only, and on adiabatic (no-heat-transfer) not isentropic (loss-less) flow. Extensions of this equation, however, to the total-to-static pressure ratio and density ratio do rely on the following isentropic relationships for a perfect gas.

Gibbs' equation for a perfect gas in simple circumstances (equation 2.16)
$T ds = du + p dv.$

Definition of enthalpy
$h \equiv u + pv,$
$dh = du + p dv + v dp.$

Therefore for a perfect gas
$dh = Cp dT = T ds + v dp.$

For an isentropic process
$ds = 0$ and $Cp dT = v dp.$

Divide both sides by the equation of state for a perfect gas (equation 2.29):

$$\frac{Cp}{R} \frac{dT}{T} = \frac{dp}{p};$$

$$\therefore \frac{Cp}{R} \ln \frac{T_2}{T_1} = \ln \frac{p_2}{p_1};$$

$$\therefore \frac{p_2}{p_1} = \left(\frac{T_2}{T_1}\right)^{Cp/R} = \left(\frac{T_2}{T_1}\right)^{\gamma/(\gamma-1)} \tag{2.40}$$

In this we used the definition of $\gamma \equiv Cp/Cv$ and equation (2.30) to derive the useful relation $Cp/R = \gamma/(\gamma - 1)$.

These relations apply to p_0/p_{st} and T_0/T_{st}, because, by definition, the total and static conditions at a point are connected by an isentrope. Therefore one-dimensional isentropic compressible-flow functions are

$$\frac{p_0}{p_{st}} = \left(\frac{T_0}{T_{st}}\right)^{\gamma/(\gamma-1)} = \left(1 + \frac{\gamma-1}{2} M^2\right)^{\gamma/(\gamma-1)}. \tag{2.41}$$

And from the equation of state

$$p_0 v_0 / p_{st} v_{st} = RT_0 / RT_{st} = (p_0/p_{st})^{(\gamma-1)/\gamma};$$

$$\therefore \frac{p_0}{p_{st}} = \left(\frac{v_0}{v_{st}}\right)^{-\gamma},$$

that is,

$$p_{st} v_{st}{}^\gamma = p_0 v_0^\gamma,$$

$$\frac{v_0}{v_{st}} = \left(1 + \frac{\gamma-1}{2} M^2\right)^{-1/(\gamma-1)} = \frac{p_{st}}{p_0}. \tag{2.42}$$

Continuity

The Mach number $M \equiv C/a$, can be introduced into the continuity equation $\dot{m} = \rho_{st} AC$:

$$\frac{\dot{m}}{A} = \rho_{st} aM. \tag{2.43}$$

For a perfect gas, from equations (2.29) and (2.35)

$$a = \sqrt{g_c \gamma R T_{st}},$$

$$p = \rho RT.$$

(By definition, $\rho \equiv 1/v$.) Therefore

$$\frac{\dot{m}}{A} = \frac{p_{st}}{RT_{st}} \sqrt{g_c \gamma RT_{st}} M,$$

$$\frac{\dot{m}\sqrt{T_{st}}}{Ap_{st}} = \sqrt{\frac{g_c \gamma}{R}} M. \tag{2.44}$$

We usually know the total, rather than the static, conditions, and it is best to convert this relation into a more useful form with equations (2.39) and (2.41):

$$\frac{\dot{m}\sqrt{T_0}}{Ap_0} = \sqrt{\frac{g_c \gamma}{R}} M \left(1 + \frac{\gamma - 1}{2} M^2\right)^{-[(\gamma+1)/2(\gamma-1)]}. \tag{2.45}$$

The cumbersome function of Mach numbers and γ in the parentheses is conveniently tabulated as a function of a hypothetical area ratio, A^*/A, in the following way. We introduce the concept (purely as a mathematical convenience) that at every point in a flowing stream the fluid should have a third set of properties. The first set are the actual static properties. The second set are hypothetical properties the fluid would have if it were brought isentropically to rest. The third set of properties, also hypothetical, are those the fluid would have if it were isentropically accelerated to sonic velocity. This hypothetical sonic velocity is known as a^* or C^* (different from both a_0 and a_{st} because of different stream temperatures in these conditions) and the other properties are p^*, T^*, u^*, and ρ^*. The conditions can be thought of as occurring at various points along an isentropic flow in a duct, although in fact the properties occur simultaneously at a point (figure 2.9).

M^* is anomalously defined:

$$M^* \equiv \left(\frac{C}{a^*}\right).$$

(The Mach number at the starred * plane is 1 by definition.)

The ratio A/A^* can be derived in terms of M and γ by starting with the continuity equation

$$\dot{m} = \rho_{st} C A = \rho^* C^* A^*,$$

$$\frac{A}{A^*} = \frac{\rho^* C^*}{\rho_{st} C} = \left(\frac{\rho^*}{\rho_0}\right)\left(\frac{\rho_0}{\rho_{st}}\right)\left(\frac{a^*}{a_0}\right)\left(\frac{a_0}{a_{st}}\right)\left(\frac{a_{st}}{C}\right)$$

$$= \frac{[1 + (M^2(\gamma - 1)/2)]^{1/(\gamma-1)}}{[1 + ((\gamma - 1)/2)]^{1/(\gamma-1)}} \times \frac{[1 + (M^2(\gamma - 1)/2)]^{1/2}}{[1 + ((\gamma - 1)/2)]^{1/2}} \times \frac{1}{M},$$

$$= \frac{1}{M} \frac{(1 + M^2(\gamma - 1)/2)^{(\gamma+1)/2(\gamma-1)}}{((\gamma + 1)/2)^{(\gamma+1)/2(\gamma-1)}}. \tag{2.46}$$

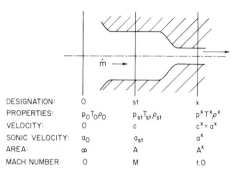

DESIGNATION:	0	st	x
PROPERTIES:	$p_0 T_0 \rho_0$	$p_{st} T_{st} \rho_{st}$	$p^x T^x \rho^x$
VELOCITY:	0	c	$c^x = a^x$
SONIC VELOCITY:	a_0	a_{st}	a^x
AREA:	∞	A	A^x
MACH NUMBER:	0	M	1.0

(a) Hypothetical isentropic duct

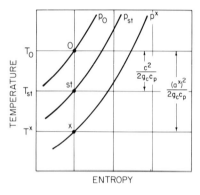

(b) Temperature – entropy diagram

Figure 2.9 Relation between hypothetical conditions at a point.

A/A^* is tabulated in *Gas Tables* as a function of M for various values of γ. The relation between M and M^* is found as follows:

$$M^* \equiv \left(\frac{C}{a^*}\right) = \left(\frac{C}{a_{st}}\right)\left(\frac{a_{st}}{a_0}\right)\left(\frac{a_0}{a^*}\right) = M\sqrt{\frac{T_{st}}{T_0}}\sqrt{\frac{T_0}{T^*}}$$

$$= \frac{[1 + ((\gamma - 1)/2)]^{1/2}}{[1 + (M^2(\gamma - 1)/2)]^{1/2}} M = M\left[\frac{(\gamma + 1)/2}{1 + (M^2(\gamma - 1)/2)}\right]^{1/2}. \qquad (2.47)$$

This is also tabulated in *Gas Tables* as a function of M for various values of γ.

The mass-flow function can then be expressed as a function of A/A^*, γ, and the molecular weight MW:

$$\frac{\dot{m}\sqrt{T_0}}{A p_0} = \frac{\sqrt{MW}\sqrt{g_c \gamma/\bar{R}}}{(A/A^*)((\gamma + 1)/2)^{(\gamma + 1)/2(\gamma - 1)}} \qquad (2.48)$$

Table 2.2
Mass-flow function

γ	$\left(\dfrac{\gamma+1}{2}\right)^{(\gamma+1)/2(\gamma-1)}\left(\dfrac{\bar{R}}{g_c\gamma}\right)^{1/2}$	
	$\dfrac{\text{N}\cdot\text{s}}{\text{kg}\sqrt{^\circ\text{K}}}$	$\dfrac{\text{lbf}\cdot\text{s}}{\text{lbm}\sqrt{^\circ\text{R}}}$
1.1	145.099	11.0297
1.2	140.586	10.6866
1.3	136.638	10.3865
1.4	133.155	10.1217
1.5	130.046	9.8854
1.6	127.265	9.6740

where $\bar{R}$ is the universal gas constant, and

$$\bar{R} = 8{,}313.219 \text{ J/kgmole-}^\circ\text{K},$$

$$= 1.98587 \text{ Btu/lbmole-}^\circ\text{R},$$

$$= 1{,}545.32 \text{ ft-lbf/lbmole-}^\circ\text{R};$$

$$MW \text{ for air} = 28.970;$$

$$R \text{ for air } (\bar{R}/MW) = 0.068549 \text{ Btu/lbm-}^\circ\text{R},$$

$$= 286.96 \text{ J/kg-}^\circ\text{K},$$

$$= 53.32 \text{ ft-lbf/lbm-}^\circ\text{R}.$$

The function of gamma is tabulated in table 2.2.

For air (MW 28.970) at low temperature, $\gamma = 1.4$, and the mass-flow function becomes

$$\frac{\dot{m}\sqrt{T_0}}{Ap_0} = \frac{0.040421}{A/A^*}\frac{\text{kg}\sqrt{^\circ\text{K}}}{s\text{N}}$$

$$= \frac{0.5318}{A/A^*}\frac{\text{lbm}\sqrt{^\circ\text{R}}}{s\text{ lbf}}. \tag{2.49}$$

The compressible-flow functions for air at low temperatures ($\gamma = 1.4$) are shown in figure 2.10.

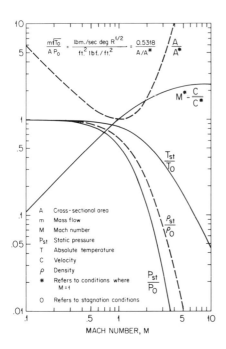

Figure 2.10 Isentropic compressible-flow functions for $\gamma = 1.4$. From an MIT Gas-Turbine-Laboratory curve.

In the figure:

$$\frac{m\sqrt{T_0}}{A\,p_0} = \frac{\text{lbm./sec deg R}^{1/2}}{\text{ft.}^2\,\text{lbf./ft.}^2} = \frac{0.5318}{A/A^*}\left/\frac{A}{A^*}\right.$$

$M^* - \dfrac{C}{C^*}$

$\dfrac{T_{st}}{T_0}$

$\dfrac{P_{st}}{P_0}$

A   Cross-sectional area
m   Mass flow
M   Mach number
P_{st} Static pressure
T   Absolute temperature
C   Velocity
ρ Density
*   Refers to conditions where $M = 1$
O   Refers to stagnation conditions

MACH NUMBER, M

Table 2.3

Steps	Method
1	$\begin{array}{c} C \\ \downarrow \\ M^* \end{array}$ $M^* = \dfrac{C}{a^*} = \dfrac{C}{a_0}\dfrac{a_0}{a^*} = \dfrac{C}{\sqrt{g_c\gamma RT_0}}\left[1 + \dfrac{\gamma - 1}{2}\right]^{1/2} = C\sqrt{\dfrac{(\gamma + 1)}{2g_c\gamma RT_0}}$
2	$\begin{array}{c} \uparrow \\ M \end{array}$ equation (2.47), or *Gas Tables* (tables 30–35)
3	$\begin{array}{c} \uparrow \\ A/A^* \end{array}$ equation (2.46), or *Gas Tables*
4	$\begin{array}{c} \uparrow \\ m\sqrt{T_0}/Ap_0 \end{array}$ equation (2.48), or table 2.2
5	$\begin{array}{c} \uparrow \\ \dfrac{m}{A} \end{array}$ use T_0 and p_0

$$\pi r^2 = \pi(12)^2 + 73.61,$$

$$r = 12.94'' \quad \text{blade height} = 0.94 \text{ in.}$$

Mach number. From table 2.3 for dry-saturated steam at 350 lbf/in^2

$$\frac{p_{st}}{p_0'} \approx 0.547;$$

$$\therefore p_0' = \frac{106}{0.547} = \frac{193.8 \text{ lbf}}{\text{in}^2}.$$

Second guess $p_{st}/p_0' = 0.5456$ for 194 lbf/in^2.
From the *h-s* chart,

$$h_0' = \frac{1{,}244 \text{ Btu}}{\text{lbm}},$$

$$h_0' - h_{st2} = \frac{54.77 \text{ Btu}}{\text{lbm}},$$

$$a_{st} = \frac{1{,}655.78 \text{ ft}}{\text{s}},$$

$$M \equiv \frac{C}{a_{st}} = 1.45.$$

2.7 Turbomachine-efficiency definitions

The overall energy efficiencies of turbomachines compare the actual work transfer with that which would occur in an ideal process. In an energy-using machine such as a pump or compressor, the work transfer in the ideal process is in the numerator. In an energy-producing machine, the ideal-process work transfer is in the denominator.

Efficiencies are seldom defined with sufficient precision. The statement that "we can deliver a compressor with an efficiency of 95 percent" is so undefined as to be almost useless for analysis. In fact competition in machine efficiencies is marked by charlatanism principally because of intentional or unintentional failure to define which of many efficiencies is being used. To take an extreme example, an intercooled compressor may have at the same time, and at the same operating point, an isothermal efficiency of 75 percent and an isentropic efficiency of over 100 percent. Or a turbine may simultaneously have isentropic total-to-static efficiencies of 90 percent and 80 percent, depending on what assumption or specification is made about the outlet plane.

General definition
The efficiency of a compressor or pump is as follows:

$$\eta \equiv \frac{\text{work transfer in ideal process}}{\text{actual work transfer}} \qquad (2.50)$$

(where the ideal process is) from inlet total pressure p_{01} and total temperature T_{01} to a defined outlet total pressure p_{02}

The efficiency of an expander or turbine is the inverse of this ratio. The definition requires precision particularly in these four aspects.

1. The ideal process must be identified. (The principal choices are isentropic, polytropic, and isothermal.)

2. The inlet and, more significantly, the outlet plane must be identified.

3. It should be stated whether the defined outlet total pressure is the actual total pressure or the "useful" total pressure.

4. The actual work transfer may be just that in the machine blading or may also include disk and seal friction and bearing losses; which of these is to be used should be specified.

Even in scientific papers it is common to identify only the ideal process. Let us discuss the implications of each of these four aspects.

Ideal processes
The two principal ideal processes used are the isentropic for adiabatic processes and the isothermal for gas compression when intercooling is (or can be) employed. In addition a variation of the isentropic efficiency is to take the limiting value at unity pressure ratio. This is termed the "small-stage" or "polytropic" efficiency, for reasons that will become obvious later.

In machines using liquids, such as pumps and hydraulic turbines, the ideal process is isentropic. It will also then be isothermal. The ideal work transfer in these various cases is found as follows.

Isentropic. The SFEE for gas turbomachinery (equation 2.9) gives $(\dot{Q} - \dot{W})/\dot{m} = \Delta h_0$ and $\dot{Q} = 0$ for an isentropic process;

$$\therefore \left(\frac{\dot{W}}{\dot{m}}\right)_s = -\Delta h_{0s} \equiv -(h'_{02s} - h_{01}), \qquad (2.51)$$

$h'_{02s} \equiv$ enthalpy after isentropic process from $h_{01}, p_{01},$ to p'_{02}, the defined outlet pressure. This is the general expression.

In the special case of a perfect gas

$$\left(\frac{\dot{W}}{\dot{m}}\right)_s = -(h'_{02s} - h_{01}) = -Cp(T'_{02s} - T_{01}) = -CpT_{01}\left(\frac{T'_{02s}}{T_{01}} - 1\right)$$

$$= -CpT_{01}\left[\left(\frac{p'_{02}}{p_{01}}\right)^{R/Cp} - 1\right] = -CpT_{01}\left[\left(\frac{p'_{02}}{p_{01}}\right)^{(\gamma-1)/\gamma} - 1\right]. \qquad (2.52)$$

In the special case of a liquid

$$\left(\frac{\dot{W}}{\dot{m}}\right)_s = -\Delta h_0 = -(\Delta u + \Delta(pv)). \tag{2.53}$$

If the liquid can be treated as incompressible, which it can for most pressure ranges, the specific volume, v, is constant. For an isentropic process in an incompressible liquid, $\Delta u = 0$,

$$\left(\frac{\dot{W}}{\dot{m}}\right)_s = -v\Delta p = -v(p_{02} - p_{01}). \tag{2.54}$$

For very-high-pressure-ratio pumps where the liquid is compressible, the general expression, equation (2.51), must be used.

Isothermal. Again, starting from the SFEE,

$$\frac{\dot{W}}{\dot{m}} = \frac{\dot{Q}}{\dot{m}} - \Delta h_0.$$

In a general reversible process

$$\frac{\dot{Q}}{\dot{m}} = T\Delta s.$$

Therefore the general expression for the work transfer in a reversible isothermal process in any fluid is

$$\left(\frac{\dot{W}_{rev}}{\dot{m}}\right)_T = T_0 \Delta s_0 - \Delta h_0$$

$$= -\int v_0 dp_0 \qquad \text{for a simple substance.} \tag{2.55}$$

For the particular case of an ideal gas, $\Delta h_0 = 0$ for $T_0 =$ constant. And from Gibbs' equation (2.28)

$$T\Delta s = \int du + \int p\,dv = \int dh - \int v\,dp.$$

Again, for $T =$ constant in a perfect gas, $dh = 0$,

$$\left(\frac{\dot{W}_{rev}}{\dot{m}}\right)_T = -\int_{01}^{02'} v\,dp\frac{RT}{pv} = -RT\int_{01}^{02'} \frac{dp}{p} = -RT\ln\left(\frac{p_{02'}}{p_{01}}\right). \tag{2.56}$$

Actual work transfer

To calculate an energy efficiency, the work transfer in the ideal process is expressed as a ratio with the actual work transfer in the process or machine.

The actual work transfer in the process is given again by the SFEE:

$$\frac{\dot{W}}{\dot{m}} = \frac{\dot{Q}}{\dot{m}} - \Delta h_0. \tag{2.57}$$

Turbomachines are generally adiabatic, in which case $\dot{Q}/\dot{m} = 0$. In some cases, however, heat transfer is significant. The two principal examples are intercooled compressors and cooled turbine expanders. Another case, a rare one, is the multistage turbine with reheat combustion or heat exchange between one or more stages. In these cases the heat transfer, or its equivalent in the case of combustion of fuel, must be included.

Equation (2.57) is the general and exact expression for the actual work transfer in the process. In other words, this will be the work transfer between the fluid and the machine's rotor(s), stator(s), and ducts. In many uses of energy efficiencies, this "process" work transfer is employed: it gives a measure of the quality of the aerodynamic and thermodynamic design.

The purchaser or user of a turbomachine is less concerned with aerodynamics than with actual energy cost, however. The purchaser wants to know how much work input is required or is produced at the shaft. Therefore "external" energy losses, those, external to the working fluid, must be added to or subtracted from the process energy.

External energy losses result from friction in bearings, labyrinths and possibly on the faces of disks ("windage"). We say "possibly" because, if the fluid causing the disk friction is bled from the working fluid and passes back into the working fluid, as is frequently the case, the energy losses appear as an increase in enthalpy (and entropy) in the working fluid and therefore are measured as an increase in the outlet enthalpy and a loss in work. The same could be true of working-fluid-lubricated bearings, for example, gas bearings, that exhaust into the machine discharge. And the same is occasionally true of labyrinth fluid.

Therefore it is not possible to identify the external energy losses in a general way. We shall simply term them $\sum$ (external losses) for identification in each individual case (see table 2.5).

2.8 Polytropic efficiency for perfect gas

The isentropic efficiency has a serious disadvantage if it is used as a measure of the quality of the aerodynamic design (or as a measure of the

Table 2.5
The "internal" efficiencies are defined by omitting the external losses

Type of efficiency	Compressor or pump	Expander or turbine
η_s isentropic, general fluid	$\left[\dfrac{h'_{02s} - h_{01}}{(h_{02} - h_{01}) + \sum(\text{ext. losses})/\dot{m}}\right]$	$\left[\dfrac{h_{01} - h_{02} - \sum(\text{ext. losses})/\dot{m}}{h_{01} - h'_{02s}}\right]$
η_s isentropic, perfect gas; equations (2.53), (2.58), or the following, may be used	$\left[\dfrac{[(p'_{02}/p_{01})^{(\gamma-1)/\gamma} - 1]}{[(p'_{02}/p_{01})^{(n-1)/n} - 1] + [\sum(\text{ext. losses})/\dot{m}CpT_{01}]}\right]$	$\left[\dfrac{[1 - (p'_{02}/p_{01})^{(n-1)/n}] - \sum(\text{ext. losses})/\dot{m}CpT_{01}}{[1 - (p'_{02}/p_{01})^{(\gamma-1)/\gamma}]}\right]$
η_s isentropic, liquid incompressible	$\left[\dfrac{+v(p'_{02} - p_{01}) + (z_2 - z_1)(g/g_c)}{\text{actual shaft work per unit mass}}\right] = \left[\dfrac{(H'_{02} - H_{01})(g/g_c)}{-\dot{W}/\dot{m}}\right]$ where $H_0 \equiv [(p_0/\rho)(g_c/g) + z]$	$\left[\dfrac{\text{actual shaft work per unit mass}}{v(p_{01} - p'_{02}) + (z_1 - z_2)(g/g_c)}\right]$
η_T isothermal, general gas	$\left[\dfrac{T\Delta s - \Delta h_0}{\Delta h_0 - \dot{Q}/\dot{m} + \sum(\text{ext. losses})/\dot{m}}\right]$	$\left[\dfrac{\dot{Q}/\dot{m} - \Delta h_0 - \sum(\text{ext. losses})/\dot{m}}{T\Delta s - \Delta h_0}\right]$
η_T isothermal, perfect gas	$\left[\dfrac{RT_0 \ln(p'_{02}/p_{01})}{Cp\Delta T_0 - \dot{Q}/\dot{m} + \sum(\text{ext. losses})/\dot{m}}\right]$	Not used for expanders

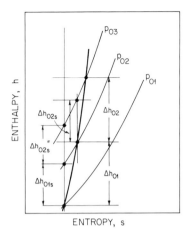

ENTROPY, s

Figure 2.12 Isentropic efficiency of a two-stage compressor.

losses) in an adiabatic machine. The value of the isentropic efficiency
is a function of pressure ratio as well as of losses, for the following reasons:

Suppose that we combine two compressors of equal isentropic efficiency
and equal enthalpy rise to make a compressor of higher pressure ratio,
(figure 2.12). The combined actual enthalpy rise will be the sum of the
individual enthalpy rises. But the sum of the isentropic enthalpy rises
will be larger than the isentropic enthalpy rise for the combined compressor, because, while

$$\Delta h_{01} = \Delta h_{02},$$

$$\eta_{s,\text{stage}} \equiv \frac{\Delta h_{01s}}{\Delta h_{01}} = \frac{\Delta h_{02s}}{\Delta h_{02}}.$$

Therefore

$$\Delta h_{01s} = \Delta h_{02s}.$$

But

$$\Delta h_{02s} > \Delta h_{02s}''$$

(see equation 2.35). Therefore

$$\eta_{s,\text{two stages}} = \frac{\Delta h_{01s} + \Delta h_{02s}''}{\Delta h_{01} + \Delta h_{02}}; \qquad (2.58)$$

∴ $\eta_{s,\text{two stages}} < \eta_{s,\text{one stage}}.$

Review of Thermodynamics

The losses are indicated by the increase of entropy in the working fluid per unit change of enthalpy: in other words, by the slope of the process line on the h-s chart. This slope is a function of the exponent $pv^n = $ constant which can be used to represent the process. (In fact this relation need only connect the end points of the process. The intermediate points will presumably be close, but the process does not have to follow the "polytropic" relation for the following to be true.)

To avoid the influence of pressure ratio on the isentropic efficiency, the limiting value of the isentropic efficiency for a given polytropic process can be used as the pressure ratio approaches unity. In steam turbines this efficiency is usually known as the "small-stage" efficiency. In gas-turbine expanders and compressors, it is known as the "polytropic" efficiency. For a perfect gas it leads to the following simple relation: the "internal" isentropic efficiency for a compressor working on a perfect gas (table 2.5) is

$$\eta_s = \frac{(p'_{02}/p_{01})^{(\gamma-1)/\gamma} - 1}{(p'_{02}/p_{01})^{(n-1)/n} - 1}.$$

Let the pressure ratio $(p'_{02}/p_{01}) \equiv 1 + \Delta$, and let $\Delta \to 0$ so that the pressure ratio approaches unity:

$$\eta_s = \frac{(1 + \Delta)^{(\gamma-1)/\gamma} - 1}{(1 + \Delta)^{(n-1)/n} - 1} = \frac{(1 + (\gamma - 1)/\gamma\Delta + \cdots + -1)}{(1 + (n - 1)/n\Delta + \cdots + -1)}.$$

As $\Delta \to 0$, Δ^2 and higher terms can be neglected.
Therefore for a compressor

$$\eta_{pc} \equiv \underset{PR \to 1.0}{\eta_s} = \left[\frac{(\gamma - 1)/\gamma}{(n - 1)/n}\right]. \qquad (2.59)$$

For an expander

$$\eta_{pe} \equiv \underset{PR \to 1.0}{\eta_s} = \left[\frac{(n - 1)/n}{(\gamma - 1)/\gamma}\right]. \qquad (2.60)$$

For perfect gases the polytropic efficiencies defined by equations (2.59) and (2.60) lead to the following useful relations:

$$\frac{T_{02}}{T_{01}} = \left(\frac{p'_{02}}{p_{01}}\right)^{((\gamma-1)/\gamma)/\eta_{pc}} = \left(\frac{p'_{02}}{p_{01}}\right)^{(R/\bar{C}p_c)/\eta_{pc}} \qquad (2.61)$$

for compressors;

$$\frac{T_{02}}{T_{01}} = \left(\frac{p'_{02}}{p_{01}}\right)^{((\gamma-1)/\gamma)\eta_{pe}} = \left(\frac{p'_{02}}{p_{01}}\right)^{(R/\bar{C}p_c)\eta_{pe}} \qquad (2.62)$$

for turbines.

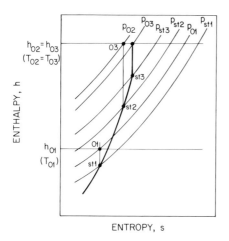

Figure 2.13 Dependence of ideal work transfer on definition of useful outlet pressure.

Definition of useful outlet pressure

These equations (2.61 and 2.62) enable the adiabatic work,

$$-\left(\frac{\dot{W}}{\dot{m}}\right)_{Q=0} = \Delta h_0 = \overline{Cp}\,\Delta T_{01} \equiv \overline{Cp}(T_{02} - T_{01}),$$

to be obtained directly from the pressure ratio. But which value of p'_{02} should be used? It is the value for which η_p is defined. A total-to-static efficiency at the outlet flange, for instance, defines the useful total pressure at this plane as the actual static pressure there.

Figure 2.13 shows four equivalent ideal polytropic adiabatic processes for different definitions of useful outlet pressure. All processes have the same T_{01} and T_{02} and therefore the same actual work. Their differing values of $(n-1)/n$ depend only on which of the four possibilities is used for p_{02}:

$$p'_{02} \equiv \begin{cases} p_{st2} & \text{rotor-outlet static pressure} \\ p_{02} & \text{rotor-outlet total pressure} \\ p_{st3} & \text{stator-outlet static pressure} \\ p_{03} & \text{stator-outlet total pressure} \end{cases} \begin{array}{l} \text{alternative} \\ \text{definitions of } p'_{02}. \end{array}$$

In a compressor the diffuser outlet and the casing flange may provide other locations for the outlet pressure to be defined. When p'_{02} is defined as a static pressure, the efficiency obtained is termed a "total-to-static" efficiency, η_{ts}. When p'_{02} is a total pressure, the efficiency obtained is "total to total," η_{tt}.

It is probable that turbomachine efficiencies are poorly defined and hence misused more because of confusion over the definition of p'_{02} than through any other factor. Here are some comments intended to clear up some of the confusion.

1. There is no "right" definition of efficiency. One may choose to use any definition one wants. It is of course essential that whichever is chosen is exactly identified.

2. There may, however, be a definition that is more appropriate in certain circumstances than others. The reason why it is more appropriate is likely to be that it saves some effort in making calculations and in accounting for energy flows and loss locations. For instance, the appropriate definition of p'_{02} for the efficiency of a compressor in a turbojet engine is the static pressure at the end of the compressor diffuser, the boundary between the compressor and the combustor, because the dynamic pressure at that point cannot be used by the combustor. There would be nothing thermodynamically wrong in choosing to use the total pressure, but doing so would have undesirable consequences. The combustor would seem to have a higher pressure drop than under the other definition, because the inlet pressure would now be listed as the total, instead of the static, pressure. The compressor would have a higher numerical efficiency under the new definition. The compressor designer would then have an incentive to cut back the diffuser, or maybe to eliminate it altogether, which would further increase the numerical value of the efficiency and further increase the allocated combustor pressure drop. At this point we would have gone beyond merely energy accounting. The actual losses would be increased if the diffuser were eliminated, so that defining the compressor efficiency in terms of a total outlet pressure could have wholly undesirable consequences.

Therefore the choice of the "appropriate" value of p'_{02} is one that must take incentives into account.

In the case of the turbine of a turbojet engines, the appropriate p'_{02} to use in the turbine-efficiency definition is the total pressure upstream of the propulsion nozzle that produces the jet. Obviously in this case the dynamic pressure in the stream is useful, and there is no need for an incentive for the turbine designer to diffuse the leaving flow.

3. Only the ideal work is affected by the choice of definition of useful outlet pressure, p'_{02}. The actual work is entirely unaffected. One has only one decision to make with regard to the actual work—whether or not to include the "external" friction losses (bearings, etc.).

4. Even if the choice of useful outlet pressure is a static pressure, it is defined as a total pressure, p'_{02}. The reason is that the ideal work is found from the steady-flow energy equation, equation (2.8), which requires the

stipulation of total enthalpies at the beginning and end of the process. We are defining a hypothetical state for a hypothetical (ideal) process. Even though it is confusing, there is nothing illogical about defining a hypothetical total pressure as being numerically equal to a known static pressure.

Shape of polytropic-process lines on *h-s* diagram
Just as the lines of constant pressure are often erroneously represented as straight on the *h-s* diagram for a perfect gas (figure 2.8), so, too, are lines of constant polytropic exponent *n*. We shall show that, in fact, these lines are curved to maintain a constant differential gradient with the constant-pressure lines.

The equation for a polytropic process is

$$pv^n = \text{constant.}$$

Differentiating and dividing give

$$\frac{1}{n}\frac{dp}{p} + \frac{dv}{v} = 0. \tag{2.63}$$

The equation of state for a perfect gas (equation 2.29) is

$$pv = RT.$$

Again, by differentiating and dividing, change this to

$$\frac{dp}{p} + \frac{dv}{v} = \frac{dT}{T}. \tag{2.64}$$

Gibbs' equation (equation 2.28) with equations (2.30 to 2.32) is

$$Tds = CpdT - vdp. \tag{2.65}$$

Combining equations (2.63), (2.64), and (2.65) gives

$$Tds = CpdT - R\frac{n}{n-1}dT,$$

$$\frac{ds}{dT} = \frac{Cp}{T}\left[1 - \frac{(\gamma-1)/\gamma}{(n-1)/n}\right]. \tag{2.66}$$

Using the definitions of compressor and expander polytropic efficiencies, equations (2.59) and (2.60), we can now derive the slope of the polytropic processes in the following simple forms:
for compression

Review of Thermodynamics

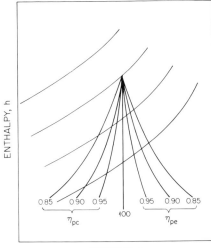

Figure 2.14 Polytropic processes.

$$\frac{dT}{ds} = \left[\frac{T}{Cp}\right]\left[\frac{1}{1 - \eta_{pc}}\right] \qquad (2.67)$$

and for expansion

$$\frac{dT}{ds} = \left[\frac{T}{Cp}\right]\left[\frac{1}{1 - 1/\eta_{pe}}\right]. \qquad (2.68)$$

These differ from the slope of the constant-pressure lines, given in equation (2.35), by a constant—the efficiency function,

$$\left(\frac{\partial T}{\partial s}\right)_p = \left[\frac{T}{Cp}\right].$$

The slope of the compression line is positive, and that of the expansion line is negative. For a polytropic efficiency of 100 percent, the slope becomes infinite: the line is vertical and is an isentropic (figure 2.14). At a temperature of absolute zero, both lines have zero slope.

The variation of curvature of the polytropic lines is of little significance for machines of small pressure ratio. However, for large pressure ratios a misrepresentation of the shape of these process lines can lead to the drawing of incorrect conclusions.

Polytropic processes for nonperfect gases
Because nonperfect gases have nonconstant values of the isentropic index connecting points on an isentropic process, the relationship just derived

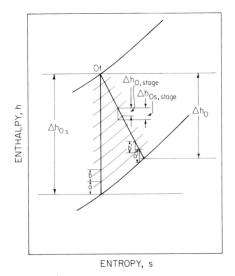

Figure 2.15 Polytropic process in a nonperfect gas.

for perfect gases cannot apply. Instead, a mean polytropic efficiency is based on a factor that represents the increased value of $(\partial h/(\partial p/p))_s$ with increased temperature. This factor is known as the reheat factor.

We combine Gibbs' equation (equation 2.28) with the definition of enthalpy (equation 2.3):

$$dh = du + pdv + vdp,$$

$$dh = Tds + vdp;$$

$$\therefore \left(\frac{\partial h}{\partial p/p}\right)_s = pv = zRT. \tag{2.69}$$

The general equation of state, $pv = zRT$, applies to nonperfect gases, with the so-called compressibility factor, z, being a slowly varying function of (reduced) pressure and temperature. R is the gas constant. Equation (2.69) shows that the isentropic enthalpy drop between lines of constant pressure increases with absolute temperature as with perfect gases, for which $z = 1$ and for which the appropriate relation was derived in equation (2.37).

Therefore, when a nonisentropic adiabatic expansion, such as that from 01 to 02 in figure 2.15, is divided into a number of actual or hypothetical small expansions, the isentropic enthalpy drop of each (for example, a', b') will be larger than the corresponding section (a, b) of the overall isentropic enthalpy drop Δh_{0s}.

That is, $a' > a$, $b' > b$, and $\sum \Delta h_{0s, \text{stage}} > \Delta h_{0s}$, where

$\Delta h_{0s, \text{stage}} \equiv$ isentropic enthalpy drop of small "stage,"

$\Delta h_{0, \text{stage}} \equiv$ actual enthalpy drop of small "stage."

However,

$\sum \Delta h_{0, \text{stage}} = \Delta h_0$.

Now the overall isentropic efficiency, η_s is

$$\eta_s \equiv \frac{\Delta h_0}{\Delta h_{0s}} = \left[\frac{\sum \Delta h_{0, \text{stage}}}{\sum \Delta h_{0s, \text{stage}}} \right] \left[\frac{\sum \Delta h_{0s, \text{stage}}}{\Delta h_{0s}} \right], \tag{2.70}$$

$\eta_s = \eta_p R_H$,

where

$$\eta_p \equiv \frac{\Delta h_{0, \text{stage}}}{\Delta h_{0s, \text{stage}}}, \quad \text{the "small-stage" isentropic efficiency,} \tag{2.71}$$

$$R_H \equiv \frac{\sum \Delta h_{0s, \text{stage}}}{\Delta h_{0s}}, \quad \text{the "reheat factor."} \tag{2.72}$$

The reheat factor is obviously a function of pressure ratio, being unity for very low pressure ratios, and of polytropic efficiency, being unity for all pressure ratios when the polytropic efficiency is unity.

Usefulness of polytropic efficiency
In removing the effect of pressure ratio, the polytropic efficiency enables machines of different pressure ratio to be compared with validity. A single-stage compressor may have an isentropic efficiency of 90 percent, total to total. If this stage is combined with others of similar performance to form a high-pressure-ratio compressor, the isentropic total-to-total efficiency of the whole compressor may be only 85 percent, making it appear to be of poor aerodynamic performance. If, however, the polytropic efficiency is used, both the single-stage and the multistage machines will have the same polytropic efficiency of about 91 percent.

The polytropic efficiency has a double advantage in analysis of Brayton power cycles, particularly when the design pressure ratio is being varied. First, as mentioned, a single value of polytropic efficiency can be entered for each compression and expansion process, instead of isentropic efficiencies that vary with pressure ratio.

Second, the equations using polytropic efficiencies are simpler than those with isentropic efficiencies, leading to a saving in computation time.

In practice the elimination of the reheat factor by the use of polytropic efficiencies does not eliminate all influence of pressure ratio from efficiency. A compressor or turbine of high pressure ratio will have higher Mach numbers, in general, and blading which is relatively longer in the low-pressure region and shorter in the high-pressure area, all factors leading to increased losses. The variation in maximum efficiency with pressure ratio is discussed in chapter 3.

Effect of leaving losses on efficiency values

The total-to-total efficiency (at machine, or diffuser, outlet) has been quoted here because of the complicating effects of the outlet kinetic energy, or "leaving losses." These leaving losses are relatively much more important for a single-stage than for a multistage machine, as will become obvious when velocity diagrams are considered. Therefore, if the more usually quoted total-to-static efficiencies had been used, some of the pressure-ratio effects of isentropic efficiency would have been compensated for by the pressure-ratio effects of leaving losses.

Actual work is unaffected by definition

From the steady-flow energy equation the actual (internal) work is the difference in total enthalpy from inlet to outlet. This work remains constant at $(h_{01} - h_{02})$ regardless of how the outlet pressure is defined.

Relations between polytropic and isentropic efficiencies for perfect gases

For compression

$$\eta_{sc} = \frac{(p'_{0,out}/p_{0,in})^{R/\bar{C}p_c} - 1}{(p'_{0,in}/p_{0,in})^{(R/\bar{C}p_c)/\eta_{pc}} - 1} \tag{2.73}$$

$$\eta_{pc} = \frac{\ln (p'_{0,out}/p_{0,in})^{R/\bar{C}p_c}}{\ln \left[\dfrac{(p'_{0,out}/p_{0,in})^{R/\bar{C}p_c} - 1}{\eta_{s,c}} + 1 \right]}. \tag{2.74}$$

For expansion

$$\eta_{s,e} = \frac{1 - (p'_{0,out}/p_{0,in})^{(R/\bar{C}p_e)\eta_{pe}}}{1 - (p'_{0,out}/p_{0,in})^{R/\bar{C}p_e}} \tag{2.75}$$

$$\eta_{p,e} = \frac{\ln [1 - \eta_{s,e}(1 - (p'_{0,out}/p_{0,in})^{R/\bar{C}p_e})]}{\ln (p'_{0,out}/p_{0,in})^{R/\bar{C}p_e}}. \tag{2.76}$$

As before, the outlet total pressure, $p'_{0,out}$, must be chosen to be consistent with the particular definition of efficiency being used.

References

Babcock & Wilcox Co. 1963. *Steam, Its Generation and Use.* New York, pp. 8–4.

Henry, R. E., M. A. Grolmes, and H. K. Fauske. 1971. *Pressure-pulse propagation in two-phase one- and two-component mixtures.* Report ANL 7792. Argonne National Laboratory, Argonne, Ill.

Hsu, Yih-Yun. 1972. *Review of critical flow rate, propagation of pressure pulse, and sonic velocity in two-phase media.* Technical note TN E-6814. NASA, Washington, D.C.

Keenan, Joseph H. 1970. *Thermodynamics.* The MIT Press, Cambridge, Mass.

Keenan, Joseph H., and Joseph Kaye. 1948. *Gas Tables.* Wiley, New York.

Reynolds, William C. 1968. *Thermodynamics.* McGraw-Hill, New York.

Problems

2.1
Does the total temperature of the working fluid rise or fall in passing through a gas-turbine expander? Why?

2.2
Does the total temperature of the water rise or fall in passing through a water turbine? Why?

2.3
By how much, and in which direction, does the total temperature of water change when it falls over a waterfall 50-m high? What is the mechanism for this change?

2.4
Sketch in your qualitative estimates of the variation of total and static enthalpy and pressure through the intercooled compressor shown in figure P2.4. Some end points are shown. (Four lines are required.)

2.5
For the subsonic axial-flow air turbine expander specified, calculate the total and static pressures and temperatures and the Mach number at the rotor-inlet plane (figure P2.5). Also find the rotor rotational speed and the nozzle-inlet blade height for constant-outer-diameter blading. Sketch the form of the complete turbine expansion on an enthalpy-entropy chart. The axial velocity will be constant at this design point.

Mass flow, $\dot{m} = 2$ kg/s.

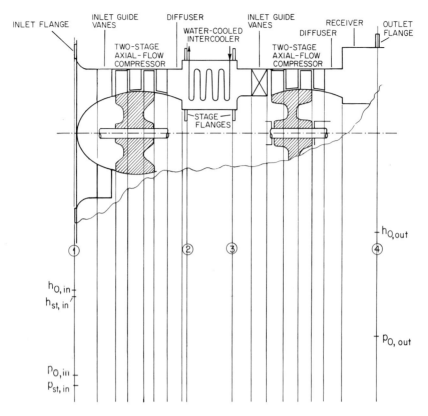

Figure P2.4

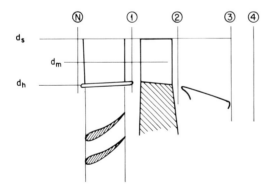

Figure P2.5

Review of Thermodynamics

Nozzle-inlet total temperature, $T_{0N} = 400$ C.

Nozzle-inlet total pressure, $p_{0N} = 3$ bars $(= 3.10^5$ N/m$^2)$ abs.

Mean diameter, $d_m = 0.25$ m.

Blade height at rotor entry, $l = 0.1d_m$ $(= 1/2\,(d_s - d_h))$.

Flow angle at nozzle exit, $\alpha_1 = 70°$ to axial direction.

Drop in total pressure in nozzle, $\Delta p_0 = 0.05$ bar.

Rotor peripheral speed at mean diameter, $u_m = 0.5$ (tangential component of nozzle outlet velocity, $C_{\theta 1}$).

2.6

Complete the table of efficiencies, table P2.6, for the turbine expander of problem 2.5, for inlet conditions 1,500 K and 8×10^5 N/m^2. The turbine total enthalpy drop is twice the mean blade speed squared $(\Delta h_0 = 2(u_m)^2)$ and the mean blade speed is 500 m/s. The flow velocity at rotor exit is axial and is 0.728 of the mean blade speed, and it is reduced by 50 percent in the diffuser. The rise in static pressure in the diffuser is 75 percent of the value it would have if the diffuser flow were isentropic. Also calculate the power, in kW, delivered by the blading. (Do not count the "external" losses—bearing friction and so forth—here or in the efficiencies.) The inlet conditions to be used for the efficiencies are the stagnation conditions at plane N.

2.7

Given that an axial-compressor stage is a row of rotor blades which act as diffusers, followed by a row of stator blades which also act as diffusers, explain (from first-law and second-law considerations) why a high Mach number is necessary if the compressor is to give a high pressure ratio.

2.8

Calculate the isentropic total-to-static internal efficiency of an intercooled air compressor (rather similar to that shown in problem 2.4) from inlet flange to outlet flange. Comment on your findings. Suppose that instead of the actual arrangement shown, two centrifugal stages are used, one upstream and one downstream of the intercooler. Each stage has a total-to-total flange-to-flange pressure ratio of $4:1$, and each has a total-to-total polytropic efficiency of 88 percent for the same planes (1 to 2 and 3 to 4). The mean velocity at the outlet flange of each stage is 30 m/s.

Table P2.6

	Outlet plane	
Efficiency	2	3
$\eta_{s,tt}$	0.95	
$\eta_{s,ts}$		
$\eta_{p,tt}$		
$\eta_{p,ts}$		

The intercooler has a total-pressure loss of 6 percent of the total pressure of the flow entering it, and it lowers the total temperature to 1.03 times the first-stage-inlet total temperature of 31.84 C, 305 K. Use a constant value of γ of 1.4. Draw a temperature-entropy diagram.

2.9
Using figure P2.9, calculate the rotor rotational speed, the static temperature and pressure at nozzle exit, and the (absolute) Mach number at that point, of a radial-inward-flow turbine expander of nozzle-outlet diameter (surface 1) of 250 mm. As an approximation, take this as the rotor-inlet diameter also. The nozzle-exit direction of the flow is 75 degrees from the radial direction, and the axial height of the blade passage is 10 percent of the rotor diameter. The turbine nozzles are supplied with 2 kg/s of air at 2 bars total pressure and 125 C total temperature. The rotor peripheral speed is 90 percent of the tangential velocity of the air at nozzle exit. The total-pressure losses of the flow through the nozzles are small enough to be neglected. Sketch the nozzle expansion on a T-s diagram.

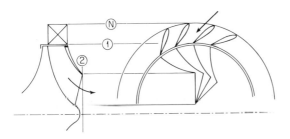

Figure P2.9

3

The thermodynamics of gas-turbine power cycles

3.1 Temperature-entropy diagrams

By definition, a gas turbine uses a gas as a working fluid. In the great majority of gas-turbine applications, the working fluid is air, or is a gas that is well removed from its liquefaction temperature in the cycle conditions chosen—for instance, hydrogen or helium. Under these conditions, it is a good approximation to treat the working fluid as a perfect gas, defined as a substance that obeys the law:

$$pv = RT,$$

where p is the pressure, v the specific volume, R is a constant, and T is the (absolute) temperature. It is convenient to perform initial analyses of cycles assuming that the working fluid is a perfect gas simply because the calculations are greatly simplified, and because the resulting ease of analysis makes it possible to gain a deeper insight into the variations that may be expected from changes in cycle conditions. Final calculations may then be made using real-gas properties in the knowledge that conditions will not be greatly changed. We shall give examples of calculations made using different assumptions for working-fluid properties.

Property diagrams are particularly useful for giving the conditions of, and relationships among, the end points of processes making up gas-turbine cycles. Perfect gases are simple substances, for which the state can be found from the value of any two properties. Therefore we could make cycle diagrams on charts with axes of p and v, or v and h, for instance.

More suitable choices for the axes of a diagram to represent the ideal gas-turbine cycle, which is known as the Brayton or Joule cycle in one form and the Ericsson cycle in another form, are T and s or h and s.

The fluid temperatures at compressor and turbine inlets are normally part of the cycle specifications. The ideal compression and expansion processes are isentropes in the Brayton cycle, and isothermals in the ideal Ericsson cycle (for compression which in practice could be approached by using many intercoolers and expansion by using many reheat combustors). Both are easily drawn on T-s diagrams. The thermodynamic or material limits for gas-turbine cycles are lines of constant temperature, again easily drawn. Atmospheric temperature is one limit, and the maximum gas temperature that can be tolerated by the expander is the other.

Atmospheric pressure is a limit for open cycles. Heat addition is carried out at a higher, constant, pressure in an ideal Brayton cycle. Constant-pressure lines have a distinct shape on a temperature-entropy diagram, a shape that helps us understand how gas-turbine cycles produce positive net power. The shape is shown in figure 2.7, and the reason why a turbine expander can give surplus power beyond that needed by the compressor is explained in section 2.5.

3.2 Actual processes

Compression and expansion processes in gas turbines are normally virtually adiabatic, or are adiabatic processes between intercoolers and reheat combustors. In real adiabatic processes the entropy must increase. The work required for compression between two pressure levels increases for an entropy-increasing process compared with that for an isentropic process, and the work obtained from expansion decreases.

Whether or not there is net power available in a real gas-turbine cycle depends on the efficiency of the nonisentropic processes (represented by the slope of the process line on the T-s or h-s diagram, as shown in figure 2.14) and on the ratio of the turbine-inlet temperature to the compressor-inlet temperature (figure 3.1).

In a real cycle there will also be pressure losses, which means that the compressor pressure ratio will be larger than the turbine expansion ratio. There may also be a leakage of mass out of the compressed-air side of the cycle, and there is usually an addition of mass with the fuel being burned. All these effects make the choice of parameters for a real gas-turbine cycle challenging and interesting.

3.3 Choice of the optimum cycle

It is easy to show that the thermal efficiency of an ideal Brayton cycle is a function only of pressure ratio. We shall not derive this result here, because it is somewhat trivial and can be misleading. One can still hear, and read in the technical literature, claims that the ability to design

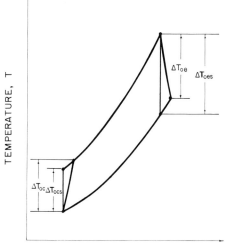

Figure 3.1 Gas-turbine cycle with nonisentropic compression and expansion.

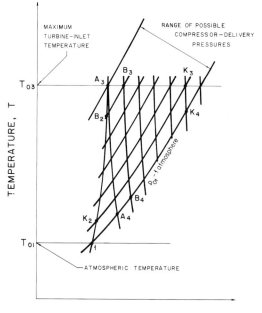

Figure 3.2 Constraints on gas-turbine cycle parameters.

higher-pressure-ratio compressors has opened the way to higher-efficiency gas turbines. This may be true for aircraft turbines but not for gas turbines where heat exchangers can be incorporated. In fact in a real gas-turbine cycle, the higher the turbine-inlet temperature, the greater the net output and the greater the cycle efficiency. Also, in real cycles, component efficiencies have a crucial effect on gas-turbine-engine viability. It will be shown that for every combination of cycle temperature ratio and component efficiencies, there exists an optimum pressure ratio for maximum thermodynamic cycle efficiency and another optimum pressure ratio for maximum specific power.

Consider the choices open to the designer of a gas-turbine cycle within the constraints shown in figure 3.2. For an open cycle the start of compression is at atmospheric temperature and pressure. A metallurgical limit fixes the expander-inlet temperature.[1] The end of expansion is also at atmospheric pressure. The best compressor technology available produces, it is supposed, the compression-process line shown. Another polytropic-process line represents the best available expansion process, which may start from different points along the T_{03} line.

Stated in these terms, the only freedom left to the designer is to choose the pressure ratio. A range of choices is shown. Let us look first at the extreme choices.

At pressure ratios close to unity, typified by the cycle 1, $K_2 K_3 K_4$, the compressor absorbs little work but the turbine expander also produces little work. Net output is vanishingly small.

The other extreme is the pressure ratio so high that compressor-delivery temperature is the turbine-inlet temperature. No heat can be added, and the flow is immediately expanded to atmospheric pressure—cycle 1 $A_3 A_4$. Turbine work is obviously less than compressor work, so that net output is negative. This cycle has no attractions.

In between these two extremes there are pressure ratios that produce net work. Therefore there is an optimum pressure ratio (figure 3.3) at which maximum net power is produced.

This same optimum pressure ratio will also apply if the designer uses a heat exchanger (as in figure 1.7), if, as a first approximation, we specify that the added heat-exchanger pressures losses will be zero. The heat-exchanger cycle is shown on a T-s diagram in figure 3.4.

At this point we would like to work an example to illustrate the existence of an optimum pressure ratio, but we have not yet developed the tools

---

1. In practice expander-inlet temperature may be influenced partly by the pressure ratio, in addition to the strong influence of the metallurgical limit. The reason is that, when an engine has a very low pressure ratio, which would be the case, for instance, if a high-effectiveness heat exchanger is used, a low blade speed may be used in the turbine expander, leading to low blade stresses and to the possibility of using a higher gas temperature for a given blade life.

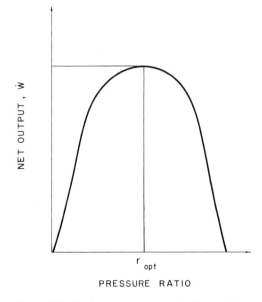

Figure 3.3 Optimum pressure ratio for maximum specific power.

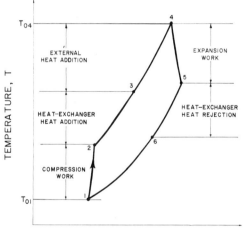

Figure 3.4 Energy transfers in a gas-turbine-engine cycle on the *T-s* diagram.

to do so easily. We resolve this minor dilemma by using, in a "plug-in" mode, a formula derived later in this chapter. Readers who prefer to wait for the formula to be developed may wish to pass over this example.

Example (calculation of pressure ratio for maximum power)

Calculate the pressure ratio which gives maximum specific power in a gas-turbine engine with the following specifications.

Compressor-inlet temperature, $T_{01} = 310$ K.

Expander-inlet temperature, $T_{04} = 1,300$ K.

Compressor polytropic efficiency, $\eta_{pc} = 0.85$.

Expander polytropic efficiency, $\eta_{pe} = 0.875$.

Total relative-pressure losses, $\sum(\Delta p_0/P_0) = 0.06$.

$R/(\overline{Cp_c}) = 0.284$.

$R/(\overline{Cp_e}) = 0.240$.

$\dot{m}_e\overline{Cp_e}/\dot{m}_c\overline{Cp_c} = 1.150$.

(Taking the last three parameters as constants is an approximation. In fact the mean specific heats would vary with pressure ratio, as would the rate of fuel addition. The effects of this approximation, however, are minor for present purposes.)

Solution We use equation (3.1), given and derived subsequently,

$$\dot{W}' \equiv \text{specific power} \equiv \frac{\dot{W}}{\dot{m}_c\overline{Cp_c}T_{01}} = \left[\frac{\dot{m}_e\overline{Cp_e}}{\dot{m}_c\overline{Cp_c}}\right]E_1 T' - C$$

with the following definitions:

$$E_1 \equiv 1 - \left[\left[1 - \sum\left(\frac{\Delta p_0}{p_0}\right)\right]r\right]^{-(R/\overline{Cp_e})\eta_{pe}},$$

$$T' \equiv \frac{T_{04}}{T_{01}},$$

$$C \equiv [r^{(R/\overline{Cp_c})/\eta_{pc}} - 1];$$

$$\therefore \dot{W}' = 1.15\{1 - [(1 - 0.06)r]^{-0.24 \times 0.875}\}\left[\frac{1,300}{310}\right] - [r^{0.284/0.85} - 1]$$

$$= 4.8226\{1 - (0.94r)^{-0.21}\} - r^{0.334} + 1.$$

At maximum specific power

$$\frac{d\dot{W}'}{dr} = 0 = -4.7603 \times (-0.21)r^{-1.21} - 0.334r^{-0.666},$$

from which $r_{opt} = 7.502$.

This optimum pressure ratio for maximum power output is unaffected by the presence or absence of a heat exchanger if we neglect, as an approximation, the effects of the additional heat-exchanger pressure drops.

The optimum pressure ratio for maximum efficiency is, however, different for simple and for heat-exchanger cycles.

3.4 Choice of optimum pressure ratio for peak efficiency

Suppose that the pressure ratio for maximum output has been selected in the way suggested in section 3.3. One can consider making small changes in pressure ratio without changing the net output, because it is at a turning point. Suppose first that we are concerned with a simple cycle. When the pressure ratio is reduced a small amount from the optimum for peak net power, the compressor-outlet temperature is reduced, and additional fuel must be supplied to make up the difference. With the net output staying constant or falling slightly, and the heat input to the cycle increasing, the efficiency falls rapidly. Conversely, if the pressure ratio is raised slightly, the compressor-outlet temperature is higher, and less fuel need be burned to bring the working fluid up to turbine-inlet temperature. The thermal efficiency is therefore increased by raising the pressure ratio above that which gives maximum output.

The maximum cycle thermal efficiency is reached at a pressure ratio only a little higher than that for maximum power. At higher pressure ratios the effect on the thermal efficiency of the fall in power output is greater than the effect of the reduced heat-addition required.

Now suppose that we are concerned with a heat-exchanger cycle. For simplicity in presenting the argument, we suppose that the heat exchanger has an effectiveness of unity, which means that the compressor-discharge air will be brought up to the temperature of the turbine discharge. In addition, as before, we assume that the heat-exchanger pressure losses are zero.

Now when the pressure ratio is raised above its maximum output level, the turbine-discharge temperature falls because of the larger expansion ratio. Therefore the temperature of the compressed air leaving the heat exchanger also falls, and more fuel must be burned to keep the turbine-inlet temperature at the desired value. When, on the other hand, the pressure ratio is dropped below that for maximum output, the expander-outlet temperature increases, and with it the temperature of the heat-exchanger compressed-air discharge to the combustor. Less fuel need be burned, and the thermal efficiency increases. The optimum pressure ratio for maximum thermal efficiency is therefore lower than the pressure ratio for maximum output. It may be considerably lower for a cycle with high-efficiency components.

All these effects can be seen in the charts of cycle thermal efficiency

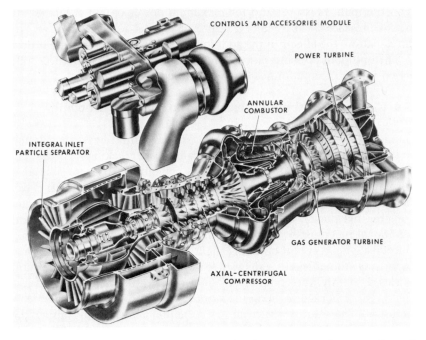

T700 shaft-power engine with axial-centrifugal compressor. Courtesy General Electric Company.

against specific power, with temperature ratio (turbine inlet to compressor inlet) and compressor pressure ratio as parameters, figures 3.5 through 3.11. Each plot is made for a set of values of component efficiencies and pressure losses.

3.5 Cycle designation

The types of cycles are indicated on the performance charts and state diagrams by symbols for the components, in the order in which they are encountered by the working fluid:

$C \equiv$ compressor,

$I \equiv$ intercooler,

$B \equiv$ heat addition from external source (e.g., burner),

$E \equiv$ expander, and

$X \equiv$ exhaust-gas-to-compressed-gas heat exchanger.

Although the working fluid passes twice through the heat exchanger, the symbol X is used in the designation only in the expander-exhaust

　　　　　　　　　　　　　　　Gas-Turbine Power Cycles

position. Thus a simple cycle is designated CBE. A heat-exchanger cycle is CBEX (not CXBEX). A cycle with an intercooled compressor and an exhaust heat exchanger is CICBEX, and so forth. The thermodynamic-cycle designation is unaffected by whether or not the components have more than one stage, or even more than one shaft. Thus the two-shaft multistage engine is designated simply as working on a CBE cycle.

3.6 Cycle calculations

The two measures of Brayton-cycle performance usually wanted in pre-liminary design are the specific power, in the form of the nondimensional parameter $(\dot{W}/\dot{m}\overline{Cp}_c T_{01})$, and the thermal efficiency η_{th}. The two most important cycle characteristics are the compressor pressure ratio, $r \equiv (p_{02}/p_{01})$, and the turbine-inlet-to-compressor-inlet temperature ratio $(T_{04}/T_{01}) \equiv T'$.

In a real, as distinct from an ideal, cycle there are many other variables that affect the specific power and the efficiency in addition to the pressure and temperature ratios. We shall account for eight: the compressor and expander polytropic efficiencies; the leakage; the total cycle pressure loss; the heat-exchanger effectiveness; and the mean specific heats for compression, expansion, and heat addition. Rather than combine all these variables into one large equation for each performance measure, we shall introduce some new variables that will simplify the task of dealing with changes in the variables.

The relations that can be used for both the simple, CBE, cycle and the heat-exchanger, CBEX, cycle are given first, and the derivations are made subsequently.

Specific power

$$\dot{W}' \equiv \frac{\dot{W}}{\dot{m}_c \overline{Cp}_c T_{01}} = \left(\frac{\dot{m}_e}{\dot{m}_c} \frac{\overline{Cp}_e}{\overline{Cp}_c}\right) E_1 T' - C. \tag{3.1}$$

Cycle thermal efficiency

$$\eta_{\text{th}} = \frac{[(\dot{m}_e/\dot{m}_c)(\overline{Cp}_e/\overline{Cp}_c)]E_1 T' - C}{[(\dot{m}_b/\dot{m}_c)(\overline{Cp}_b/\overline{Cp}_c)][T'(1 - \varepsilon_x(1 - E_1)) - (1 + C(1 - \varepsilon_x))]} \tag{3.2}$$

where $\dot{W} \equiv$ power output,

$\dot{m}_c \equiv$ mass flow through compressor,

$\dot{m}_e \equiv$ mass flow through expander (including effects of fuel addition and working-fluid leakage),

$\dot{m}_b \equiv$ mass flow through burner,

$\overline{Cp_c} \equiv$ mean specific heat during compression,

$\overline{Cp_e} \equiv$ mean specific heat during expansion,

$\overline{Cp_b} \equiv$ mean specific heat during heat addition,

$$E_1 \equiv (T_{04} - T_{05})/T_{04}$$
$$= 1 - \{[(1 - \sum(\Delta p_0/p_0))r]^{(R/\overline{Cp_e})\eta_{pe}}\}, \tag{3.3}$$

$\sum(\Delta p_0/p_0) \equiv$ total cycle relative pressure losses,

$$r \equiv (p_{02}/p_{01}),$$

$$R \equiv \text{gas constant},$$

$\eta_{pe} \equiv$ expander polytropic efficiency,

$$T' \equiv T_{04}/T_{01},$$

$$C \equiv (T_{02} - T_{01})/T_{01} = [r^{(R/\overline{Cp_c})/\eta_{pc}} - 1], \tag{3.4}$$

$\eta_{pc} \equiv$ compressor polytropic efficiency,

$\iota_x \equiv$ heat-exchanger effectiveness $=$
$$(T_{03} - T_{02})/(T_{05} - T_{02}). \tag{3.5}$$

For the simple CBE cycle the specific power is unchanged (equation 3.1). In the efficiency relation the heat-exchanger effectiveness, ε_x, vanishes, and the expression becomes

$$\eta_{th,CBE} = \frac{[(\dot{m}_e/\dot{m}_c)(\overline{Cp_e}/\overline{Cp_c})]E_1 T' \quad C}{[(\dot{m}_b/\dot{m}_c)(\overline{Cp_b}/\overline{Cp_c})][T' - C - 1]}. \tag{3.6}$$

In preliminary design fuel addition and working-fluid leakage are often ignored, and calculations are made with a single value of mean specific heat. Then equation (3.6) becomes

$$\eta_{th,CBE} = \frac{E_1 T' - C}{T' - C - 1}. \tag{3.6a}$$

Equations (3.1) and (3.2) can be derived as follows.

The steady-flow energy equation (equation 2.9), $(\dot{Q} - \dot{W})/\dot{m} = h_{02} - h_{01}$, can be used to find the work transfer in the (adiabatic) compressor and expander, and the heat transfer in the heat exchanger and burner:

Net power $\equiv \dot{W} \equiv$ expander power + compressor power

$$= \dot{m}_e(h_{04} - h_{05}) - \dot{m}_c(h_{02} - h_{01}) \tag{3.7}$$

$$= \dot{m}_e\overline{Cp_e}(T_{04} - T_{05}) - \dot{m}_c\overline{Cp_c}(T_{02} - T_{01}). \tag{3.8}$$

Specific power $\equiv \dfrac{\dot{W}}{\dot{m}_c \overline{Cp}_c T_{01}}$

$$= \left[\frac{\dot{m}_e}{\dot{m}_c} \frac{\overline{Cp}_e}{\overline{Cp}_c}\right]\left[\frac{T_{04} - T_{05}}{T_{04}}\right]\left[\frac{T_{04}}{T_{01}}\right] - \left[\frac{T_{02} - T_{01}}{T_{01}}\right]$$

$$= \left[\frac{\dot{m}_e}{\dot{m}_c} \frac{\overline{Cp}_e}{\overline{Cp}_c}\right] E_1 T' - C, \tag{3.9}$$

which is equation (3.1). The parameters $E_1 T'$ and C were defined earlier. The parameter C can be obtained from the pressure ratio using equation (2.61):

$$\left(\frac{T_{02}}{T_{01}}\right) = \left(\frac{p_{02}}{p_{01}}\right)^{(R/\overline{Cp}_c)/\eta_{pc}};$$

$$\therefore \frac{T_{02} - T_{01}}{T_{01}} \equiv C = \left[r^{(R/\overline{Cp}_c)/\eta_{pc}} - 1\right].$$

Similarly, the parameter E_1 can be obtained using equation (2.62):

$$\left(\frac{T_{04}}{T_{05}}\right) = \left(\frac{p_{04}}{p_{05}}\right)^{(R/\overline{Cp}_e)\eta_{pe}}.$$

But the expansion ratio $p_{04}/p_{05} = (p_{02}/p_{01})(1 - \sum(\Delta p_0/p_0))$;

$$\therefore \frac{T_{04} - T_{05}}{T_{04}} \equiv E_1 = 1 - \frac{1}{(T_{04}/T_{05})} = 1 - \left[r\left(1 - \sum\left(\frac{\Delta p_0}{p_0}\right)\right)\right]^{(-R/\overline{Cp}_e)\eta_{pe}}.$$

The specific power is the numerator of the expression for thermal efficiency, equation (3.2). It is unaffected by the presence or absence of a heat exchanger except for the influence of any additional pressure drop on E_1. The denominator is the heat addition, derived as follows for CBE and CBEX cycles:

$$\text{Heat addition} = \dot{m}_b(h_{04} - h_{03}) \tag{3.10}$$

$$= \dot{m}_b \overline{Cp}_b(T_{04} - T_{03}). \tag{3.11}$$

We substitute for T_{03} from the definition of heat-exchanger effectiveness, and we divide by the same terms used to make the specific power non-dimensional. We call this the specific heat addition, $\dot{Q}'$:

$$\dot{Q}' \equiv \left[\frac{\text{Heat addition}}{\dot{m}_c \overline{Cp}_c T_{01}}\right] = \left[\frac{\dot{m}_b \overline{Cp}_b}{\dot{m}_c \overline{Cp}_c}\right]\left[\frac{T_{04}}{T_{01}} - \varepsilon_x \frac{T_{05} - T_{02}}{T_{01}} - \frac{T_{02}}{T_{01}}\right]$$

$$= \left[\frac{\dot{m}_b \overline{Cp}_b}{\dot{m}_c \overline{Cp}_c}\right]\left[T' - \varepsilon_x\left[\frac{T_{05} - T_{04}}{T_{04}}\right]\left[\frac{T_{04}}{T_{01}}\right] - \varepsilon_x\left[\frac{T_{04}}{T_{01}}\right]\right.$$

$$+ \varepsilon_x\left[\frac{T_{02}}{T_{01}}\right] - \left[\frac{T_{02}}{T_{01}}\right]$$

$$= \left[\frac{\dot{m}_b \overline{Cp}_b}{\dot{m}_c \overline{Cp}_c}\right]\left\{T' + \varepsilon_x E_1 T' - \varepsilon_x T' + \varepsilon_x(1 + C)\right.$$

$$\left. - (1 + C)\right\};$$

$$\therefore \left[\frac{\text{Heat addition}}{\dot{m}_c \overline{Cp}_c T_{01}}\right] = \left[\frac{\dot{m}_b \overline{Cp}_b}{\dot{m}_c \overline{Cp}_c}\right]\{T'[1 - \varepsilon_x(1 - E_1)\}$$

$$- (1 + C)(1 - \varepsilon_x)]. \tag{3.12}$$

3.7 Efficiency versus specific-power charts

The thermal efficiency and specific power of CBE and CBEX Brayton cycles can be calculated by using equations (3.1) and (3.2), and by inserting the specified or expected component efficiencies, pressure losses, mass flows, and heat-exchanger effectiveness. Wolf (1983) developed a computer program and plotting routine to produce the charts shown in figures 3.5, 3.6, and 3.9 through 3.11.[2]

The mean specific heats used were those for air, with the compressor-inlet temperature at 300 K. The property data for air and for combustion products, and the required mass flow of a standard hydrocarbon fuel, were calculated from the method and correlations of Chappell and Cockshutt (1974) summarized in appendix A1. However, the charts would be reasonably accurate for gases of other molecular weights and fuels.

The charts have been calculated for conditions of no leakage of compressed air to the atmosphere or exhaust, as could occur through compressed-air-buffered labyrinth oil seals, or through leakage past the seals and in the matrix of a periodic-flow heat exchanger. Compressed air from the compressor discharge may also be diverted past the combustion system and used for cooling the disks and blades of high-temperature

---

2. Thomas Wolf and André By (personal communications as part of thesis research work at MIT).

turbines. One method of allowing for this compressed-air usage in cycle calculations is by estimating a reduced polytropic efficiency of expansion.

The charts are applicable to open or closed cycles, and to those with single or multiple stages of compression and expansion, and with single or multiple shafts. Such construction details affect the part-load working points of given gas-turbine engines. Now equations (3.1) and (3.2) apply equally well to part-load conditions as to design-point conditions, but they require that all the inputs to the equations be known. Obtaining these inputs from iterative calculations of the off-design performance of all the components is too complex to be fully covered in this book, although aspects of the part-load performance of different components are discussed in later chapters. This point is made to avoid confusion in the interpretation of the efficiency-specific-power charts. The lines on the charts connect different cycles having the same temperature ratio, going from 4.0 to 7.0 reading from left to right in all charts. The pressure ratios are marked as points on the temperature-ratio curves. These charts do not show the off-design performance of a single engine cycle.

Charts for the "simple" CBE cycle

Only two charts are shown for the simple CBE cycle, figures 3.5a and 3.5b. The reason for this brief treatment is that the CBE is not a high-efficiency cycle. This is not to claim that it is anything but the optimum choice for most aircraft-propulsion systems; its light weight and small size make it attractive, not its high efficiency.

Even so, thermal efficiencies of over 0.5 can be seen in figure 3.5a for temperature ratios of 6 and 7, the two right-hand curves. The pressure ratios are marked as points that must be counted from the right-hand end of each curve starting at 5 and ending at 80 in increments of 5. For the lowest temperature ratio, 4, the highest-pressure-ratio point shown as giving a thermal efficiency above 0.2, the cutoff point on the graph, is 60. The pressure ratios for maximum thermal efficiency are about 25 and 55 for temperature ratios of 4 and 5, and over 80 for the two higher temperature ratios.

The pressure loss for all the CBE cycles shown is that of the combustion system and ducting and is taken at the typical value of 0.04. The polytropic efficiencies of compressor and expander are both taken as constant 0.9. This is an easily attainable value at pressure ratios up to 10 in the larger machines (which can be defined as having a compressor-inlet blading diameter of over one meter), but this efficiency is, with present technology, unrealistic for the higher pressure ratios. Therefore the upper parts of the curves must be regarded as optimistic in terms of practicable thermal efficiencies. In other words, thermal efficiencies of over 0.5 cannot be reached with CBE cycles using present technology.

A more pessimistic view of the potential for CBE cycles is given by

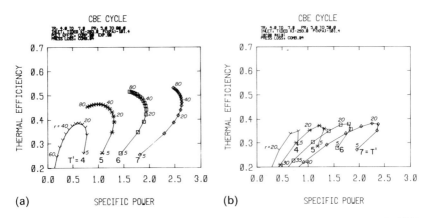

Figure 3.5 Thermal-efficiency versus specific-power charts for "simple" CBE cycles: (a) with constant polytropic efficiencies (compressor and expander, 90 percent; combustor pressure drop 4 percent); (b) with compressor and expander losses increasing with pressure ratio.

figure 3.5b. Here the compressor and turbine efficiencies are made functions of pressure ratio. Simple linear relations are used for the losses (defined as (1 − polytropic efficiency))·

for compressor losses

$$(1 - n_{pc}) = 0.04 + (r - 1)/150$$

and for turbine losses

$$(1 - n_{pe}) = 0.03 + (r - 1)/180,$$

where r is the compressor pressure ratio. These relations, which are intended to represent the best attainable performances with present technology, are probably sufficiently accurate for pressure ratios in the range of 10 to 30 but give losses that are too high for the higher pressure ratios. The qualitative effects are obvious from figure 3.5b: for each temperature-ratio curve the peak thermal efficiency is reduced from that in figure 3.5a by the increased losses that inevitably result to some degree at higher pressure ratios. The benefit of using high temperature ratios in these circumstances is principally to increase specific power. Pressure ratios giving peak thermal efficiencies are between 12 and 15, the range used for most industrial simple-cycle shaft-power gas-turbine engines, so that to this extent at least these two loss assumptions are partly confirmed as being reasonable. "Multispool" (independently rotating sets of stages of compressor and turbine) aircraft engines have losses at high pressure ratio lower than those assumed and could be expected to have an intermediate performance between those shown in figures 3.5a and 3.5b.

Charts for heat-exchanger cycles

The first CBEX chart, figure 3.6a, is for high-efficiency (0.925) compressor and turbine and for high-effectiveness (0.98) heat exchanger, but with high pressure losses (0.16). The pressure ratio for maximum thermal efficiency is between 2 and 3 at the lower temperature ratios, and between 3 and 4 at the higher temperature ratios. At a temperature ratio of 6, representative of the highest of current aircraft engines, a thermal efficiency of over 60 percent is indicated. This should be reduced slightly by bearing, labyrinth, and windage losses to arrive at a practical shaft efficiency, and by the effects of air bled from the compressor and discharged into the turbine, lowering its efficiency, if air-cooling of the blading is used.

In figures 3.6b, c, and d a slightly lower heat-exchanger effectiveness (0.975) and a lower total-pressure loss (0.12) are specified, and the compressor and turbine efficiencies are specified as varying with pressure ratio. The losses are somewhat arbitrarily ascribed as being appropriate to gas-turbine engines having small radial turbomachinery (figure 3.6b), large radial turbomachinery (figure 3.6c), and large axial turbomachinery (figure 3.6d) according to the following relations.

For small radial (figure 3.6b)

$$(1 - \eta_{pc,ts}) = 0.15 + (r - 1)/90 \quad \text{(compressor)},$$

$$(1 - \eta_{pe,ts}) = 0.10 + (r - 1)/180 \quad \text{(turbine)}.$$

For large radial (figure 3.6c)

$$(1 - \eta_{pc,ts}) = 0.10 + (r - 1)/90 \quad \text{(compressor)},$$

$$(1 - \eta_{pe,ts}) = 0.07 + (r - 1)/180 \quad \text{(turbine)}.$$

For large axial (figure 3.6d)

$$(1 - \eta_{pc,ts}) = 0.04 + (r - 1)/150 \quad \text{(compressor)},$$

$$(1 - \eta_{pe,ts}) = 0.03 + (r - 1)/180 \quad \text{(turbine)}.$$

In all cases the pressure ratio for maximum thermal efficiency is close to 3. For the large machine with its higher component efficiencies, "gross" thermal efficiencies of 0.6 or higher are indicated for temperature ratios of 5 and higher, well within current technology. As will be explained in chapter 10, it is theoretically possible to design heat exchangers even of

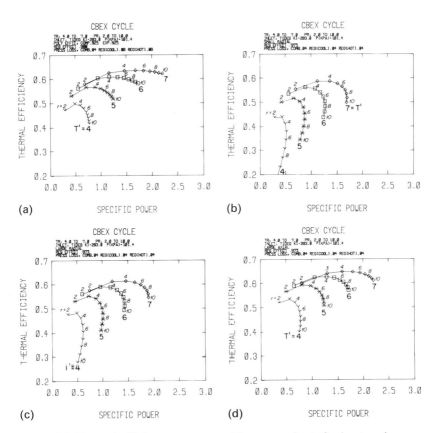

Figure 3.6 Thermal-efficiency versus specific-power charts for heat-exchanger, CBEX, cycles: (a) with constant polytropic efficiencies (compressor and expander, 92.5 percent; regenerator effectiveness, 98 percent; sum of total-pressure losses 16 percent); (b) with losses for "small radial" components; (c) with losses for "large radial" components; (d) with losses for "large axial" components.

very high effectiveness, as is specified here, with very low pressure losses, so that even higher efficiencies than those shown in figure 3.6d, at lower optimum pressure ratios, are theoretically possible. A low-pressure-ratio cycle intrinsically has a good part-load performance, and engines designed to such specifications should be superior in fuel economy to any other heat engine, including the compression-ignition engine, which is itself undergoing rapid development in the direction of even higher efficiencies.

The cycle calculations are made for total pressure losses in the combustor and in connecting ducts of 4 percent, to which are added typical pressure losses in the two "sides" of the heat exchanger and the ducting. Wherever a pressure loss occurs in a gas-turbine cycle, its effect is to reduce the pressure ratio across the expander. If each pressure loss is expressed as a proportion of the local total pressure, all the proportions (or percentages) can be simply added and used to determine the amount by which the compressor pressure ratio should be reduced to give the expander pressure ratio.

Let us examine cycles with two temperature ratios that are at the low and the high ends of the range of current (1983) industrial gas-turbine engines—4.0 and 5.0. With compressor-inlet temperatures of 80 F, (540 R, 300 K), these give expander-inlet temperatures of 1,700 F (or 927 C) at the low end, and 2,240 F (or 1,227 C) at the high end. (Figure 1.10 shows that in fact much higher expander-inlet temperatures are already in aircraft service, and very much higher temperatures are being tested.)

If we examine the curves for the same temperature ratios, 4.0 and 5.0, as we looked at for the simple cycle, we find that the trends and the actual values confirm the earlier discussion. The optimum pressure ratios for maximum specific power, and the values of specific power themselves, are almost identical to those for the simple cycle with similar component efficiencies. There is a small difference resulting from the increased pressure losses in the heat-exchanger cycle.

The pressure ratios for maximum efficiency are, as stated earlier, much lower for a CBEX cycle than for a corresponding CBE cycle. Values obtained from a detailed calculation of a typical CBEX cycle are shown in table 3.1.

Obviously, the number of charts needed to give one the ability to determine the combined effects of many changes would be enormous. The effects of small changes in the efficiency of one component are linear over a small but useful range. Table 3.2 shows calculations made for three "base" cases for heat-exchanger cycles rather similar to those already examined, but allowing for leakage of compressed air after heat has been added—in other words, at a point just before expander entry.

Some of the influence coefficients shown in table 3.3 demonstrate that a high-efficiency heat-exchanger gas-turbine engine is far from being a "Micawber engine" as described in chapter 1. The early low-efficiency engines were shown to be precarious balances between turbine output and compressor demands, so that a small change in component efficiency would have a very large effect on cycle efficiency. In contrast, table 3.3 shows that in some cases a fall of 1 percent in expander efficiency can cause less than a 1 percent fall in cycle efficiency. The reason is that the lower-efficiency turbine will have a higher outlet temperature, which will be transferred by a high-effectiveness heat exchanger to the compressed air. Less fuel will then be needed.

This example also explains why the largest influence coefficients are for drops in heat-exchanger effectiveness.

Table 3.1
Effect of cycle temperature ratio on optimum pressure ratios in the CBEX cycle (typical)

Temperature ratio	4.0	5.0
Pressure ratio for maximum specific power	7.0	10.0
Maximum specific power	0.62	1.02
Pressure ratio for maximum thermal efficiency	3.5	4.5
Maximum thermal efficiency	0.445	0.525

Table 3.2
Influence coefficients for heat-exchanger cycles

Compressor pressure ratio		3.0	3.5	4.0
Specifications of base cases				
Compressor polytropic efficiency		0.9	0.9	0.9
Expander polytropic efficiency		0.9	0.9	0.9
Temperature ratio		4.0	4.0	5.0
Heat-exchanger effectiveness		0.9	0.9	0.9
Sum of total-pressure losses		0.05	0.05	0.06
Leakage		0.05	0.05	0.05
Base-case performance				
Thermal efficiency, η_{th}		0.4375	0.4363	0.4842
Specific work M J/kg air		0.1664	0.1826	0.2532
Effect on thermal efficiency of				
1% additional leakage	$\Delta\eta_{th}$	0.0084	0.00866	0.00853
	$(\Delta\eta_{th}/\eta_{th})$	0.0176	0.0180	0.0176
1% drop in effectiveness	$(\Delta\eta_{th}/\eta_{th})$	0.0489	0.01222	0.0254
	$\Delta\eta_{th}$	0.00710	0.00586	0.0123
1% fall in expander efficiency	$\Delta\eta_{th}$	0.00429	0.00438	0.006
	$(\Delta\eta_{th}/\eta_{th})$	0.00899	0.00914	0.0124
1% fall in compressor efficiency	$\Delta\eta_{th}$	0.00249	0.00276	0.0054
	$(\Delta\eta_{th}/\eta_{th})$	0.00522	0.00575	0.01115
1% increase in pressure losses, 0.05 to 0.06	$\Delta\eta_{th}$	0.00437	0.00389	0.0053
	$(\Delta\eta_{th}/\eta_{th})$	0.00917	0.00811	0.011
1% fall in temperature ratio	$\Delta\eta_{th}$	0.0028	0.0031	0.0188
	$(\Delta\eta_{th}/\eta_{th})$	0.0064	0.0071	0.0352

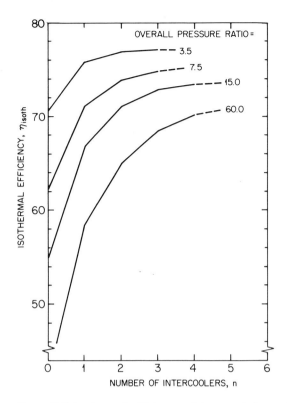

Figure 3.7 Isothermal efficiency of intercooled centrifugal compressors. From Wilson 1965.

Intercooled cycles

We showed earlier that compression work was directly proportional to absolute temperature. A large pressure ratio will, in an adiabatic machine, produce a relatively large increase in temperature, so that the work required to accomplish the latter part of the compression will be much larger than for the low-pressure part.

It is therefore attractive to consider the possibility of breaking the compression process into several parts and cooling the compressed gas between the stages or groups of stages. The power required can, even with quite inefficient compressor stages, be less than for an ideal adiabatic process, so that polytropic and isentropic efficiencies of well over 100 percent can be obtained. Accordingly, the efficiencies of intercooled compressors for the process industries are normally given in terms of the ideal isothermal process, figure 3.7 and 3.8.

When the compressor power is reduced to such an extent in a gas-turbine cycle, the net output increases to high levels. Without a heat

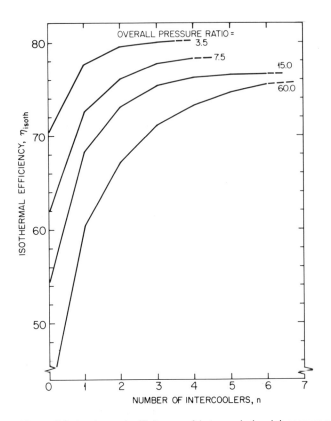

Figure 3.8 Isothermal efficiency of intercooled axial compressors. From Wilson 1965.

exchanger, however, the thermal efficiency would be low because of the large amount of additional fuel required to heat up the compressed gas from the low compressor-delivery temperature. This disadvantage is completely overcome when a heat exchanger is incorporated. The performance of some heat-exchanger intercooled cycles (CICBEX) is shown on figures 3.9 through 3.11. On each of the charts of figure 3.9 the compressor and expander polytropic efficiencies are constant, as are the heat-exchanger effectivenesses, but these values increase from figure 3.9a to 3.9c. In figures 3.10a, b, c, and 3.11a, b, and c the compressor and expander losses are those listed in figure 3.6 for small radial, large radial, and large axial components. In figure 3.10 the effectiveness of the heat exchanger is 0.75 and of the intercooler 0.9, and the total pressure losses are 0.10; in figure 3.11 the same values become 0.975, 0.975, and 0.15.

There have been several attempts to harvest the benefits of the intercooled cycle. Rolls Royce produced the RM 60, with two intercoolers

Gas-Turbine Power Cycles

and three separate rotating shafts. This engine was used to power naval vessels, in which the ready availability of cooling water made it advantageous. Ford developed an automobile engine with a rather similar cycle. In this case the intercoolers had to be air-cooled, giving higher losses and reduced intercooling. Both engines were too complex for the then-current state of the art. The intercooled cycle will undoubtedly find a place in the future. Even for a high-temperature-ratio, highly regenerated CBEX cycle having a low optimum pressure ratio of 3.5, the addition of one intercooler will increase the compressor efficiency by about seven points (figure 3.8), which would produce a significant increase in thermal efficiency and specific power.

Reheat cycles

Similar gains to those for the intercooled cycle can theoretically be realized by breaking the expansion process into several steps and reheating the gas to the maximum allowable temperature between each step. Developing combustors to operate with very high inlet temperatures, as well as high outlet temperatures, is extremely difficult, however. Reheat cycles may become more practicable when ceramic combustors become reliable.

3.8 Performance calculation for cycles

Intercooled reheat cycle

The formulas for calculating the specific work and the thermal efficiency of an intercooled reheated heat-exchange cycle are given in this section. The derivation is similar to that for equations (3.1) and (3.2) and will not be given in detail.

The cycle considered has three compressors of equal pressure ratio r, and three expanders also of equal pressure ratio. The pressure ratio of each expander will then equal the compressor pressure ratio less one third of the total cycle pressure losses $\sum(\Delta p_0/p_0)$.

The first heat-addition combustor, 3-4a in figure 3.12, and the two reheat combustors, 5a-4b and 5b-4c, take the working fluid to the same maximum temperature: $T_{04a} = T_{04b} = T_{04c}$.

The compressors, however, have different inlet and outlet temperatures. The two intercoolers have been specified to have the same effectiveness, ε_i, and to reject heat to the ambient at T_{01a}. Therefore $T_{01} > T_{01b} > T_{01a}$, and $T_{02c} > T_{02b} > T_{02a}$.

Specific power $\dot{W}' \equiv \dot{W}/\dot{m}\overline{Cp}_c T_{01a}$ = work from 3 expanders − work to 3 compressors:

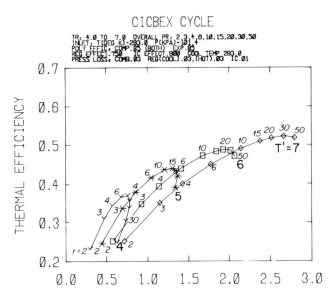

Figure 3.9 (a)

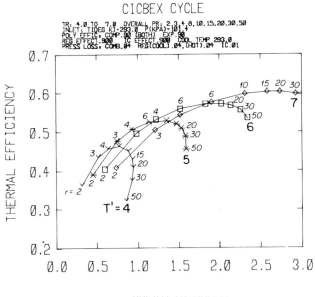

Figure 3.9 (b)

Gas-Turbine Power Cycles

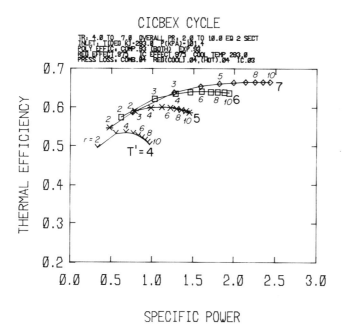

Figure 3.9 (c)

Figure 3.9 Thermal-efficiency versus specific-power charts for intercooled heat-exchanger (CICBEX) cycles: (a) low component efficiencies (compressor and expander polytropic efficiencies 85 percent; heat-exchanger effectiveness 75 percent; sum of total-pressure drops 10 percent); (b) good component efficiencies (compressor and expander polytropic efficiencies 90 percent; heat-exchanger effectiveness 90 percent; sum of total-pressure drops 13 percent); (c) high component efficiencies (compressor and expander polytropic efficiencies 93 percent; heat-exchanger effectiveness 97.5 percent; sum of total-pressure drops 15 percent).

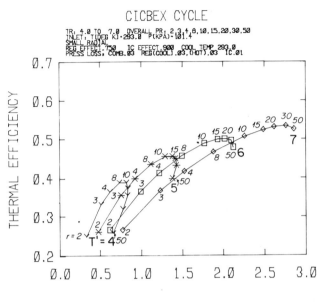

Figure 3.10 (a)

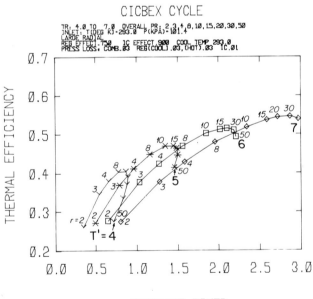

Figure 3.10 (b)

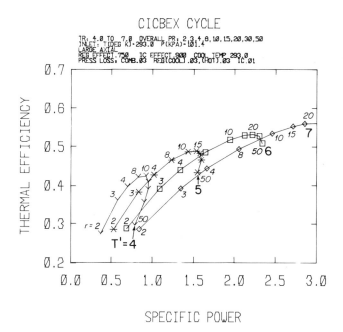

Figure 3.10 (c)

Figure 3.10 Thermal-efficiency versus specific-power charts for intercooled heat-exchanger (CICBEX) cycles with low (75 percent) heat-exchanger effectiveness and component efficiencies varying with pressure ratio: (a) efficiencies for "small radial" components; (b) efficiencies for "large radial" components; (c) efficiencies for "large axial" components.

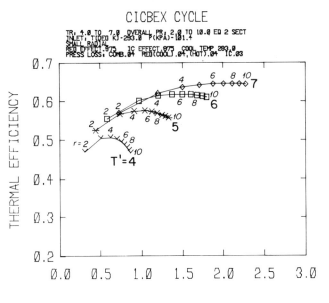

Figure 3.11 (a)

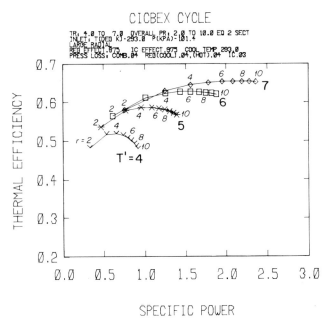

Figure 3.11 (b)

Gas-Turbine Power Cycles

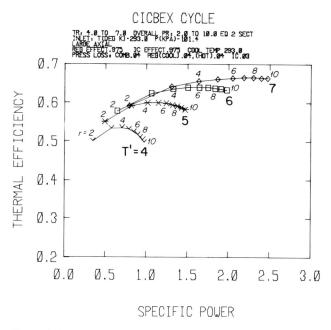

Figure 3.11 (c)

Figure 3.11 Thermal-efficiency versus specific-power charts for intercooled heat-exchanger (CICBEX) cycles with high (97.5 percent) heat-exchanger effectiveness and component efficiencies varying with pressure ratio: (a) efficiencies for "small radial" components; (b) efficiencies for "large radial" components; (c) efficiencies for "large axial" components.

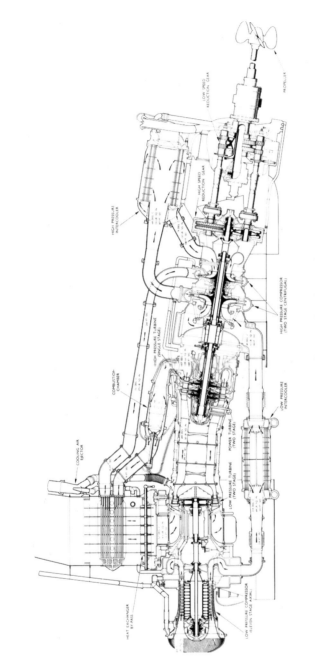

Rolls-Royce RM60 marine engine with two intercoolers and exhaust heat exchanger. Courtesy Rolls-Royce Limited.

Gas-Turbine Power Cycles

$$\frac{\dot{W}}{\dot{m}\overline{Cp}_c T_{01a}} = \left[\frac{\dot{m}_e}{\dot{m}_c}\frac{\overline{Cp}_e}{\overline{Cp}_c}\right] T' \times 3E_3$$

$$- \underbrace{C\{1}_{} + \underbrace{[C(1-\varepsilon_i)+1]}_{} + \underbrace{[(C(1-\varepsilon_i)+1)(1+C)(1-\varepsilon_i)+\varepsilon_i]}_{}\}.$$

First   Second   Third compressor
compressor compressor   (3.13)

Specific heat addition $\dot{Q}' \equiv \dot{Q}/\dot{m}\overline{Cp}_c T_{01a}$:

$$\frac{\dot{Q}}{\dot{m}\overline{Cp}_c T_{01a}} = \left[\frac{\dot{m}_b}{\dot{m}_c}\frac{\overline{Cp}_b}{\overline{Cp}_c}\right] T'$$

$$\times \left\{\underbrace{\left[1 - \varepsilon_x(1-E_3) + (1-\varepsilon_x)(C(1-\varepsilon_i)+1)(1+C)^2(1-\varepsilon_i)\frac{1}{T'}\right]}_{} + 2E_3\right\}.$$

First combustor   Second
                                                                                       and third
                                                                                       combustors
                                                                                              (3.14)

The cycle thermal efficiency is the ratio of the specific power (equation 3.13) to the specific heat addition (equation 3.14):

$$\eta_{th} = \frac{[\dot{W}/\dot{m}\overline{Cp}_c T_{01a}]}{[\dot{Q}/\dot{m}\overline{Cp}_c T_{01a}]}.$$ (3.15)

The symbols are as defined in section 3.6, except that here

$$E_3 \equiv \frac{T_{04c} - T_{05c}}{T_{04c}} = \left\{1 - \left[\left(1 - \frac{1}{3}\sum\left(\frac{\Delta p_0}{p_0}\right)\right) r\right]^{-(R/\overline{Cp}_c)\eta_{pe}}\right\},$$

$$r \equiv \frac{p_{02a}}{p_{01a}} = \frac{p_{02b}}{p_{01b}} = \frac{p_{02c}}{p_{01c}},$$

$$C \equiv \frac{T_{02a} - T_{01a}}{T_{01a}} = \frac{T_{02b} - T_{01b}}{T_{01b}} = \frac{T_{02c} - T_{01c}}{T_{01c}} = [r^{(R/\overline{Cp}_c)/\eta_{pc}} - 1],$$

$$\varepsilon_x \equiv \frac{T_{03} - T_{02c}}{T_{05} - T_{02c}},$$

$$\varepsilon_i \equiv \frac{T_{02a} - T_{01b}}{T_{02a} - T_{01a}} = \frac{T_{02b} - T_{01c}}{T_{02b} - T_{01a}},$$

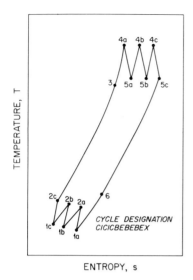

Figure 3.12 Diagram of a multiply intercooled reheated heat-exchanger cycle.

$$T' \equiv \frac{T_{04}}{T_{01}}.$$

The relative magnitudes of these terms may be seen from an example.

Example

Find the specific power, the thermal efficiency, and the individual contributions of the different compressors, expanders and combustors to these performance characteristics, for a double-intercooled-reheated heat-exchanger cycle, as shown in figure 3.12, with the following specifications:

$\left.\begin{array}{l} T_{01a} = 300 \text{ K} \\[2mm] T_{04a} = T_{04b} = T_{04c} = 1{,}500 \text{ K} \end{array}\right\} \therefore T' = 5.0.$

$\left.\begin{array}{l} p_{01a} = 0.95 \text{ at} \\[4mm] \dfrac{p_{02a}}{p_{01a}} = \dfrac{p_{02b}}{p_{01b}} = \dfrac{p_{02c}}{p_{01c}} = 2.90 \end{array}\right\} \therefore C = 0.40.$

$\left.\begin{array}{l} \eta_{pc} = 0.9 = \eta_{pc} \\[2mm] \sum\left(\dfrac{\Delta p_0}{p_0}\right) = 0.15 \end{array}\right\} \therefore E_3 = 0.1933 \text{ (after iterating for } \overline{Cp_e}).$

$\varepsilon_x = \varepsilon_i = 0.9$

The leakage and fuel addition are such that $[(\dot{m}_e/\dot{m}_c)(\overline{Cp_e}/\overline{Cp_c})] = 1.0$.

Solution These parameters are inserted into equations (3.13) through (3.15).

Contribution of the first expander to the specific power $= 0.1933 \times 5 = +0.967$.
Contribution of the second expander to the specific power $= 0.1933 \times 5 = +0.967$.
Contribution of the third expander to the specific power $= 0.1933 \times 5 = +0.967$.
Contribution of the first compressor to the specific power $= 1.0 \times (-0.4) = -0.40$.
Contribution of the second compressor to the specific power $= 1.04 \times (-0.4) = -0.416$.
Contribution of the third compressor to the specific power $= 1.0456 \times (-0.4) = -0.418$.

Net specific power $\equiv \dfrac{\dot{W}}{\dot{m}_c \overline{Cp}_c T_{01}} = 1.661$.

Contribution of the first combustor to the heat added $= 0.278 \times 5 = 1.390$.
Contribution of the second combustor to the heat added $= 0.1933 \times 5 = 0.967$.
Contribution of the third combustor to the heat added $= 0.1933 \times 5 = 0.967$.

Total heat added $= \dfrac{\dot{Q}}{\dot{m}_c \overline{Cp}_c T_{01a}} = 3.324$.

Thermal efficiency $= 0.500$.

Performance calculations for other cycles

With appropriate changes, equations (3.13) and (3.14) may be used for other cycles.

Single-intercooled-reheated heat-exchanger cycle (CICBEBEX cycle) (figure 3.13)

Specific power $\equiv \dfrac{\dot{W}}{\dot{m}_c \overline{Cp}_c T_{01a}}$

$$= \left[\frac{\dot{m}_e \ \overline{Cp}_e}{\dot{m}_c \ \overline{Cp}_c} \right] T' \times 2E_2 - C[2 + C(1 - \varepsilon_i)], \qquad (3.16)$$

where the symbols are defined as in section 3.3, except that

$$E_2 \equiv 1 - \frac{T_{05}}{T_{04}} = \left\{ 1 - \left[\left(1 - \frac{1}{2} \sum \left(\frac{\Delta p_0}{p_0} \right) \right) r \right]^{-(R/\overline{Cp}_e)\eta_{pe}} \right\}.$$

Specific heat addition

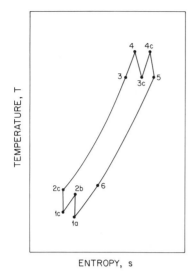

Figure 3.13 Intercooled-reheated (CICBEBEX) cycle.

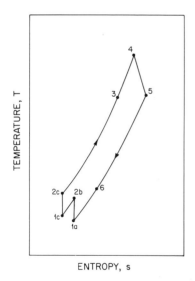

Figure 3.14 Intercooled heat-exchanger (CICBEX) cycle.

Gas-Turbine Power Cycles

$$\frac{\dot{Q}}{\dot{m}_c \overline{Cp}_c T_{01a}} = \left[\frac{\dot{m}_b \overline{Cp}_b}{\dot{m}_c \overline{Cp}_c} \right]$$

$$\times T' \left\{ (1 + E_2) + \varepsilon_x (1 - E_2) - \frac{(1 + C)}{T'} (1 - \varepsilon_x)[(1 + C)(1 - \varepsilon_i) + \varepsilon_i] \right\}.$$

$$(3.17)$$

The cycle thermal efficiency is the ratio (specific power/specific heat added).

Single-intercooled heat-exchanger cycle CICBEX (figure 3.14)

$$\text{Specific power} \equiv \frac{\dot{W}}{\dot{m}_c \overline{Cp}_c T_{01a}}$$

$$= \left[\frac{\dot{m}_e \; \overline{Cp}_e}{\dot{m}_c \; \overline{Cp}_c} \right] T' E_3 - C\{2 + C(1 - \varepsilon_i)\}, \quad (3.18)$$

where the symbols are as defined in section 3.3 except that

$$E_3 \equiv \frac{T_{04} - T_{05}}{T_{04}} = \left\{ 1 - \left[\left(1 - \sum \left(\frac{\Delta p_0}{p_0} \right) \right) r^2 \right]^{-(R/\overline{Cp}_e)\eta_{pe}} \right\}.$$

$$\text{Specific heat addition} \equiv \frac{\dot{Q}}{\dot{m}_c \overline{Cp}_c T_{01a}} = \left[\frac{\dot{m}_b \overline{Cp}_b}{\dot{m}_c \overline{Cp}_c} \right]$$

$$\times T' \left\{ 1 - (1 - E_3)\varepsilon_x - \frac{(1 + C)}{T'} (1 - \varepsilon_x) \times [1 + C(1 - \varepsilon_x)] \right\}. \quad (3.19)$$

The cycle thermal efficiency is the ratio of equations (3.18) and (3.19) in the same way as before.

3.9 Cryogenic cycles

A gas-turbine engine normally produces power from heat while rejecting heat to a cold sink. Engines can also be designed to produce power from cold, while absorbing heat ("rejecting cold") to a (relatively) warm sink. Figure 3.15 shows, for instance, a temperature-entropy diagram of a hydrogen closed cycle that could be used for evaporating liquefied natural gas and simultaneously recovering a proportion of the power used at the gas field in liquefying it. Currently, natural gas is "deliquefied" by the expenditure of heat, often from combustion of the fuel itself, an obviously wasteful process.

In the cycle shown, which is similar to the cycle of figure 3.12, the

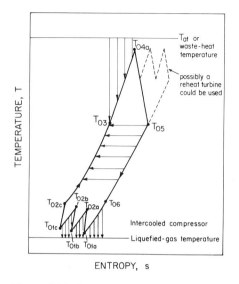

Figure 3.15 Cryogenic intercooled reheated cycle.

latent heat of the liquefied gas would be given up in cooling the hydrogen to compressor-inlet temperature after leaving the heat exchanger, from T_{06} to T_{01a}, and in recooling the hydrogen in one or more intercoolers, T_{02a} to T_{01b} and T_{02b} to T_{01c}. After the final compression stage the hydrogen would be heated in the high-pressure side of the heat exchanger, $T_{02c} \rightarrow T_{03}$, by internal heat transfer, and would receive its final increment of temperature $T_{03} \rightarrow T_{04a}$, by absorbing heat from the atmosphere or from, for instance, a hot-water power-plant cooling-water discharge.

The intercooled cycle (figures 3.9 to 3.11) could also be used for this duty. These charts can be used as a guide to the efficiency and specific power. They have been calculated on the basis of air properties, and the similar value of γ for hydrogen should produce similar optimum pressure ratios and specific-power values. Equations (3.18) and (3.19) can be used for more exact values for the intercooled heat-exchanger cycle. More complex cycles may also be used at low temperatures.

This cryogenic application is one case where the use of reheaters between expansion stages would be easily practicable and would increase the power output and the efficiency. Many stages of compression and expansion are needed to produce a change of temperature sufficient for the use of intercoolers or reheaters when hydrogen is used, because of the very high specific heat. Whether or not the high additional cost of intercoolers and reheaters would be justified would depend on the value of the additional power produced.

3.10 The inverted Brayton cycle

The ideal Brayton cycle involves an enclosed working fluid being taken through a succession of processes. Although there are some closed-cycle gas-turbine engines that operate on an approximation to this cycle, most gas turbines use a so-called "open cycle" in which the atmosphere substitutes for the constant-pressure cooling process. (The atmosphere also regenerates the exhaust gases so that "internal" combustion can be used.)

When air is used as the working fluid, the closed cycle can be "opened" to the atmosphere at any point. In the usual open cycle the turbine exhausts to, and the compressor draws from, the atmosphere. The inverted cycle is simply the name given to the type of engine cycle in which the pressure (but not necessarily the temperature) at inlet to the turbine expander is approximately atmospheric. The inverted Brayton cycle as an independent prime mover consists of atmospheric air being drawn through a source of heat, such as a combustor; then being expanded from high temperature and approximately atmospheric pressure to sub-atmospheric pressure; then being cooled at approximately constant, low, pressure; and finally being recompressed to atmospheric pressure and being discharged (figure 3.16).

The thermal efficiency and specific power output of inverted cycles were evaluated as functions of temperature and pressure ratios by Hodge (1955). So far as is known, no inverted-cycle gas-turbine engine using this principle has ever been built and run. The reasons are obvious. All the components, and particularly the turbine and compressor, are much larger than those of a gas turbine working on the direct open cycle with high-pressure combustion. Expressed another way, a given compressor-turbine combination will produce many times the power operating with compression from atmospheric pressure than can be achieved when expanding from atmospheric pressure. Moreover, although the thermal efficiency for the given combination of temperature and pressure ratio should be identical in the two cases, the inverted cycle adds a component, the cooler, which is not necessary in the simple direct cycle and which adds a penalizing pressure drop; therefore the inverted-cycle thermal efficiency is lower.

The unattractiveness of the inverted cycle would seem to have been firmly established, and there would seem to be no reason to give it any further consideration.

There are, however, two mitigating factors. First, the inverted gas-turbine should be considered for use only when the heat input is free. In other words, it can be considered when there is a flow of hot gas at atmospheric pressure which is going to waste, such as the exhaust of a

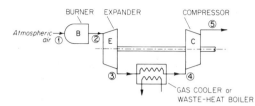

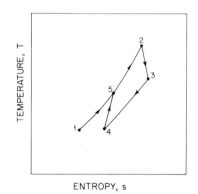

Figure 3.16 The simple inverted (ILC) cycle.

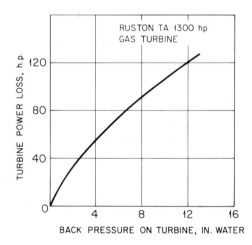

Figure 3.17 Effect of back pressure on gas-turbine-engine power output.

Gas-Turbine Power Cycles

gas-turbine engine. A conventional way of using engine exhaust is to lead the gas to a waste-heat boiler and, if power is desired, to incorporate this boiler in a steam-power cycle. This is sometimes called a "bottoming" cycle. The inverted Brayton cycle is an alternative bottoming cycle. In certain cases it can be advantageous.

The second mitigating factor is that, when the inverted cycle is added to an arrangement where the waste heat from a conventional gas turbine is utilized in a heating boiler, the power output can be increased. The attractive potential of this form of the inverted cycle was brought to light in the following way.

Many conventional gas turbines are sold in a so-called "total-energy" package in which a waste-heat boiler is incorporated to remove sensible heat from the turbine exhaust.

A performance penalty must normally be accepted when one adds an exhaust-heated boiler to a gas turbine. The penalty is usually given in the form shown in figure 3.17, which is for an early version of the Ruston and Hornsby TA 1,250 kW gas turbine (Forbes 1956). The addition of a pressure drop to the turbine exhaust produces a power-output reduction which is linear in the region of interest.

A Canadian purchaser of a TA turbine wrote to Ruston and Hornsby to inform them that their performance penalty seemed to be in error. The reason for his statement was that he had decided to add a waste-heat boiler to his existing TA turbine, but he could not allow the turbine power output to decrease by the 100 hp that was predicted. He therefore decided to install an electrically driven induced-draft fan after the waste-heat boiler so that the pressure at the turbine-exhaust flange would be restored to atmospheric (figure 3.18). He found that he could do this with a fan absorbing approximately 60 hp. How could he obtain a power increase of 100 hp with the expenditure of 60 hp?

One way of answering this question is to state that the fan had to compress a smaller volume of gas since it had been cooled than expanded in the turbine. Therefore its power absorption was less than the restored power gain of the turbine. Another way of describing the same phenomenon is that an inverted cycle had been added to the existing gas turbine (figure 3.19).

Some further questions of interest immediately arise. If restoring the turbine-exhaust pressure to atmospheric produces a net power gain over the case where the boiler has no induced-draft fan, does decreasing the turbine-exhaust pressure to below atmospheric produce a greater gain? It can be shown that it does. By how much must the waste-heat boiler be increased in size to meet the same design-point conditions? Can the added induced draft fan pay for itself in different circumstances? These questions were examined by Dunteman (1970). The results of

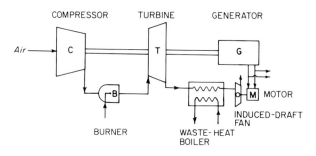

Figure 3.18 Use of induced-draft fan with waste-heat boiler.

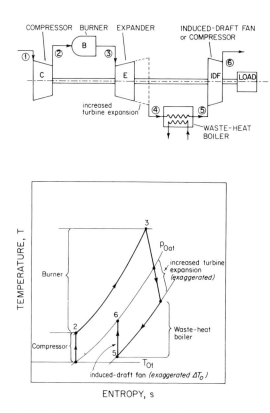

Figure 3.19 Inverted cycle added to gas-turbine engine.

Gas-Turbine Power Cycles

Table 3.3
Inverted-cycle economics

	Case 1	Case 2
Gas-turbine exhaust temperature	1,500 R	1,200 R
I.D. compressor temperature rise	264°R	173°R
Return on investment (1970 energy and borrowing costs)	34%	21.9%

economic optimizations show that in certain circumstances the inverted cycle shows very attractive rates of return.

For the inverted cycle considered as an add-on device (waste-heat boiler and induced-draft fan) for a standard power-producing gas turbine, the return on investment seems to be of the order of 30 percent (1970 conditions) because of the significant power increase brought about in the gas-turbine plant with only the addition of the induced-draft fan. This large return on investment is calculated on the assumptions that the gas turbine would be exhausting into the waste-heat boiler in any event and would suffer a power loss as a result of the back pressure. Accordingly, the only capital investment necessary is for the induced-draft fan or compressor and its associated drive. Two fairly typical cases that were calculated are shown in table 3.3.

References

Chappell, M. S., and E. P. Cockshutt. 1974. *Gas-turbine cycle calculations: thermodynamic data tables for air and combustion products for three systems of units.* National Research Council of Canada Aeronautical Report LR-579, NRC no. 14300. Ottawa.

Dunteman, N. Richard. 1970. A new look at the competitive position of the inverted-cycle gas turbine for waste-heat utilization and other applications. MSME thesis. MIT, Cambridge, Mass.

Forbes, S. M. 1956. TA turbine performance data. T. D. report 125. Ruston & Hornsby Ltd., Lincoln, England.

Hodge, James. 1955. *Gas-turbine Cycles and Performance Estimation.* Butterworths, London.

Vermeulen, Bert Martin. 1981. Computer-aided visualization for design decisions. BS/MSME thesis. MIT, Cambridge, Mass.

Wilson, D. G. 1965. The specification of high-mass-flow intercooled air compressors for process applications. Paper 65-WA/FE-25. ASME, New York.

Problems

3.1

The diagram in figure P3.1–P3.2 is of a pumped-storage scheme in which an air compressor can be electrically driven during the night to pump air into an underground cavern. What is the isentropic efficiency of the air compressor, total to total, from the total inlet conditions shown at 1 to the total conditions at outlet from the cavern, shown at 3? The total-to-total polytropic efficiency from the inlet of the compressor, 1, to the outlet of the machine, 2, is 90 percent. The conditions at the various stations are shown on the diagram, and the air may be assumed to be a perfect gas of $\gamma = 1.4$; $Cp = 1,005$ J/kg-°K. State other assumptions you need to make.

3.2

For the complete cycle shown in figure P3.1–P3.2 the air from the underground cavern can be heated in a combustor B and expanded through a turbine E, and this can be done during the day with the compressor declutched (not driven).

(a) What is the qualitative effect on the cycle efficiency of the pressure loss from point 2 to point 3?

(b) What is the qualitative effect on the cycle thermal efficiency and the specific work (per kg of air compressed) of the cooling of the compressed air to 15 C in the cavern?

(c) What would be the effect on the cycle thermal efficiency and the specific work of adding a heat exchanger (turbine exhaust to burner inlet)?

(d) How does the size of the alternator/motor compare with an alternator for a single-shaft simple-cycle gas-turbine engine (CBE) using the same compressor and turbine?

(e) What is the approximate effect of a 1 percent drop in alternator/motor efficiency on the overall cycle thermal efficiency?

(f) Would it be useful to make a second cavern into which the air from an intermediate stage of the compressor could be delivered and returned? Should the walls of this cavern and of the storage cavern after the last stage be good insulators or good conductors?

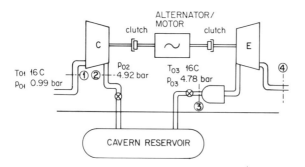

Figure P3.1–P3.2

3.3

Find the maximum percentage of air that must be taken from the compressor delivery to cool a high-temperature turbine expander for the overall CBE cycle efficiency to drop by not more than five percentage points from the value calculated for no usage of compressed air. Here are the CBE cycle specifications:

Compressor pressure ratio, 20 : 1.

Compressor inlet temperature, 38 C.

Turbine inlet temperature, 1,094 C.

Turbine exhaust pressure $p_{04} = p_{01} =$ atmospheric.

Combustor pressure drop, 5 percent of inlet p_0.

Compressor and turbine polytropic efficiencies, $\eta_{pc} = \eta_{pe} = 0.90$ (based on given pressures).

Assume that neither the compressed air added to the turbine for cooling nor the mass of the fuel added in the burner affect the turbine power. For simplicity in calculation use $\gamma = 1.4$ throughout the cycle.

3.4

(a) Find the pressure ratio for maximum efficiency for a nonregenerative (CBE) simple-cycle gas turbine by calculating the overall efficiency at three pressure ratios and attempting to draw a curve through the efficiencies obtained. The cycle specifications are as follows:

Inlet temperature to compressor, $T_{01} = 38$ C.

Inlet temperature to the turbine, $T_{03} = 983$ C.

Compressor and turbine polytropic efficiencies, $\eta_{pc} = \eta_{pe} = 0.88$.

Combustor pressure drop 5 percent of the inlet total pressure, $\sum (\Delta p_0/p_0)$. $\dot{m}_e/\dot{m}_c = 1.025$.

Calculations using mean air tables of constant specific heats can be used and the effects of fuel addition on the gas properties can be ignored for simplicity.

(b) Then determine how the optimum pressure ratio and the efficiency at that optimum changes when intercooling is added. The intercooling should divide the compressor into two compressors having equal pressure ratios and should cool the air to 36 C. There will be a 4 percent loss in total pressure through the intercooler.

3.5

(a) Find the cycle thermal efficiency for an open-cycle gas-turbine engine with a heat exchanger (CBEX cycle) having an inlet air temperature of 27 C, and turbine-inlet temperature of 927 C, a pressure ratio of 4 : 1 in the compressor; a compressor polytropic efficiency of 92.5 percent; a turbine polytropic efficiency of 92.5 percent; a heat-exchanger effectiveness of 98 percent of the maximum theoretically possible; and pressure drops in the air and gas streams through each side of the heat exchanger and through the combustor of $5\frac{1}{3}$ percent of the inlet pressure in each case.

(b) These specifications are those of figure 3.6a. Calculate the power output per kg per second air flow and compare your values of power output and efficiency with those given by the curve. Then assume that there is a leakage of 2 percent of the entering air in the heat exchanger, going directly to atmosphere; or that there is a 6 percent pressure drop in the combustor; or that there is both leakage and increased combustor pressure drop together. Hence discuss the sensitivity of this cycle to leakage and pressure drop.

(c) These calculations can be made using the data in appendix A for air and gas enthalpies or mean values of specific heats. However, make one complete cycle calculation using gas tables and compare the results.

3.6
(a) The temperature ratio of a simple-cycle (CBE) gas-turbine engine could be increased with blade cooling, but compressed air from the compressor delivery would be needed to cool the turbine blades; this cooling air would produce no useful work in the turbine expansion. Also it is possible that the turbine itself would have a lower polytropic efficiency because of the thicker blade profiles necessary. In the two cases of blade cooling given in table P3.6 find the maximum amount of compressor air that may be used for cooling if the specific power of the gas turbine is to be the same as for the base case.

(b) The mean specific heat in compression is 1,005 J/kg°K and in expansion 1,160 J/kg°K. The universal gas constant R is 8,313 J/kg mole°K; the molecular weight is 28.9; and the inlet total temperature and pressure are 5 C and 0.95 bar. Effects of fuel mass addition may be neglected. Will the cycle thermodynamic efficiencies be higher or lower than for the base case?

3.7
Investigate the thermal efficiency and the power output per kg per second of the following heat-exchanger open-cycle (CBEX) gas-turbine engines and comment on their possible attractiveness for automotive, merchant-ship, and naval use. Compare the results for the new cycles with the nearest equivalents in figure 3.6.

Common values

Inlet temperature, $T_{01} = 15$ C.
Inlet pressure, $P_{01} = 0.98$ bar.
Compressor = turbine polytropic efficiency = 0.90.
Regenerator thermal effectiveness, $\varepsilon_x = 0.90$.

Table P3.6

	Base case (BC)	Blade cooling (A)	Blade cooling (B)
Temp. ratio, T'	4	5	5
Pressure ratio, r	36	36	36
Compressor η_{pc}	0.90	0.90	0.90
Turbine η_{pe}	0.90	0.90	0.87
Cooling air [$\delta\dot{m}/\dot{m}_c$]	0	?	?
$\sum(\Delta p_0/p_0)$	0.14	0.14	0.14

Expander-inlet temperature, T_{04}, and pressure ratio, r (compressor).

For varying values

$T_{04} = 1{,}227$ C, $r = 6$,
$T_{04} = 1{,}077$ C, $r = 5$,
$T_{04} = 927$ C, $r = 4$.

3.8
Find the cycle thermal efficiency and the hydrogen mass flow in kg per second for the cycle shown, if it is to evaporate 30 kg/s of liquid methane. The heat absorption is from furnace waste heat. The hydrogen may be treated as a perfect gas. Property values and cycle conditions are shown on the h-s diagram in figure P3.8. Make and state such other assumptions or approximations as seem necessary or desirable.
    In figure P3.8

$p_{04}/p_{05} = 0.85(p_{02}/p_{01})$.
Hydrogen properties
$MW = 2.0$,
$\overline{Cp} = 14{,}317$ J/kg-°K,
$\quad \gamma = 1.41$,
$\quad R = 4{,}125.6$ J/kg-°K.

Liquid-methane properties
$MW = 16.0$.
Boiling point at atmospheric pressure $= 111.11$ K.
Latent heat of evaporation $= 514.52$ kJ/kg.

3.9
We want to find the effects of various component losses on the net output (specific power) and the thermal efficiency of simple (CBE) and of heat-exchanger (CBEX)

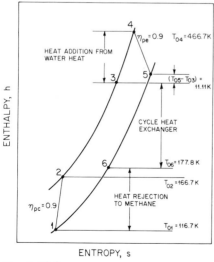

Figure P3.8

gas-turbine cycles. Choose one pressure ratio for the CBE cycle from those listed and another pressure ratio for the CBEX cycle. Calculate the thermal efficiency in each case. Then successively make the compressor and turbine efficiencies to be 100 percent, and then eliminate the pressure drops. Then, for the heat-exchanger cycle, make the heat-exchanger effectiveness 100 percent. Find the thermal efficiency at each step. We want (from a group of students) to produce a plot of base-cycle thermal efficiencies against pressure ratio and to give the effects of component losses.

Simple-cycle pressure-ratio choices: 5, 10, 24, 48, 64.

Heat-exchanger-cycle pressure-ratio choices: 2.5, 5, 10, 20.

Total-to-total (effective) polytropic efficiency, compression: 0.87 expansion: 0.85.

Compressor-inlet temperature: 288.2 K; pressure: 1.0 at.

Expander-inlet temperature: 1,673.0 K.

Pressure losses (reducing the expansion ratio): 10 percent for the simple cycle; 15 percent for the HX cycle.

Heat-exchanger effectiveness: 75 percent.

Ignore leakage. Preferably the correlations in appendix A should be used to calculate fuel mass flow and air and gas properties.

3.10
(a) Find the effect on the overall thermal efficiency, the waste heat recovered, and the cycle mass flow of designing a heat-exchanger for waste-heat recovery (or "waste-heat boiler," WHB) on a CBE gas-turbine engine with and without an induced-draft fan or compressor. This is the component between points 5 and 6 in figure 3.19. If employed, it would have a pressure ratio of 1.8 and an isentropic efficiency of 85 percent and would thereby increase both the pressure ratio to be designed for in the expander and the power extracted by the compressor-drive shaft.

The design shaft power for this marine power unit is to be 15,000 kW. The main compressor pressure ratio is 4.5 and the turbine-inlet temperature is 1,027 C. The compressor-inlet temperature is 16 C, and the atmospheric pressure at the inlet (upper-deck level) is 1 bar. The pressure losses as proportions of local total pressure are as follows:

Inlet duct, 0.03,

Burner, 0.06,

Turbine duct, 0.02,

Waste-heat boiler, 0.04,

Exhaust duct, 0.03.

In both cases the waste-heat boiler is designed for an outlet temperature T_{05} of 150 C. The expander isentropic efficiency, which includes the effects of leakage, is 86 percent and that of the compressor is 87 percent.

(b) Assume that the expander and WHB mass flow is increased by 1.05 percent in both cases to account for fuel addition (even though it will differ somewhat in the two cases). Use Keenan et al. Gas Tables, table 1, for the air side and table 4 (400 percent excess air) for the products of combustion, taking the molecular weight of the products to be 28.9, or appendix A. Give your calculations in the form of a table. Briefly discuss your results. Indicate how you forecast your results would change (qualitatively) if there were a heat exchanger (with the turbine exhaust heating the compressor-delivery air) before the waste-heat boiler.

4

Diffusion and diffusers

The design of turbomachinery is dominated by diffusion—the conversion of velocity or "dynamic" head into stream pressure. Every blade row in a typical axial compressor is a collection of parallel diffusers. In most centrifugal compressors both the rotor and the radial diffuser are limited by the diffusion capabilities of the flow channels. And the performance of turbines of all types is enhanced, sometimes markedly, by an efficient exhaust diffuser. A large proportion of the aerodynamic and hydro-dynamic losses (apart from kinetic-energy leaving losses) in turbomachines is due to local or general areas of boundary-layer separation resulting from a local or general degree of diffusion that is too large for the boundary layer to overcome. It was principally the lack of understanding of diffusion that delayed the arrival of efficient pumps and compressors until some decades into this century, and of viable gas-turbine engines until the mid-1930s.

The analysis of diffusion is extremely complicated. Only a small part of the multidimensional matrix of diffusion parameters has been investi-gated experimentally. Nevertheless, we shall propose a very useful rule to be employed in preliminary design: that a simple mainstream velocity ratio, (W_2/W_1), should be kept above a minimum value if losses are to be low.

Diffusion can occur on an isolated surface (figure 4.1a) or in a duct (figure 4.1b). In either case, for the desired velocity reduction from W_1 to W_2 to occur, the boundary layers must remain attached. Where separation occurs, the main flow forms a jet that dissipates into turbulence.

The data quoted and analyzed here will be for ducts, but we shall apply the findings to diffusion in general, including that found on isolated or nearly isolated bodies (such as axial-compressor blades).

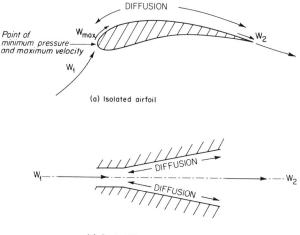

(a) Isolated airfoil

(b) Duct diffuser

Figure 4.1 Examples of diffusion: (a) isolated airfoil; (b) duct diffuser.

4.1 Diffusion in ducts

Diffusion can occur because of area change or because of flow curvature. Subsonic diffusers have increasing area. Figure 4.2 shows some different types of symmetrical, straight-axis diffusers.

Curved diffusers

Diffusion occurs locally when flow approaches a bend in a constant-area duct (figure 4.3). Potential (incompressible, inviscid) flow, which is a good model of flow away from areas dominated by viscosity, requires that the flow speed up on the inside of a bend and slow down on the outside. Therefore diffusion occurs approaching the outside of a bend and leaving the inside of a bend, and these are locations where separation is possible. Also boundary layers are less stable on concave than on plane surfaces.

When the axis of a diffuser is curved, the overall diffusion is added to the local diffusion. For separation not to occur, less overall diffusion must obviously be attempted. In most cases separation has to occur in only one place for complete flow breakdown.

Influence of inlet boundary layer

Flows with laminar boundary layers at inlet, or with thick turbulent layers, will not withstand as much diffusion without separation as will thin turbulent layers.

A model or analogy of a flow undergoing diffusion is a line of people on skateboards on a walled track approaching a hill (figure 4.4).

Suppose that the skaters in the middle have enough velocity to get over

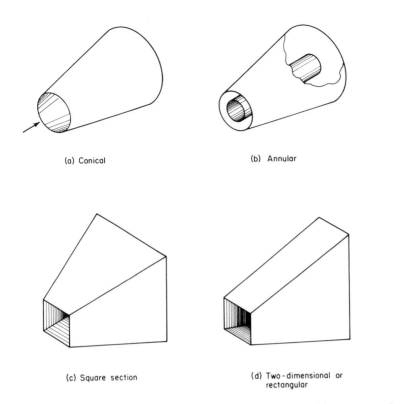

(a) Conical

(b) Annular

(c) Square section

(d) Two-dimensional or rectangular

Figure 4.2 Axial-flow diffusers: (a) conical, (b) annular, (c) square section, (d) two-dimensional or rectangular.

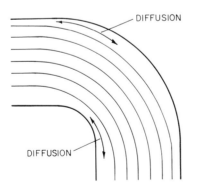

Figure 4.3 Diffusion in a constant-area bend.

Figure 4.4 Skate-boarding analogy to diffusion.

the hill. The skaters on the outside are brushing against the walls, and their velocity has been reduced so much that they have insufficient kinetic energy to surmount the hill. If they come to rest half way up the hill and begin sliding back, any following lines of skate-boarders will have to leave the wall in an analogy of separation.

The only way for these skate-boarders at the sides to get over the hill (if we assume that they are allowed only to roll and are not to use their muscles) is for their neighbors who are going at higher speeds to give them a hand—to drag them up the hill. In laminar boundary layers the (viscous) drag forces between layers is small, and this is why laminar flows cannot diffuse far without separation. In turbulent flows the inner strata of the boundary layers are given energy by exchange of "packets" of fluid from the higher-energy regions. In the analogy skate-boarders far from the wall, going at high speed, would exchange places with slow skate-boarders near the wall. The bumping that would occur during these exchanges would ensure that no skate-boarders could go very slowly in conditions of moderate diffusion or moderate slope. The energy losses during this bumping process would, of course, be larger than for the laminar model, but the losses are always small compared to those that occur if separation is allowed to take place.

Thus the inlet boundary-layer condition has a very significant influence on the performance of any diffuser.

Separation
The term "separation" is, then, short for "separation of the flow from the guiding surface." It occurs when the main flow is encountering a pressure rise (or rate of diffusion) severe enough to bring the boundary layer to rest. Usually, a vortex forms just downstream of this stationary point, with some of the flow reversing direction against the wall. This vortex effectively presents another bounding surface to the flow. Sometimes the new streamlines of the main flow result in a pressure distribution sufficiently modified (in the direction of reducing the severity of the adverse pressure gradient) for the main flow to "reattach" to the wall. The backward-turning vortex then forms a "separation bubble" which acts rather as does a roller bearing, and overall flow losses are small. We describe this phenomenon, and its effects on the pressure and heat-transfer distributions on a gas-turbine blade, in chapter 10.

Frequently, however, the flow does not reattach but continues as a high-velocity jet flow away from the wall. The jet dissipates in turbulent mixing, and there is little further pressure recovery. This type of separation is to be avoided wherever possible. Where it is unavoidable, it should be limited in extent. For instance, in a typical aircraft-engine layout the compressor will deliver air at around 200 m/s to a coaxial annular diffuser, which can be designed to reduce the velocity efficiently to a little under

100 m/s. The combustion system needs air flowing at a much lower velocity, but to continue to attempt to diffuse will invite a jet-type separation which might then occur further upstream than is necessary. By placing some type of break in the smooth wall, the separation can be located to yield maximum diffusion.

Separation as described is a steady-flow phenomenon. Frequently, however, it is unsteady. Sometimes separation that rapidly travels up and down a wall without ever resulting in gross jetlike separation yields maximum diffusion. On the other hand, the type of unsteadiness where a jet oscillates from one wall of a rectangular diffusing passage to the other will give high losses in the diffuser. Such a flow can also produce losses in, for instance, a downstream blade row, designed to accept a fully attached flow. This problem produces arguments among component designers, with the person responsible for a downstream diffuser or blade row blaming poor performance on the velocity profile delivered by the upstream component. We often cannot resolve such arguments satisfactorily, because we do not know how to characterize an acceptable velocity profile except to state that gross jet-type separation, steady or unsteady, is usually (with combustion systems being necessarily an exception) unacceptable.

Separation that is steady in an upstream component, for instance, a rotor, may appear to be unsteady in the form received by a downstream component. Radial-flow compressors of high pressure ratio generally experience a region of separation near the rotor circumference on the suction side of the blades (chapter 9). In those cases where a circular row of diffuser vanes is used downstream (outside) of the rotor, it appears to be desirable to allow a relatively short radial distance, of about 5 percent of the rotor radius, for the jets and wakes to mix out sufficiently for the diffuser row to be capable of operating efficiently. On the other hand, many studies of the effect of varying the axial distance between rotor and stator rows in axial-flow compressors have failed to show that there is any advantage to increasing the spacing to more than that necessary for purely mechanical reasons. These different experimental conclusions may simply reflect the difference in the extent of the unavoidable separation in radial-flow and axial-flow compressors.

Some of the cases of separation just discussed have been two-dimensional—the jet flow oscillating from one side to the other of a rectangular-section passage—and some have been three-dimensional, as occurs in a conical or annular diffuser. We do not know enough about the interaction of a separation from one wall of a rectangular diffuser with the flow in the corner and the influence on the flow on the neighboring walls to do more than refer the reader to the experimental results given here and elsewhere, and to sound a warning about applying results from two-dimensional diffuser tests to three-dimensional cases.

Diffuser flow regimes

Kline and his co-workers at Stanford University have identified the significant flow regimes of diffusers. Figure 4.5 shows these regimes for straight-walled two-dimensional diffusers with thin turbulent inlet boundary layers. They found that maximum pressure recovery occurred when a condition of transitory stall existed. This is a condition where short-duration flow reversal propagates up and down the diffuser walls, but general flow separation does not occur.

4.2 Performance measures

Pressure-rise coefficient

The purpose of diffusers is to convert dynamic pressure at inlet to added static pressure at outlet. Therefore a useful measure of performance is the ratio of these two parameters:

$$\text{pressure-rise coefficient, } Cpr \equiv \frac{p_{st2} - p_{st1}}{p_{01} - p_{st1}}, \tag{4.1}$$

where $(p_{01} - p_{st1})$ is also given the symbol q_1, the dynamic pressure at diffuser inlet. Ideally, the value of Cpr could reach unity. In practice, only in very favorable circumstances can a pressure-rise coefficient of over 0.80 be obtained.

The pressure-rise coefficient in a compressible-flow lossy diffuser

The pressure-rise coefficient is defined as

$$Cpr \equiv \frac{p_{st2} - p_{st1}}{p_{01} - p_{st1}},$$

as before. The pressures, temperatures, and velocities are shown on a $T - s$ diagram in figure 4.6.

$$
\begin{aligned}
\frac{p_{st}}{p_0} &= \left[\frac{T_{st}}{T_0} \right]^{Cp/R} \\
&= \left[\frac{T_0 - \dfrac{C^2}{2g_c Cp}}{T_0} \right]^{Cp/R} \\
&= \left[1 - \frac{C^2}{2g_c Cp T_0} \right]^{Cp/R}.
\end{aligned}
\tag{4.2}
$$

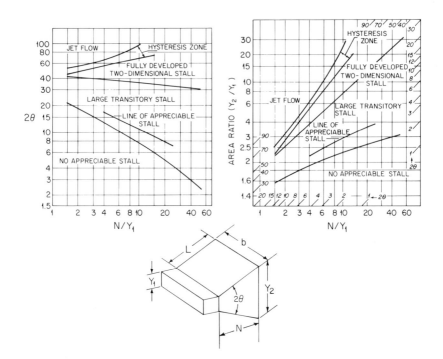

Figure 4.5 Flow regimes in straight-wall two-dimensional diffusers. From Fox and Kline 1962.

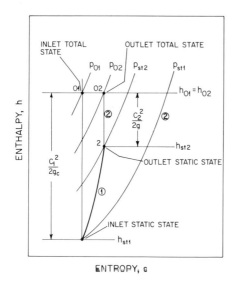

Figure 4.6 Diffusion on the *h-s* diagram.

This is the exact solution, which can also be given in the form of equation (2.41):

$$\frac{p_{st}}{p_0} = \left[1 + \frac{\gamma - 1}{2} M^2\right]^{-\gamma/(\gamma-1)}.$$ (4.3)

An approximate solution can be obtained by using the first two terms only of the binomial expansion.

$$\frac{p_{st}}{p_0} \approx 1 - \frac{C^2}{2g_c R T_0} = 1 - \frac{\rho_0 C^2}{2g_c p_0},$$

$$p_{st} \approx p_0 - \frac{\rho_0 C^2}{2g_c}.$$ (4.4)

This is similar to the frequently used incompressible relation for the dynamic head. For it to be a fair approximation to compressible-flow conditions, the total density, ρ_0, not ρ_{st}, must be used.

The term including the velocity, C, in equation (4.2) is a Mach number, as can be seen from the form of the relation given in equation (2.41).

The approximate relation, equation (4.4), can also be obtained in terms of the local Mach number by substitution of equation (2.39):

$$\left(\frac{p_{st}}{p_0}\right) \approx 1 - \frac{C^2}{2g_c R T_0} = \left[1 + \frac{\gamma}{(2/M^2) - 1}\right]^{-1}.$$ (4.5)

To show the degree of approximation involved in using equation (4.5) instead of the exact form, equation (4.2), values for each, and for the ratio of the approximate to the exact values, are listed in table 4.1 for $\gamma = 1.4$, as a function of Mach number.

The errors involved in using the incompressible form for the dynamic head are less than 1 percent up to a Mach number of 0.4, and are still less than 2.5 percent at $M = 0.6$.

The substitution of equation (4.4) into (4.1) leads to

$$Cpr \approx \frac{p_{02} - p_{01}}{\rho_{01} C_1^2/2g_c} + \left[1 - \frac{\rho_{02}}{\rho_{01}}\left(\frac{C_2}{C_1}\right)^2\right]$$ (4.6)

for low Mach numbers. The total temperature does not change in an adiabatic diffuser, so that we can put $\rho_{02}/\rho_{01} = p_{02}/p_{01}$ for perfect-gas flow. For lossless flow,

$$p_{02} = p_{01} \quad \text{and} \quad Cpr_{th} = 1 - \left(\frac{C_2}{C_1}\right)^2.$$ (4.7)

Diffusion and Diffusers

Table 4.1
Density-ratio approximation for $\gamma = 1.4$

	Ratio of static pressure to total pressure		
	Exact form	Approximate form	Approximate/ exact
Velocity relation	$\left[1 - \dfrac{C^2}{2g_c CpT_0}\right]^{Cp/R}$	$\left[1 - \dfrac{\rho_0 C^2}{2g_c p_0}\right]$	
Mach-number relation	$\left[1 + \dfrac{\gamma - 1}{2}M^2\right]^{\gamma/(\gamma-1)}$	$\left[1 + \dfrac{\gamma}{2/M^2 - 1}\right]^{-1}$	
Mach number			
0.2	0.97250	0.97222	0.99971
0.4	0.89561	0.89147	0.99538
0.5	0.84302	0.83333	0.98851
0.6	0.78400	0.76493	0.97567
0.7	0.72093	0.68761	0.95379
0.8	0.65602	0.60284	0.91893

This is the theoretical pressure-rise coefficient. The velocity ratio (C_2/C_1) is more often given here as (W_2/W_1), where W is a relative velocity, so that Cpr_{th} can be applied to moving as well as to stationary ducts.

The diffuser velocity ratio has been found to be a good guide to permissible diffusion levels in nonisentropic and compressible flows. The ratio (W_2/W_1) is sometimes known as the de Haller number, after the Swiss engineer who suggested that this ratio should not be less than 0.72 for compressor cascades. The relation between this value and Cpr_{th} is shown in table 4.2 and figure 4.7.

This number agrees well with the upper limit of 0.5 for the value of Cpr_{th}, which is often used for compressor cascades. If the cascades have to work in conditions where the blade surfaces become roughened, such as by industrial pollution or salt deposits, a lower value of Cpr_{th} should be used, such as 0.45 or 0.40. A lower value should also be used for blade rows that are operated over a wide range of incidences, such as the first stages in a high-pressure-ratio compressor (see chapter 8). An upper limit of 0.4 would be appropriate for these first stages for many situations.

4.3 Theoretical Cpr_{th} or $\Delta p/q$ as a design guide

The theoretical value of Cpr_{th} (or $\Delta p/q$, which is identical, but Cpr_{th} is more frequently used in blade-row and stage design) chosen for any particular diffuser or blade row may not necessarily be related closely to the

Table 4.2
Velocity ratio and Cpr_{th}

(W_2/W_1)	Cpr_{th}
0.6	0.64
0.7	0.51
0.75	0.44
0.8	0.36
0.85	0.28
0.9	0.19

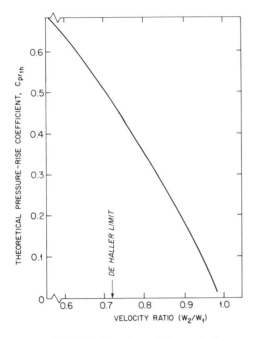

Figure 4.7 Velocity ratio and theoretical-pressure-rise coefficient.

Cpr actually achieved. It is simply a design guide to the value of outlet relative velocity to be specified. Thus, if the inlet relative velocity is W_1, the diffuser outlet relative velocity W_2 may be chosen from equation (4.8).

$$W_2 = W_1(1 - Cpr_{th})^{1/2}. \tag{4.8}$$

4.4 Diffuser effectiveness

There is in fact a performance measure that is defined as the ratio between the actual and the theoretical pressure-rise coefficients. The ratio of the actual to the theoretical is defined as the diffuser effectiveness:

$$\eta_{\text{diff}} \equiv \frac{(\Delta p/q)_{\text{ac}}}{(\Delta p/q)_{\text{th}}} = \frac{Cpr}{Cpr_{\text{th}}}. \tag{4.9}$$

The effectiveness is sometimes cross-plotted on diffuser-performance charts. It is related to the actual pressure-rise coefficient and to the area ratio, typically as shown in figure 4.8a and b (Reneau et al. 1967). The variation of the actual to the theoretical pressure-rise coefficients for different inlet-boundary-layer thicknesses is plotted by Nebo in figure 4.8c.[1]

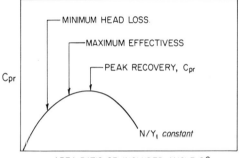

Figure 4.8 (a)

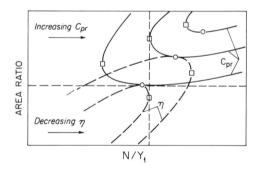

Figure 4.8 (b)

---

1 A. C. Nebo (1975), response to assignment in course 2.275, MIT, Cambridge, Mass.

Diffusion and Diffusers

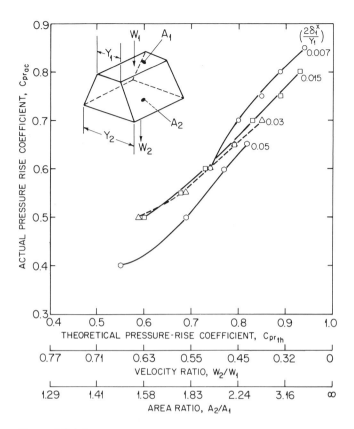

Figure 4.8 (c)

Figure 4.8 Relationship of diffuser effectiveness to actual pressure-rise coefficient: (a) Representative locations of several optima of performance at constant N/Y_1. From Reneau, Johnston, and Kline 1967. (b) Locations of optima of performance at constant length (square symbols) and constant area (round symbols). From Reneau, Johnston, and Kline, 1967. (c) Actual pressure-rise coefficient versus theoretical pressure-rise coefficient, velocity ratio and area ratio for straight two-dimensional diffusers. From Nebo 1975.

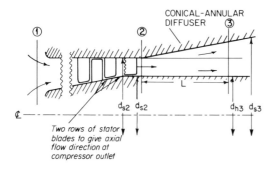

CONICAL-ANNULAR DIFFUSER ③

d_{s2} d_{s2} d_{h3} d_{s3}

$\mathcal{L}$

Two rows of stator
blades to give axial
flow direction at
compressor outlet

L

Figure 4.9 Conical-annular diffuser at outlet of axial compressor.

Example (axial-compressor diffuser design)

Design a conical-annular diffuser, shown diagramatically in figure 4.9, for an axial compressor to have a theoretical pressure-rise coefficient, (Cpr_{th}) of 0.55. Then calculate a one-dimensional value for $Cpr \equiv [(p_{st3} - p_{st2})/(p_{02} - p_{st2})]$, and compare it with the theoretical value. Also, locate a possible annular diffuser from figure 4.16. Assume these conditions:

Mass flow, $\dot{m} = 30$ kg/s.

Diffuser-entrance total pressure, $p_{02} = 500$ kN/m².

Diffuser-entrance total temperature, $T_{02} = 450$ K.

Diffuser-entrance velocity, axial, $C_{x2} = 200$ m/s.

Diffuser total-pressure loss, $(p_{02} - p_{03}) = 0.1 (p_{02} - p_{st2})$.

Compressor-inlet total temperature, $T_{01} = 288$ K.

Compressor-inlet total pressure, $p_{01} = 100$ kN/m².

$(d_{h2}/d_{s2}) = 0.90$; $d_{h2} = d_{h3}$; $2L/(d_{s2} - d_{h2}) = 5.0$.

Use a mean specific heat based on the mean static temperature of the diffuser. Find d_{h2}, d_{s2}, and d_{s3} for one-dimensional-flow conditions.

Summary of calculations The dimensions were calculated to be

$d_{h2} = d_{h3} = 0.4854$ m,

$d_s = 0.5393$ m,

$d_{s3} = 0.5604$ m,

$L = 0.1348$ m.

The actual value of the pressure-rise coefficient was calculated to be 0.442, in comparison with the theoretical (incompressible inviscid) value of 0.55. The actual value is lower than the theoretical value, first, because there is a frictional drop in total pressure leading to a smaller increase in static pressure than would otherwise occur and, second, because the calculations include approximations.

Calculations The theoretical inviscid pressure-rise coefficient is

$$Cpr_{th} = 1 - \left(\frac{C_3}{C_2}\right)^2,$$

where $C_3 = C_{x3}$,
$\qquad C_2 = C_{x2}$.

(The radial components of velocity may be neglected.)
For $Cpr_{th} = 0.55$,

$$\frac{C_{x3}}{C_{x2}} = 0.671;$$

$$\therefore \; C_{x3} = 0.671 \times 200 = 134.20 \text{ m/s}.$$

To calculate p_{st2} and p_{st3},

$$T_{st2} = T_{02} - \frac{C_{x2}^2}{2g_c \overline{Cp}} \quad \text{(first guess at } \overline{Cp} = 1{,}000 \text{ J/kg-}^\circ\text{K)}$$

$$= 450 \text{ K} - \frac{200^2}{2 \times 1{,}000} = 450 - 20 = 430 \text{ K},$$

$$T_{st3} = 450 \text{ K} - 0.45 \times 20 = 450 - 9 = 441 \text{ K}.$$

Diffuser mean static temperature $= 435.5$ K. By interpolation in table A.2 (appendix A),

$$\overline{Cp} = 1{,}018.8 \text{ J/kg-}^\circ\text{K}.$$

A second iteration with this closer value of $\overline{Cp}$ leads to

$$T_{st2} = 430.37 \text{ K},$$

$$T_{st3} = 441.2 \text{ K}.$$

Then

$$\left(\frac{p_{02}}{p_{st2}}\right) = \left(\frac{T_{02}}{T_{st2}}\right)^{Cp/R}$$

$$= \left(\frac{450}{430.35}\right)^{3.5464}$$

$$= 1.1716$$

$$\therefore \; p_{st2} = 426.78 \text{ kN/m}^2;$$

$$(p_{02} - p_{st2}) = 73.221 \text{ kN/m}^2$$

$$(p_{02} - p_{03}) = 0.1(p_{02} - p_{st2}) = 7.322 \text{ kN/m}^2$$

$$\therefore \; p_{02} = 492.68 \text{ kN/m}^2$$

$$\frac{(p_{st3} - p_{st2})}{(p_{02} - p_{st2})} = 0.442.$$

And

$$\left(\frac{p_{03}}{p_{st3}}\right) = \left(\frac{T_{03}}{T_{st3}}\right)^{Cp/R}$$

$$= \left(\frac{450}{441.2}\right)^{3.5464}$$

$$= 1.0730;$$

$$\therefore p_{st3} = 459.17 \text{ kN/m}^2$$

$$(p_{st3} - p_{st2}) = 32.389 \text{ kN/m}^2.$$

Sizing We must calculate first the total and then the static densities at inlet and outlet of diffuser:

$$\rho_{02} = \frac{p_{02}}{RT_{02}} = \frac{500,000 \text{ N/m}^2}{286.96 \text{ J/kg-K } 450 \text{ K}} = 3.8720 \text{ kg/m}^3,$$

$$\frac{\rho_{02}}{\rho_{st2}} = \left(\frac{T_{02}}{T_{st2}}\right)^{(\overline{C}p/R)-1} = \left[\frac{450}{430.35}\right]^{2.550} = 1.1204;$$

$$\therefore \rho_{st2} = 3.4558 \text{ kg/m}^3.$$

$$\rho_{03} = \frac{p_{03}}{RT_{03}} = \frac{492,680 \text{ N/m}^2}{286.96 \text{ J/kg-K } 450 \text{ K}} = 3.8153 \text{ kg/m}^3,$$

$$\left(\frac{\rho_{03}}{\rho_{st3}}\right) = \left(\frac{T_{03}}{T_{st3}}\right)^{(\overline{C}p/R)-1} = \left[\frac{450}{441.15}\right]^{2.550} = 1.0519;$$

$$\therefore \rho_{st3} = 3.6271 \text{ kg/m}^3.$$

Continuity

$$\dot{m} = \rho A C,$$

$$A_2 = \frac{\pi d_{s2}^2}{4}[1 - 0.9^2] = \frac{30 \text{ kg/s}}{3.4558 \text{ kg/m}^2 \ 200 \text{ m/s}} = 0.0434 \text{ m}^2;$$

$$\therefore d_{s2} = 0.5393 \text{ m}.$$

$$d_{h2} = 0.4854 \text{ m},$$

$$A_3 = \frac{\pi}{4}[d_{s3}^2 - 0.4854^2] = \frac{30 \text{ kg/s}}{3.6271 \text{ kg/m}^3 \ 134.2 \text{ m/s}} = 0.0616 \text{ m}^2;$$

$$\therefore d_{s3} = 0.5604 \text{ m}.$$

$$\frac{d_{s2} - d_{h2}}{2} = \text{blade height} = 26.95 \text{ mm.}$$

$$L = 5\left(\frac{d_{s2} - d_{h2}}{2}\right) = 134.75 \text{ mm.}$$

Not all the input information was required.

4.5 Diffuser performance data

The very simple design rules for selecting values of $(\Delta p/q)_{\text{th}}$ for different diffuser applications are one alternative open to the designer. The other is to find experimental data for the actual configuration and conditions under study. We report some representative test data for different configurations and conditions.

This alternative approach can provide only occasional guidelines in a multidimensional space. There are too many variables for the whole field to have been more than spotted with experiments. These are the principal variables.

Diffuser cross section (circular, rectangular, elliptical, annular, etc.).

Ratio of length to inlet hydraulic diameter.

Size and shape of inlet boundary layers.

Mach number of inlet flow.

Direction of inlet flow (axial, at an incidence, swirling, log spiral, etc.).

Steady, periodic, or fluctuating flow.

Straight axis or curved.

Smooth walls or rough, or shaped (e.g., ribbed).

Downstream conditions (plenum, tailpipe, transverse wall, etc.).

With such a vast array of possible combinations of conditions, the designer is unlikely to find test results that exactly match the conditions of interest. The data, however, can be extrapolated or interpolated in many cases, and the area of uncertainty in which the designer's judgment must be relied upon will thereby be reduced.

Conical-diffuser performance

The following performance data for conical diffusers were taken principally from work performed by Dolan and Runstadler (1973) for NASA.

The variables were length-to-inlet-diameter ratio L/d_{ht}; inlet (throat) Mach number M_t; throat boundary-layer blockage B_t; throat Reynolds number ($\mathrm{Re}_t \equiv W_t d_{ht} \rho / \mu$); and total-pressure level p_0. The fluid was air; the flow was axial, steady, and nonswirling. The diffusers had straight axes, smooth walls, and discharged into a plenum. The throat boundary-layer blockage is the ratio of the sum of the boundary-layer displacement thickness (δ^*) (or equivalent stream thickness) to the total throat width. The performance is given as the pressure-recovery coefficient, Cpr, which is identical to ($\Delta p / q$).

The optimum divergence angle for a fixed L/d_{ht} ratio (8) and midrange values of other conditions is shown to be between 7 and 8 degrees in figure 4.10. This optimum angle gives a maximum pressure-recovery coefficient of 0.65.

At a fixed divergence angle of 4 degrees, the pressure-recovery coefficient Cpr rises as the diffuser is made longer, until at $L/D = 25$, Cpr is about 0.76 for the conditions for which figure 4.11 is plotted.

Correlations of Cpr against area ratio and L/d_{ht} with divergence angle as a parameter are shown in figure 4.12 for different inlet conditions. It can be seen that for all conditions the optimum divergence angle decreases from about 8 degrees for short diffusers (L/d_{ht} below 5) to between 4 and 6 degrees for long diffusers (L/d_{ht} over 25).

In figure 4.13 the throat Mach number is shown to have a surprisingly small effect on Cpr. The performance improves with increase of Reynolds number at low Mach numbers (figure 4.12a and b) but not, apparently, at high Mach numbers (figure 4.12c and d).

Square-diffuser performance
A comparison of conical- (circular) and rectangular-diffuser (square throat) performance by Dolan and Runstadler (figure 4.14) shows only a slight falloff for the rectangular diffusers. Therefore the data for conical diffusers may be used for square diffusers, with an appropriate correction for large inlet boundary layers.

Rectangular-diffuser performance
Correlations of peak pressure recovery at constant L/d_{ht} ratios by Reneau, Johnston, and Kline are shown in figure 4.15. Values of the inlet boundary-layer blockage are given for various test points. The optimum divergence angles (figure 4.15b) can be seen to be about twice those for conical diffusers at equivalent L/d_{ht} ratios, as would be expected.

Annular-diffuser performance
Figure 4.16a shows that first stall in annular diffusers occurs at a smaller area ratio for a given value of N/d_{ht} and N/h (where h is the annular height) than for a conical diffuser. Values of Cpr for three straight-core (constant inner diameter) annular diffusers of divergence angles (2θ) 10, 20, and 30 degrees and for a range of Reynolds numbers are shown in figure 4.16b

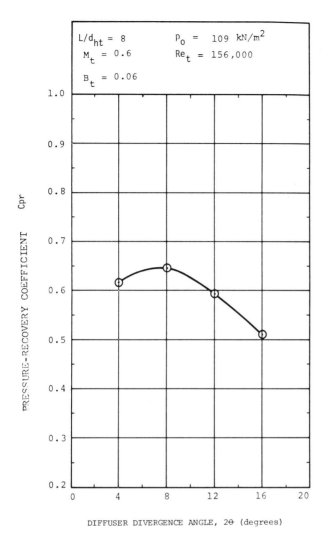

Figure 4.10 Pressure recovery versus divergence angle. From Dolan and Runstadler 1973.

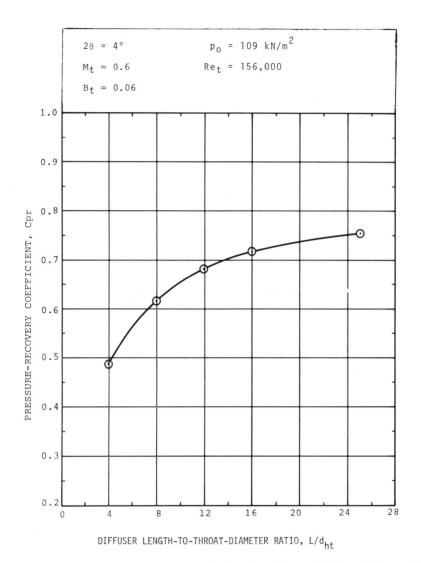

Figure 4.11 Pressure recovery versus diffuser length. From Dolan and Runstadler 1973.

Diffusion and Diffusers

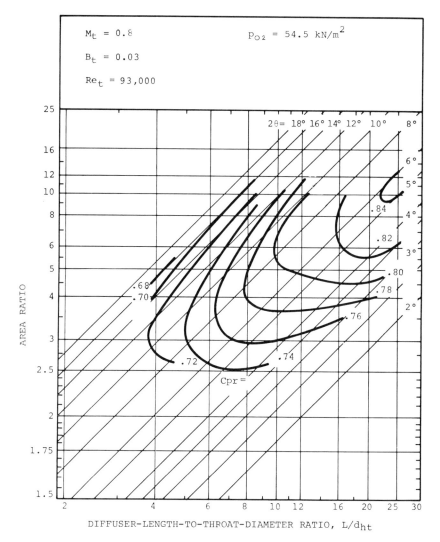

$M_t = 0.8$ $P_{O_2} = 54.5 \ kN/m^2$

$B_t = 0.03$

$Re_t = 93,000$

$2\theta = $ 18° 16° 14° 12° 10°   8°

6°

5°

4°

.84

3°

.82

.80

2°

.78

.76

.68

.70

.74

.72

$C_{pr} = $

AREA RATIO

DIFFUSER-LENGTH-TO-THROAT-DIAMETER RATIO, L/d_{ht}

Figure 4.12 (a)

Diffusion and Diffusers   167

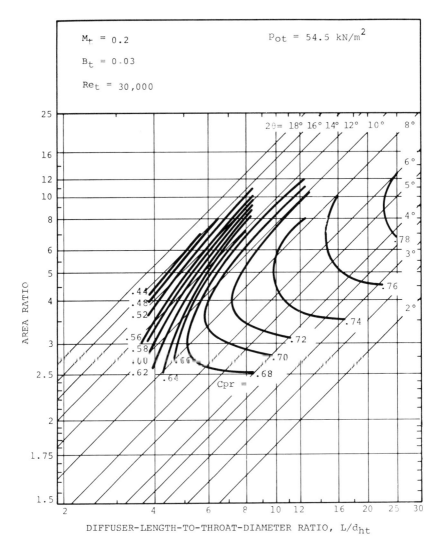

Figure 4.12 (b)

Diffusion and Diffusers

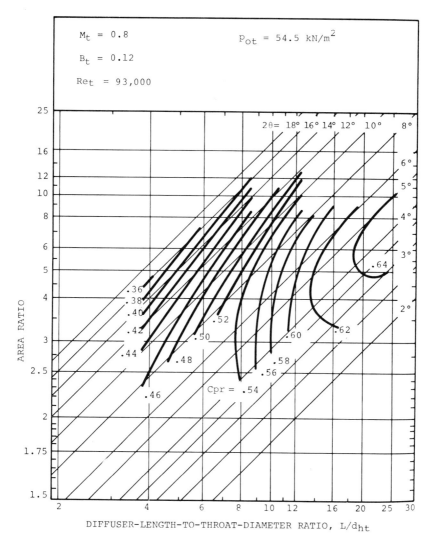

Figure 4.12 (c)

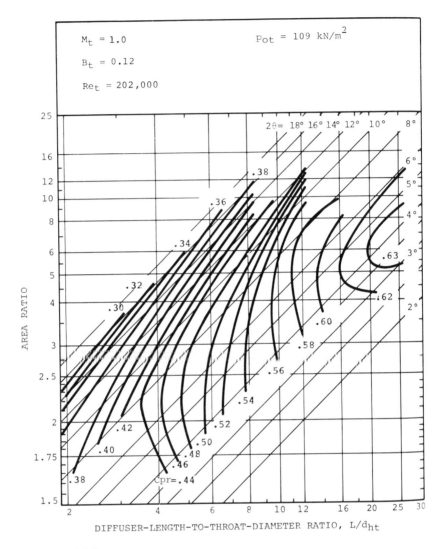

Figure 4.12 (d)

Figure 4.12 Conical-diffuser performance map. From Dolan and Runstadler 1973. (a) $M_t = 0.8$, $B_t = 0.03$; (b) $M_t = 0.2$, $B_t = 0.03$; (c) $M_t = 0.8$, $B_t = 0.12$; (d) $M_t = 1.0$, $B_t = 0.12$.

Diffusion and Diffusers

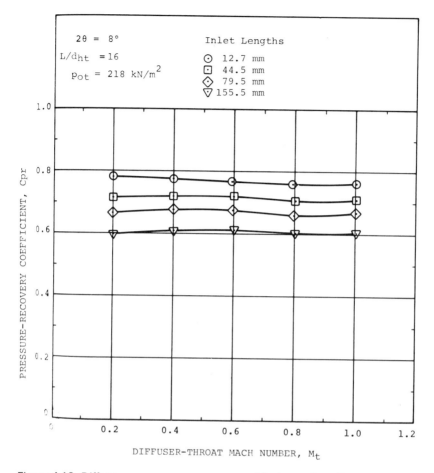

Figure 4.13 Diffuser pressure recovery versus Mach number. From Dolan and Runstadler 1973.

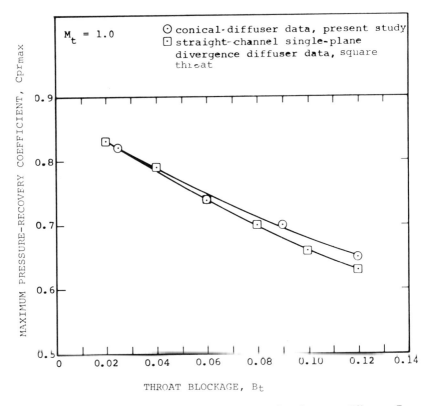

Figure 4.14 Maximum pressure recovery of conical and square diffusers. From Dolan and Runstadler 1973.

and c. It is evident that in this configuration (figure 4.16d) the 20-degree angle is optimum for big diffusers $((N/h) > 2.5)$ and the 30-degree angle is better for shorter diffusers. The test points are shown in relation to the Cpr contours in figure 4.16e.

Tailpipe effects
When a rectangular diffuser discharges into a straight constant-area tailpipe, instead of into a plenum as is used for most tests, an increase of pressure recovery normally occurs. Typical test results of Kelnhofer and Derick are shown in figure 4.17. The parameter for the curves is (L_d/Y_1), where L_d is the length of added tailpipe and Y_1 the inlet or throat width.

Wall contouring: splitter vanes
Substitution of a ribbed or finned surface (figure 4.18a) for a smooth surface has been shown by several investigators to give improved pressure recovery. Also the use of a bell-shaped wall contour (figure 4.18b) rather

Diffusion and Diffusers

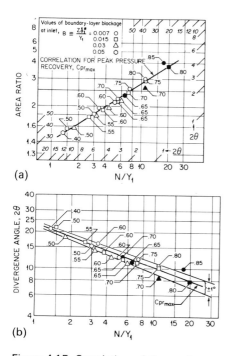

Figure 4.15 Correlation of the peak pressure recovery at constant (N/Y_1): (a) area ratio versus (N/Y_1); (b) divergence angle versus (N/Y_1).

than a straight configuration has been found to give slightly improved performance, or shorter diffusers at the same performance.

Shorter diffusers can also be obtained by using splitter vanes to produce a number of small-angle diffusers in parallel (figure 4.18c).

In none of these approaches have design data adequate to cover general cases been produced. Designers wishing to take advantage of potential improvements must resort to test programs.

4.6 The effects of swirl on axial-diffuser performance

A small degree of swirl in the inlet flow of a diffuser with its axis generally in line with the flow (as distinct from a radial diffuser) may cause a sharp drop in performance and perhaps complete flow breakdown. As little as 10 degrees may be sufficient; see figure 4.19. The flow tries to set up a free vortex, with the tangential component of velocity inversely proportional to the radius. Such a flow pattern would result in infinite tangential velocities at the core. Shear forces prevent this occurring, and the central part of such a flow rotates as a solid body. The energy that has been dissipated leaves a total-pressure deficit, and the outer rotating flow produces a low static pressure at the core. With only small degrees of swirl, high-

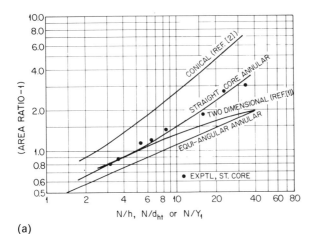

(a)

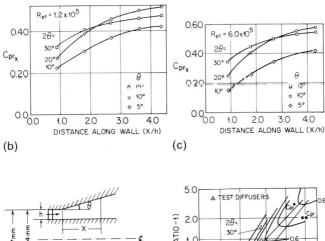

(b)   (c)

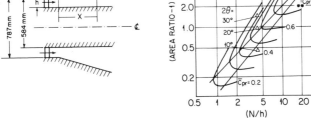

(d)

Figure 4.16 Performance of annular diffusers: (a) Lines of first stall. From Howard, Henseller, and Thornton-Trump 1967. (b) Axial static-pressure distribution for three diffusers at Re = 1.2 × 10⁶. From Adenubi 1975. (c) Axial static-pressure distribution for three diffusers at Re = 6.0 × 10⁶. From Adenubi 1975. (d) Test geometries in relation to Cpr contours of Sovran and Kemp. From Adenubi 1975.

Diffusion and Diffusers

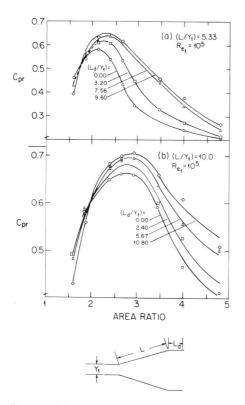

Figure 4.17 Straight-walled diffuser pressure-recovery coefficient with effect of tail-pipe addition. From Kelnhofer and Derick 1971. (a) $(L/Y_1) = 5.33$; (b) $(L/Y_1) = 10.0$.

pressure downstream flow will reverse to flow into this low-pressure core and to produce general flow breakdown.

Flow-straightening vanes are desirable upstream of the throat of turbomachine diffusers. Some compressor designs incorporate three rows of straightening vanes. Turbines almost never have straightening vanes, despite the much larger kinetic energy being discharged. It has to be concluded that turbine diffusers seldom recover much pressure over much of their operating ranges, because, in off-design conditions, swirl angles will be high. (One very successful engine that incorporates straightening vanes in the turbine exhaust is the Pratt & Whitney JT9.)

Radial diffusers

In contrast to axial diffusers, radial diffusers can diffuse flows with large amounts of swirl. They are used typically downstream of radial-flow compressor and pump rotors. The swirl angles may be 70 degrees to the

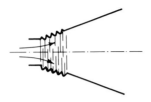

(a) Ribbed diffuser (diagrammatic)

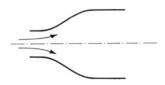

(b) Bell-shaped diffuser

(c) Diffuser with splitter vanes

Figure 4.18 Alternative forms of axial diffuser: (a) ribbed diffuser; (b) bell-shaped diffuser; (c) diffuser with splitter vanes.

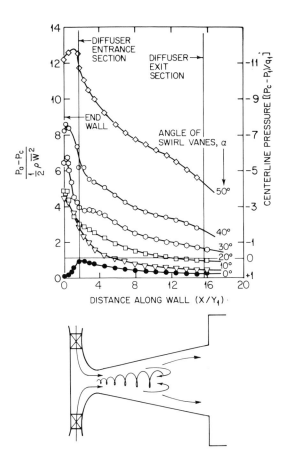

Figure 4.19 Centerline static-pressure distribution for swirling flow in conical diffusers. From So 1967.

radial direction, and in compressors the inlet Mach number may be well above unity.

It is useful to consider the two components of the entering flow separately.

The radial-flow component diffuses because the area increases, and because this component has a low Mach number even when the Mach number of the total velocity is above unity. In axisymmetric flow the pressure gradient is in the radial direction and acts on this radial component.

The tangential-velocity component diffuses because in axisymmetric flow, and in the absence of turning vanes and wall friction, the tangential velocity is inversely proportional to the radius. It is unaffected by area change. It can pass from supersonic to subsonic flow without shock (because it can adjust to direction changes from signals received through the subsonic radial component).

When the radial diffuser is used downstream of a typical compressor or pump, the tangential component of momentum will typically be over twice the radial component (the flow direction will be at more than 65 degrees to the radius). The radial component has to surmount the radial pressure gradient, however. When reverse flow of the radial boundary layer occurs, it is not possible for the tangential component to continue diffusing in axisymmetric flow, because this would imply that there would be a pressure increase in one component and not in the other. Therefore breakdown must occur nonsymmetrically. In fact rotating jets are found, producing flow spirals similar to those of spiral nebulae in space. We call this form of flow breakdown a rotating stall.

The best available guide to the geometric and flow conditions which will cause breakdown has been given by Jansen (1964), whose stability charts are reproduced as figure 4.20.

These charts are for radial diffusers bounded by plane parallel walls. Some designers have attempted to obtain more diffusion by increasing the axial width between the diffuser walls as the radius is increased. Such wall contouring is likely to have an adverse effect on pressure recovery, because the sluggish radial momentum would then break down earlier. There is much to be said for the opposite approach—of decreasing the wall spacing with the increase in radius. This is especially the case at the diffuser entrance, where a narrowing of the spacing downstream of the rotor will tend to energize the boundary layers in the radial direction but have no effect on the tangential component wherein is most of the momentum.

A wide range of design data for the pressure-rise coefficients of vaneless diffusers is not available. The pressure rise would be expected to be somewhat lower than that for an axial diffuser of equivalent area ratio.

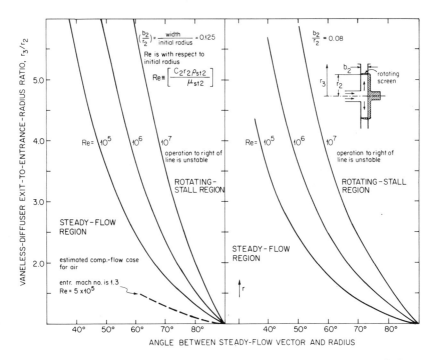

Figure 4.20 Stable operating range of vaneless diffusers. From Jansen 1964.

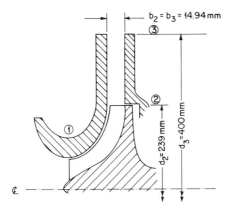

Figure 4.21 Diagram of vaneless radial diffuser for stability analysis.

Diffusion and Diffusers

179

Example (estimation of radial-diffuser stability)

Using figure 4.20, estimate whether or not the radial diffuser in figure 4.21 will be stable. The specified data are

$p_{01} = 100 \text{ kN/m}^2$.

$p_{02} = 653.7 \text{ kN/m}^2$.

$C_2 = 508.23 \text{ m/s}$.

$p_{03} = 641.23 \text{ kN/m}^2$.

$T_{01} = 288.5 \text{ K}$.

$T_{02} = 537.8 \text{ K}$.

$\alpha_2 = 65°$ (assume uniform).

$\dot{m} = 5.139 \text{ kg/s}$.

$N = 40{,}000 \text{ rev/min}$.

$p_{st2} = 251.78 \text{ kN/m}^2$.

$p_{st3} = 486.16 \text{ kN/m}^2$.

$\alpha_3 = 73.72°$.

$\rho_{st2} = 2.1333 \text{ kg/m}^3$.

Summary of results The diffuser would be stable, using Jansen's criterion literally as in figure 4.20.

Calculations To enter figure 4.20, we need (b_2/r_2), α_2, and Re_2 (the Reynolds number):

$$b_2 = 14.94 \text{ mm},$$

$$r_2 = \frac{239}{2} \text{ mm};$$

$$\therefore \left(\frac{b_2}{r_2} \right) = 0.125.$$

Therefore figure 4.20 can be used directly:

$$\alpha_2 = 65° \quad \text{as specified},$$

$$Re_2 \equiv \frac{\rho_{st2} C_2 (d_2/2)}{\mu_{st2}}.$$

We use table A.2 first to find T_{st}, by iterating $Cp_2 : T_{st2} = 410.6 \text{ K}$. Then, by interpolation, $\mu_{st2} = 2.328 \times 10^{-5} \text{ kg/ms}$;

$$\therefore Re_2 = \frac{2.1333 \text{ kg/m}^3 \times 508.23 \text{ m/s} \times 0.239/2 \text{ m}}{2.328 \times 10^{-5} \text{ kg/ms}}$$

$$= 5.56 \times 10^6.$$

By interpolation on figure 4.20a for the Reynolds number, stability would be given for an inlet flow angle of 65° up to a radius ratio of about 2.75. Here the radius

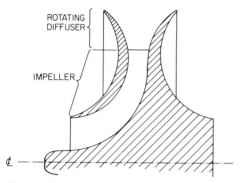

Figure 4.22 Rotating vaneless diffuser.

ratio is 1.67; therefore the radial diffuser under consideration appears to have good stability.

Only a small part of the input information provided was needed for this calculation.

Rotating vaneless diffusers

The concept of attaching the vaneless diffuser to the compressor rotor (figure 4.22) was patented in Germany by the Neu Company. The principle is that centrifugal action energizes the radial component of the boundary layer and enables it to overcome larger pressure gradients. Accordingly, such compressor rotors were frequently profiled to give increasing diffuser-wall spacing with radius. Critics of the approach maintained that the additional fluid friction on the outside of the rotating walls took away more than was gained in increased pressure recovery. For whatever reason, the concept does not appear to have survived. However, Rodgers and Mnew (1974) at Solar have tested a freely rotating vaneless diffuser with apparently promising results.

Vaned radial diffusers

To achieve a given pressure rise, a vaneless diffuser requires a considerable radial space, and it incurs fairly high frictional losses because of the long path length of the spiral flow. Vanes can be used to direct the flow more in the radial direction, to increase the rate of diffusion, and to decrease the losses.

Vaned diffusers may be of many forms, some of which are illustrated in figure 4.23. They become similar to axial-flow diffusers with square, rectangular, or occasionally circular cross sections. The inlet conditions are very complex, however, even when the machine is running at its design

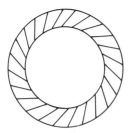

(a) Straight vanes

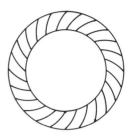

(b) Curved vanes

(c) Wedge vanes

(d) Multiple – trumpets

(e) Single scroll

Figure 4.23 Configurations of vaned radial diffusers.

Diffusion and Diffusers

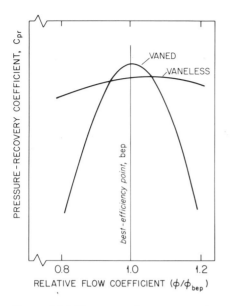

Figure 4.24 Representative pressure rise versus range for vaned and vaneless radial diffusers.

point when the flow should enter without incidence. Even in this condition the flow is highly three-dimensional. At off-design conditions incidence is added to complicate further the flow pattern and to increase the likelihood of flow separation and stalling. (The stall is likely to rotate, as it does in vaneless diffusers.)

No general design data are available for vaned diffusers. Even if the design of a reportedly successful diffuser is scrupulously copied, the use of a different rotor can give quite different inlet conditions and therefore different performance. Development testing must normally be resorted to. However, in general, it can be said that a vaned diffuser can give a higher pressure rise than a vaneless diffuser, but this will be found over a smaller range (figure 4.24).

4.7 The risk factor in diffuser design

Large gains in efficiency and in specific power may be made in turbomachinery design by increasing the diffusion of kinetic energy to pressure energy. To be greedy, however, by trying to get too much diffusion is to invite disaster. This is why.

Suppose we have a rectangular diffuser with two diverging walls between two plane walls (figure 4.25). We start with a small amount of diffusion and fully attached flow, and we gradually increase the amount of diffusion

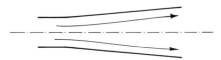

(a) Small area ratio: moderate pressure recovery

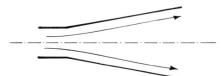

(b) Moderate area ratio: higher pressure recovery

(c) Higher area ratio: no pressure recovery

Figure 4.25 The risk factor in diffuser design: (a) Small area ratio—moderate pressure recovery; (b) Moderate area ratio—higher pressure recovery; (c) Higher area ratio—no pressure recovery.

either by increasing the length of the diverging walls at constant divergence angle, or by increasing the angle at constant length. The diffusion will increase up to a certain point, when separation will occur. Now, unfortunately, separation in such a case does not take place just at the maximum previous area ratio: it propagates back through the previously attached boundary layer and causes the whole flow to be separated. The last increment of diffusion demanded becomes the straw that breaks the camel's back.

An analogy to the diffuser-design risk is the problem of selling a used car with one advertisement. If one judges that the car is worth at least $500 and possibly $750, one can be absolutely sure of a sale by advertising the car at $450. One can probably make more money by putting the figure at $500, or $600, or even $700, but the chances of no one responding increase. If one advertises at $800, it is almost certain that no one will answer the advertisement. The extra greed has lost the whole enterprise.

References

Adenubi, S. O. 1975. *Performance and flow regime of annular diffusers with axial-turbomachine-discharge inlet conditions.* Paper 75-WA/FE 5. ASME, New York.

Dolan, Francis X., and Peter W. Runstadler, Jr. 1973. *Pressure-recovery performance of conical diffusers at high subsonic Mach numbers.* Report CR-2299, (July). NASA, Washington, D.C.

Fox, R. W., and S. J. Kline, 1962. Flow-regime data and design methods for curved subsonic diffusers. *Trans. ASME J. Basic Eng.*, D 84 (September).

Howard, J. H. G., H. J. Henseller, and A. B. Thornton-Trump. 1967. *Performance and flow regimes for annular diffusers.* Paper 67-WA/FE-21. ASME, New York.

Jansen, W. 1964. Rotating stall in a radial vaneless diffuser. *Trans. ASME J. Basic Eng.*, 86: 750–758.

Kelnhofer, William J., and Charles T. Derick. 1971. *Tailpipe* effects on gas-turbine-diffuser performance with fully developed inlet conditions. *Trans. ASME J. Eng. Power* 93 (January): 57–62.

Reneau, L. R., J. P. Johnston, and S. J. Kline. 1967. Performance and design of straight, two-dimensional diffusers, *Trans. ASME J. Basic Eng.* 89 D, (March): 141–150.

Rodgers, C., and H. Mnew. 1974. Experiments with a model free-rotating vaneless diffuser. *Trans. ASME J. Eng. Power* (paper 74-GT-58. ASME, New York).

So, Kwan L. 1967. Vortex phenomena in a conical diffuser. *AIAA Journal*, 5 (June): 1072–1078.

Problems

4.1
Write down in equations, relations, or words the steps you would take to find the outlet diameter of a plane-walled vaneless diffuser in figure P4.1 which is to give a rise in static pressure of 70 kN/m² in assumed conditions of isentropic, asisymmetric, steady flow in radial equilibrium. At the inlet, diameter 300 mm, the air is at 350

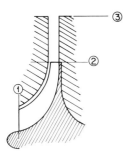

Figure P4.1

kN/m², 44.5 K, total pressure and temperature, and flows at 70° to the radial direction with a radial component of velocity of 55 m/s. (These figures are given only to indicate the degree of compressibility, etc., and are not intended to be used for calculations.)

4.2
Determine the rise in static pressure across the blade row shown in figure P4.2 when it is passing 33.93 kg/s air (subsonically), assuming uniform one-dimensional flow (ignoring boundary layers, etc.). Use the Mach-number charts figure 2.11 or tables.

4.3
Compare the actual pressure coefficient Cpr with the so-called theoretical incompressible Cpr_{th} as defined here for the row of axial-compressor blades sketched in figure P4.3. Comment on any difference.

$$Cpr \equiv \frac{p_{st2} - p_{st1}}{p_{01} - p_{st1}}, \quad Cpr_{th} \equiv 1 - \left(\frac{C_2}{C_1}\right)^2$$

where C_2 and C_1 are mean velocities at inlet and outlet. Use the *Gas Tables* or compressible-flow charts for gamma = 1.4, or appendix A data. The loss in total pressure between planes 1 and 2 is 15 percent of the inlet dynamic head ($p_{01} - p_{st1}$).

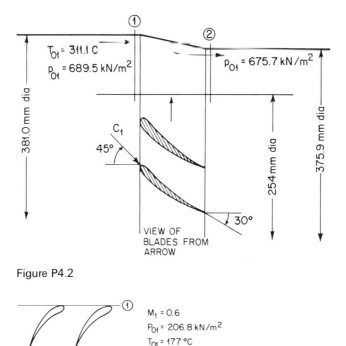

Figure P4.2

$M_1 = 0.6$
$P_{01} = 206.8 \text{ kN/m}^2$
$T_{01} = 177 \text{ °C}$
$\gamma = 1.4$
$A_2 = \Sigma A_1$ actual flow areas

Figure P4.3

Diffusion and Diffusers

4.4

If, for experimental convenience, you use air instead of water at similar velocities to develop a diffuser for a water pump, would you expect the actual pressure-rise coefficient to be higher or lower than when the diffuser is run in water? Why?

5

Energy transfer in turbomachines

In this chapter we shall first apply Newton's law to turbomachines and thereby derive Euler's equation. Then we shall use this to examine velocity diagrams for turbomachines and the various parameters that can be used to define velocity diagrams.

5.1 Euler's equation

Consider the flow of a single streamtube of fluid that enters the rotor of a turbomachine at one radius with one velocity and leaves at another radius with another velocity (figure 5.1). The change of momentum between the flow entering and leaving the rotor can be used to calculate the force on the rotor and the equal and opposite force on the fluid.

For our purposes it is helpful to consider the three principal components of this force: axial, radial, and tangential. The axial and radial components are important for the design of bearings and for the analysis of vibration excitations, for instances. But these two components cannot contribute to the work transfer between the working fluid and the rotor. Only the

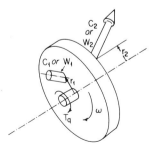

Figure 5.1 Flow through a rotor.

tangential component of the force can produce a change in enthalpy through a work transfer. This work can be computed from the forces and the torques as follows.

The flow filament enters the rotor at a radius r_1, with a tangential component of the absolute velocity $C_{\theta 1}$, positive in the direction of rotor rotation ω.

Tangential force on rotor from entering fluid $= \dot{m}C_{\theta 1}/g_c$.

Torque on rotor $= \dot{m}r_1 C_{\theta 1}/g_c$.

Net torque, Tq, on rotor from entering and leaving flows $= \dot{m}[r_1 C_{\theta 1} - r_2 C_{\theta 2}]/g_c$ assuming steady flow.

Energy transfer $= Tq\omega = \dot{m}[\omega(r_1 C_{\theta 1} - r_2 C_{\theta 2})]/g_c$.

Now $\omega r_1 = u_1$, the rotor peripheral speed at the radius of entry of the streamtube. And $\omega r_2 = u_2$;

$$\therefore + \dot{W} = Tq\omega = \frac{\dot{m}}{g_c}(u_1 C_{\theta 1} - u_2 C_{\theta 2})$$

or

$$\frac{\dot{W}}{\dot{m}} = \frac{1}{g_c}(u_1 C_{\theta 1} - u_2 C_{\theta 2}). \tag{5.1}$$

This is known as Euler's equation. Its origin is set in context in the brief history at the beginning of the book. Positive work means that work is delivered from the turbomachine shaft: that is, it is a turbine rather than a pump, fan, or compressor.

Euler's equation is generally applicable. The flow maybe compressible or incompressible, idealized or frictional. The one restriction of steady flow is relatively unimportant, because of the near impossibility of storing fluid within a rotor.

No torques other than those given by the fluid during passage through the rotor have been considered. Frictional torques on the disks or bearings or from the casing must be accounted for. Usually this is straightforward. Interaction between the casing (the stationary shroud) and the blade tips in an axial machine—where the flow in a streamtube may flow in a spiral, first performing useful work and then contributing purely to friction losses, or may do both simultaneously—is difficult to account for accurately but is normally relatively small.

For an adiabatic rotor in the absence of external torques, or large changes in elevation, a combination of the steady-flow energy equation with Euler's equation gives

$$g_c(h_{02} - h_{01}) = u_2 C_{\theta 2} - u_1 C_{\theta 1}$$

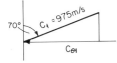

Figure 5.2 Vector diagram of flow leaving nozzles.

or

$$g_c \Delta_1^2 h_0 = \Delta_1^2 (u C_\theta). \tag{5.2}$$

This is the most useful single relation in turbomachinery design. It can be further simplified in many cases of axial- and radial-flow machines, as follows.

In the preliminary design of axial-flow machines, the change of radius of the mean flow can often be ignored, so that a more restricted version of Euler's equation becomes

$$g_c \Delta_1^2 h_0 = u \Delta_1^2 C_\theta. \tag{5.3}$$

In radial-flow compressors and pumps the flow is usually led to the rotor without "swirl"—the tangential component of the inlet velocity, $C_{\theta 1}$, is zero:

$$\therefore g_c \Delta_1^2 h_0 = u_2 C_{\theta 2}. \tag{5.4}$$

Example (use of the Euler equation for a steam-turbine stage)

What is the power output (kW) of a single-stage steam-turbine expander which takes 6 kg/s of dry saturated steam at 550 C, passes the flow through nozzles, from which it leaves at a direction 70° from that of the axial, at a velocity of 975 m/s, as diagramed in figure 5.2, and discharges it from the rotor blading without swirl ($C_{\theta 2} = 0$)? The mean diameter of the expander blading is 1 m, and the shaft speed is 10,000 rev/min. The turbine has an isentropic total-to-total stage efficiency, to the rotor outlet, of 75 percent.

Solution Using equation (5.2),

$$g_c(h_{02} - h_{01}) = u_2 C_{\theta 2} - u_1 C_{\theta 1} = -u_1 C_{\theta 1},$$

$$u_1 = 10,000 \times \frac{2\pi}{60} \times \frac{1}{2} = 523.60 \, \frac{\text{m}}{\text{s}},$$

$$C_{\theta 1} = C_1 \sin 70° = 916.20 \, \frac{\text{m}}{\text{s}};$$

$$\therefore \; g_c(h_{02} - h_{01}) = -479{,}721 \; \frac{m^2}{s^2},$$

$$g_c = 1.0 \quad \text{in SI units.}$$

The simple form of the steady-flow energy equation, equation (2.9), can be used to find the power output, $\dot{W}$:

$$\frac{\dot{Q} - \dot{W}}{\dot{m}} = \Delta_1^2 h_0.$$

The turbine can be treated as adiabatic, so that $\dot{Q} = 0$:

$$\therefore \; -\dot{W} = \dot{m}\Delta_1^2 h_0 = -6 \; \frac{kg}{s} \times 479.721 \; \frac{m^2}{s^2},$$

$$\dot{W} = 2.878 \; \text{MW.}$$

Comments

1. This result gives the true power delivered by the expander-turbine blading. To obtain the shaft power, one would have to deduct only the effect of the various friction torques, principally bearings, seals, and disk friction.

2. The steam conditions were not needed. In fact the same result would have been obtained had the fluid been hydrogen, water, or molasses.

3. We do not need to know the turbine efficiency.

Example (application of Euler's equation to a radial-flow air compressor)

Calculate the pressure ratio (outlet static pressure/inlet total pressure) of a radial-flow air compressor having a rotor of 300 mm diameter, turning at 20,000 rev/min, and having 15 radial blades. The air (0.9 kg/s) enters the rotor without swirl, at 15 C and 100 kN/m², and leaves the rotor with 90 percent of the rotor's tangential velocity. The compressor polytropic efficiency, inlet total to outlet static conditions, is 80 percent.

Solution We use the version of Euler's equation for zero inlet swirl:

$$C_{\theta 1} = 0 \quad \text{(equation 5.4),}$$

$$g_c \Delta_1^2 h_0 = u_2 C_{\theta 2}$$

$$= 0.9 u_2^2$$

$$u_2 = 20{,}000 \times \frac{2\pi}{60} \times \frac{0.3}{2} = 314.61 \; \frac{m}{s};$$

$$\therefore g_c \Delta_1^2 h_0 = 88{,}826 \ \frac{\text{m}^2}{\text{s}^2},$$

$$g_c = 1.0 \quad \text{for SI units.}$$

We could use tables to find the inlet air enthalpy, h_{01}, and we could add the isentropic enthalpy rise (which would require converting the polytropic efficiency to the isentropic efficiency) and so find the outlet pressure. Let us use, however, the perfect-gas model, defined by equations (2.29) through (2.34):

$$\Delta_1^2 h_0 = \overline{Cp}(T_{02} - T_{01}) = T_{01}\overline{Cp}_{12}\left[\frac{T_{02}}{T_{01}} - 1\right];$$

$$\therefore \frac{T_{02}}{T_{01}} = \frac{88.826}{\overline{Cp}_{12}T_{01}} + 1.$$

We guess at $\overline{Cp}_{12}$ and iterate.

Guess $\overline{Cp}_{12} = 1{,}012$ J/kg-°K. Then

$$T_{02} - T_{01} = \frac{88.826 \ \text{m}^2/\text{s}^2}{1{,}012 \ \text{J/kg-°K}} = 87.77°\text{K};$$

$$\therefore \overline{T}_{012} = T_{01} + \left[\frac{T_{02} - T_{01}}{2}\right] = 331.89 \ \text{K},$$

at which $Cp = 1{,}007.8$ J/kg-°K (table A.2).

We use this value for $\overline{Cp}_{12}$ as being close enough (even though the value of Cp at the arithmetic mean temperature is not the mean Cp, because the variation of Cp with T is nonlinear in this region). Then we use equation (2.66) to find the pressure ratio:

$$\frac{p_{02'}'}{p_{01}} = \left(\frac{T_{02}}{T_{01}}\right)^{(\overline{Cp}_{12}/R)\eta_{pc}} = \left[1 + \frac{88{,}826 \ \text{m}^2/\text{s}^2}{1{,}007.8 \ \text{J/kg-°K} \times 288°\text{K}}\right]^{(1{,}007.8/286.96)0.8}$$

$$= 2.117.$$

Comments

1. The most confusing aspect of this result will undoubtedly be that a ratio of total pressure, $p_{02'}'/p_{01}$, has been obtained, whereas that was asked for was the ratio of the outlet static pressure, p_{st2}, to the inlet total pressure, p_{01}.

The answer to this difficulty is found in the general definition of a turbomachine efficiency, equation (2.51), where the outlet conditions include the *defined* outlet total pressure, p_{02}'. For a total-to-static efficiency, η_{pcts}, the (useful) outlet total pressure p_{02}' is defined as being equal to the actual outlet static pressure, p_{st2}.

Therefore $p_{02}'/p_{01} = p_{st2}/p_{01}$, and the answer is as desired.

2. The mass-flow rate is not required in the calculation. Nor is the number of blades or the inlet pressure.

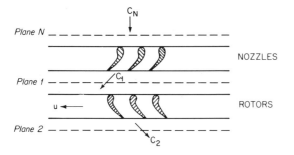

Figure 5.3 Turbine-stage blade rows.

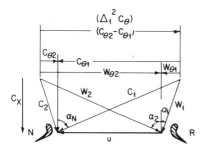

Figure 5.4 Turbine-stage velocity diagram.

5.2 Velocity diagrams and the parameters that describe them

The velocity vectors of fluid flow through blade rows of turbomachines can be combined to form velocity diagrams. The velocity vectors shown in such diagrams are conventionally those for the mean flow at a specified radius at entry to, or at exit from, a rotor or a stator or both. The flow directions within stator and rotor passages and around blades are not shown in velocity diagrams. We shall show that the use of Euler's equation in interpreting velocity diagrams enables the relative enthalpy change obtainable in any given diagram to be estimated on sight.

As an example of the construction of a velocity diagram we show in figures 5.3 and 5.4 the blading cross section and the velocity diagram of a turbine "stage": a combination of a stator and a rotor (not necessarily in that order, although turbine stages are almost inevitably analyzed in this way). The type of velocity-diagram construction that we shall use is based on the rotor peripheral velocity at flow exit, u_2. In this and several later examples of axial-flow machines we show a "simple" velocity diagram, defined as one having a constant axial flow velocity, C_x, from inlet to outlet and a constant rotor peripheral speed, u (constant streamtube diameter), from inlet to outlet.

Absolute and relative velocities

In the velocity diagram we use the convention of giving "absolute" velocities—velocities in the reference frame of the stators or nozzles—the designation C, and of giving "relative" velocities—velocities relative to moving surfaces, the rotor blades—the designation W.

The axial velocity, C_x, is, in this simple diagram, identical to the velocity entering the nozzles, C_N. This stage could be followed by another, in which the second-stage nozzles would receive the flow vector C_2 at an angle of α_2 to the axis.

Angle convention

We give angles with the axial direction. Another convention, used, for instance, in steam-turbine engineering, gives flow and blade angles relative to the plane of the blade rows.

Approximate blade shapes

This form of velocity diagram enables the approximate blade shapes to be sketched in. In general, blade angles are not equal to flow angles. The flow usually enters a blade row at an angle of "incidence" and leaves with an angle of "deviation," each being a usually small angle between the flow direction and the tangent to the blade centerline at inlet and exit, respectively. Blades are designed from the velocity diagrams (see chapters 7 and 8), rather than vice versa.

Change in tangential velocity and stage work

The absolute and relative tangential components of velocity (C_θ and W_θ) are shown in figure 5.4, as is the change in tangential velocity (also known as the change in "whirl" or "swirl") across the rotor, $\Delta C_\theta = \Delta W_\theta$. For simple diagrams (those with streamtubes or streamsurfaces entering and leaving the rotor at the same diameter, so that $u_1 = u_2$) the specific work done in (by) the stage is equal to $-u\Delta_1^2 C_\theta = -g_c\Delta_1^2 h_0$ for that streamtube or streamsurface. The work coefficient, the first of three parameters that can completely specify a stage velocity diagram of the simple kind, is then defined as follows.

The work or loading coefficient, ψ

$$\psi \equiv \frac{-g_c\Delta_1^2 h_0}{u^2} = -\left[\frac{\Delta_1^2(uC_\theta)}{u^2}\right]_{ad} = -\left[\frac{\Delta_1^2 C_\theta}{u}\right]_{ad,simple} \tag{5.5}$$

(ψ is positive for turbines, negative for pumps, compressors, fans, and so forth. In some places the work coefficient is defined as twice this value.) The approximate value of the work coefficient can be determined immediately from the stage velocity diagram. In the example shown in figure 5.4,

$\psi = 1.36$. It is sometimes known as the "loading" coefficient. In turbines values of ψ above 1.5 would lead to them being called "highly loaded" or "high-work" turbines, or turbine "sections" (if the diagram refers to one radial position along a long blade). Values of ψ below 1.0 are for "low-work" or "lightly loaded" turbine stages. In compressor stages highly and lowly loaded stages would have values of above 0.5 and below 0.3, respectively.

The work coefficient alone is insufficient to specify the aerodynamic or boundary-layer loading on the blade rows and inner and outer walls. With the same value of ψ and a high or low value of C_x, the required flow deflection through any blade row could be low or high, respectively. We therefore define the second velocity-diagram parameter as the flow coefficient.

The flow coefficient, ϕ

$$\phi \equiv \frac{C_x}{u}. \tag{5.6}$$

In a simple velocity diagram the flow coefficient is constant and refers to the stage as a whole. In the general case, where both C_x and u vary through the stage for any streamsurface, there will be a different flow coefficient at rotor inlet and at rotor outlet, and the flow coefficient varies with radius.

These two parameters, the work coefficient and the flow coefficient, fix a large part of the velocity diagram. But the geometrical relationship of the difference of tangential velocity to the blade peripheral speed has yet to be specified. This is done by a quantity known as the "reaction."

The stage reaction, R_n

The strict definition of reaction is the ratio of the change in static enthalpy to the change in total enthalpy of the flow passing through the rotor:

$$R_n \equiv \left(\frac{\Delta h_{st}}{\Delta h_0} \right)_{ro} = \frac{\Delta h_{st,ro}}{\Delta h_{0,stage}}, \tag{5.7}$$

(because $\Delta h_{0,stage} = \Delta h_{0,ro}$). The resulting ratio is often referred to as a percentage: for instance, a 50-percent-reaction turbine. The exact value of reaction is not very significant, however, and turbines with reaction above perhaps 20 percent are often referred to just as "reaction" turbines.

A less-precise definition of reaction substitutes pressures for enthalpies in this relation. It serves as a useful benchmark to remember that zero-reaction (or "pure-impulse") turbines have no change of static pressure through the rotor.

The precise definition of reaction can be developed in terms of velocity-

Energy Transfer in Turbomachines

diagram values to show its geometric influence on the final specified form of the diagram.

For either the expansion or compression processes shown in figures 5.5 and 5.6,

$$\Delta h_{st} = \Delta h_0 + \frac{C_1^2}{2g_c} - \frac{C_2^2}{2g_c},$$

where $\Delta h_{st} \equiv h_{st2} - h_{st1}$,

$$\Delta h_0 \equiv h_{02} - h_{01} = u_2 C_{\theta 2} - u_1 C_{\theta 1},$$

$C_1 \equiv$ absolute velocity into rotor,

$C_2 \equiv$ absolute velocity out of rotor;

$$\therefore R_n \equiv \frac{\Delta h_{st}}{\Delta h_0} = 1 - \frac{(C_2^2 - C_1^2)/2}{u_2 C_{\theta 2} - u_1 C_{\theta 1}}. \tag{5.8}$$

This expression is valid for compressors and turbines and for general velocity diagrams (those with u_1 in general not equal to u_2, and $C_{x1} \neq C_{x2}$).

For a simple velocity diagram, $u_1 = u_2$ and $C_{x1} = C_{x2}$. Also in general,

$$C_1^2 = C_x^2 + C_{\theta 1}^2,$$

$$C_2^2 = C_x^2 + C_{\theta 2}^2,$$

as can be seen from figure 5.4. Therefore for a simple velocity diagram

$$R_n = 1 - \frac{1}{2} \frac{[C_x^2 + C_{\theta 2}^2 - C_x^2 - C_{\theta 1}^2]}{u(C_{\theta 2} - C_{\theta 1})}$$

$$= 1 - \frac{1}{2} \frac{(C_{\theta 2} - C_{\theta 1})(C_{\theta 2} + C_{\theta 1})}{u(C_{\theta 2} - C_{\theta 1})}$$

$$= 1 - \frac{C_{\theta 1} + C_{\theta 2}}{2u}. \tag{5.9}$$

Thus the reaction of a simple velocity diagram is the ratio of the mean tangential-velocity vector, $(C_{\theta 1} + C_{\theta 2})/2$, to the peripheral velocity, u, deducted from 1.0.

Zero-reaction diagrams

In zero-reaction turbines the mean tangential-velocity vector, $(C_{\theta 1} + C_{\theta 2})/2$, is identical to the peripheral velocity, u. Zero-reaction diagrams are normally used in turbines, not in compressors, where the flow swirl after rotors and stators would be very high. (The German company Junkers tried zero-reaction compressors at an early phase of development.) A

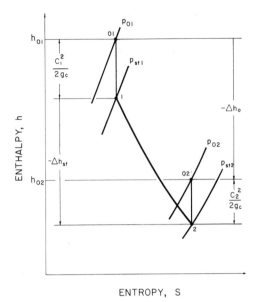

Figure 5.5 Enthalpy-entropy (h-s) diagram for a turbine expansion.

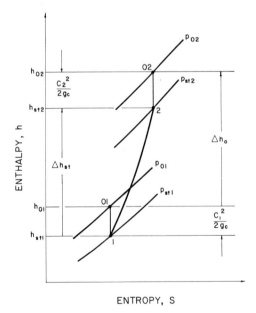

Figure 5.6 Enthalpy-entropy (h-s) diagram for a compression process.

Energy Transfer in Turbomachines

zero-reaction turbine is called an "impulse" turbine because there is no expansion or acceleration of the flow through the rotor blades, and the rotor torque comes wholly from the "impulse" of the nozzle stream. With no pressure drop across the rotor-blade row, pressure seals (which in reaction turbines usually take the form of a ring or shroud attached to the blade tips and carrying a labyrinth) are unnecessary.

The full-line impulse diagram in figure 5.7 is used a great deal because the stage entry and exit flows can be axial in direction (no swirl) and because the turbine stage has a high work coefficient: 2.0.

A zero-reaction compressor stage has no intrinsic advantages and is to be avoided.

Fifty-percent-reaction, or symmetrical, diagrams

For the reaction to be 50 percent, the mean tangential-velocity vector $(C_{\theta 1} + C_{\theta 2})/2$ must equal $u/2$, which is why the diagrams become symmetrical.

Such diagrams are frequently favored for turbines because they have accelerating flow to an equal extent in rotor- and stator-blade passages, which leads to low losses. The "rectangular-diagram" turbine shown in the center of figure 5.8 has the additional advantage of having axial flow at stage inlet and outlet. Also tests show that this diagram gives the peak polytropic efficiency for turbine stages (see chapter 7).

When used in compressors, rotors and stators have equal diffusion coefficients Cpr_{th}, and approximately equal—and minimum—relative Mach numbers for a given work coefficient.

High-reaction diagrams

A diagram with 100 percent reaction has $C_{\theta 1} = C_{\theta 2}$, so that the mean C_{θ} vector coincides with $u = 0$ (figure 5.9). Such a diagram has no intrinsic advantages for turbines or compressors. However, diagrams having axial flow at inlet and outlet and high reactions do have applications.

The rotor-stator compressor diagram at the top of figure 5.10 has the advantage of a low diffusion coefficient, Cpr_{th}, for a given work coefficient and the disadvantage of a high rotor Mach number. This diagram also applies to fans that operate without stators.

The central diagram has over 100 percent reaction, the stators having accelerating flow and therefore losing some of the static-pressure rise produced by the rotor row.

The lower diagram is similarly over 100 percent reaction, but when applied to a turbine, this becomes the rotor-only diagram of a windmill. In this case C_2 is the vector of the leaving flow.

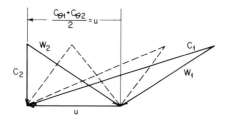

Figure 5.7 Zero-reaction diagrams for turbines.

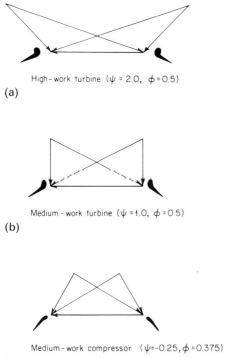

High-work turbine ($\psi = 2.0$, $\phi = 0.5$)

(a)

Medium-work turbine ($\psi = 1.0$, $\phi = 0.5$)

(b)

Medium-work compressor ($\psi = -0.25$, $\phi = 0.375$)

(c)

Figure 5.8 Symmetrical (50-percent-reaction) diagrams: (a) high-work turbine ($\psi = 2.0$, $\phi = 0.5$); (b) medium-work turbine ($\psi = 1.0$, $\phi = 0.5$); (c) medium-'ork compressor ($\psi = -0.25$, $\phi = 0.375$).

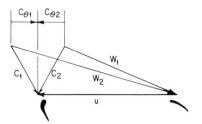

Figure 5.9 100-percent-reaction compressor diagram.

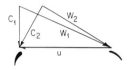

Rotor-stator compressor or fan
under 100% reaction ($\psi = -0.31$, $\phi = 0.5$)

(a)

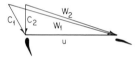

Stator-rotor compressor or fan
over 100% reaction ($\psi = -0.25$, $\phi = 0.37$)

(b)

Windmill rotor – no stator
over 100% reaction ($\psi = 0.1$, $\phi = 0.17$)

(c)

Figure 5.10 High-reaction diagrams: (a) rotor-stator compressor or fan under-100-percent reaction ($\psi = -0.31$, $\phi = 0.5$); (b) stator-rotor compressor or fan over-100-percent reaction ($\psi = -0.25$, $\phi = 0.37$); (c) windmill rotor—no stator, over-100-percent reaction ($\psi = 0.1$, $\phi = 0.17$).

Example (calculation of pump velocity diagrams)

Calculate and draw two mean-diameter "simple" velocity diagrams suitable for use in a town fire department's pump to pass 0.08 m³/s (80 liter/s) water at a delivery total head of 150 m:

(a) Axial-flow stator-rotor pump, where work coefficient $\psi = -0.4$, flow coefficient $\phi = 0.4$, and axial outlet flow from the rotor.

(b) Axial-flow rotor-stator pump, where work coefficient $\psi = -0.3$, flow coefficient $\phi = 0.3$, and axial outlet flow from the stator.

Give the peripheral velocity, u, the relative velocity into the rotor, W_1, and calculate the degree of reaction, R_n, and the rotor and stator diffusion ratio (relative outlet velocity/relative inlet velocity, W_2/W_1). Make an initial guess of the blading efficiency (theoretical work/actual work) at 0.85. Comment on the suitability of each as a single-stage pump.

Solution We first use the steady-flow energy equation, applied to incompressible fluids, equation (2.12):

$$\frac{\dot{Q} - \dot{W}}{\dot{m}} = \Delta_1^2 \left[\frac{H_0 g}{g_c} + u + \frac{Zg}{g_c} \right]$$

The flow through the pump will be adiabatic, so $\dot{Q} = 0$. When a delivery head across a pump is specified, it is implied that the measuring stations are at the same height, that is $Z_1 = Z_2$:

$$\therefore \quad -\frac{\dot{W}}{\dot{m}} = (H_{02} - H_{01})\frac{g}{g_c} + (u_2 - u_1)$$

$$= (150 \text{ m})\frac{g}{g_c} + (u_2 - u_1).$$

The losses are represented by $(u_2 - u_1)$. The ideal or theoretical work, in a pump with no losses, is

$$-\frac{\dot{W}_{th}}{\dot{m}} = (H_{02} - H_{01})\frac{g}{g_c} = -0.85\frac{\dot{W}}{\dot{m}}.$$

Therefore the actual work required to be given by the blades is

$$-\frac{\dot{W}}{\dot{m}} = \frac{150 \text{ m}}{0.85}\frac{g}{g_c}.$$

This can be used in the basic Euler equation, (5.1):

$$-\frac{\dot{W}}{\dot{m}} = \frac{1}{g_c}[u_2 C_{\theta 2} - u_1 C_{\theta 1}] = \frac{150 \text{ m}}{0.85}\frac{g}{g_c},$$

Therefore for a simple velocity diagram

$$u(C_{\theta 2} - C_{\theta 1}) = \frac{150 \text{ m}}{0.85} g = 1{,}730.65 \frac{\text{m}^2}{\text{s}^2}.$$

Now we can draw the two velocity diagrams (figure 5.11) and calculate the various velocities from the specified parameters.

Stator-rotor pump

From equation (5.4), $\psi \equiv -\dfrac{\Delta C_\theta}{u} = -\left(\dfrac{C_{\theta 2} - C_{\theta 1}}{u}\right)$:

$C_{\theta 2} = 0$;

$\therefore C_{\theta 1} = -0.4u.$

Rotor-stator pump

$C_{\theta 1} = 0$;

$\therefore C_{\theta 2} = +0.3u.$

Inserting this into the Euler equation enables the peripheral velocity to be found in each case:

$$u(0 + 0.4u) = 1{,}730.65 \frac{\text{m}^2}{\text{s}^2};$$

$$\therefore u = 65.78 \frac{\text{m}}{\text{s}};$$

$$W_1 = \sqrt{(0.4u)^2 + (1.4u)^2}$$

$$= 95.78 \frac{\text{m}}{\text{s}}.$$

$$u(0.3u + 0) = 1{,}730.65 \frac{\text{m}^2}{\text{s}^2};$$

$$\therefore u = 75.95 \frac{\text{m}}{\text{s}};$$

$$W_1 = \sqrt{(0.3u)^2 + u^2}$$

$$= 79.29 \frac{\text{m}}{\text{s}}.$$

For a simple diagram the degree of reaction is given by equation (5.8):

$$R_n = 1 - \frac{C_{\theta 1} + C_{\theta 2}}{2u},$$

$$R_n = 1 - \frac{(-0.4u + 0)}{2u};$$

$$\therefore R_n = 1.2.$$

The reaction is therefore 120 percent. The stator has accelerating flow.

$$R_n = 1 - \left(\frac{0 + 0.3u}{2u}\right);$$

$$\therefore R_n = 0.85.$$

The reaction is 85 percent. The stator diffusion ratio is C_x/C_2:

$$\frac{C_x}{C_2} = \frac{0.3u}{\sqrt{(0.3u)^2 + (0.3u)^2}} = \frac{1}{\sqrt{2}} = 0.707,$$

which is acceptable for favorable conditions.

The rotor diffusion ratio, W_2/W_1, is

$$\frac{W_2}{W_1} = \sqrt{\frac{(0.4u)^2 + u^2}{(0.4u)^2 + (1.4u)^2}} = 0.74,$$

which is acceptable for most duties.

The rotor diffusion ratio, W_2/W_1, is

$$\frac{W_2}{W_1} = \sqrt{\frac{(0.3u)^2 + (0.7u)^2}{(0.3u)^2 + u^2}} = 0.73,$$

which is acceptable.

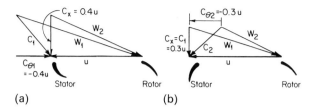

Figure 5.11 Alternative pump velocity diagrams: (a) stator-rotor pump; (b) rotor-stator pump.

Comment The stator-rotor pump has the advantage of accelerating stator blades, which will tolerate a disturbed inlet flow, and a moderate degree of diffusion in the rotor blades ($W_2/W_1 = 0.74$). But the rotor-inlet relative velocity is very high, (95.78 m/s) which could lead to cavitation problems. The rotor-stator design has a much lower inlet relative velocity, (79.29 m/s), but a dangerous degree of diffusion is required in the stators. The rotor-stator design with a slightly lower (magnitude) work coefficient—say $\psi = -0.25$, to increase (C_x/C_2) somewhat—seems preferable.

One must keep in mind also that, if the mean-diameter velocity diagram is marginal in any respect, conditions at the hub or shroud are likely to be unacceptable. This problem is treated in the next chapter.

5.3 Axial-compressor and pump velocity diagrams

Velocity diagrams for axial compressors and pumps can be specified by the three variables described in the previous section: flow coefficient, work coefficient, and reaction. However, other forms of specifications are usually preferred, for the following reasons.

Axial compressors and pumps are composed of series of rotating and stationary blade rows, each of which has a high relative velocity at entry and a lower relative velocity at exit. (This is strictly true only for machines with reactions between 0 and 100 percent as may be seen from the example of a stator-rotor pump in section 5.2.) Each blade row is therefore a group of parallel diffusers. We should expect the rules governing diffuser design to apply to axial compressors and pumps, and this is the case. The principal rule, as was stated in chapter 4, is that the outlet-to-inlet relative velocity ratio (the de Haller number) should be above some minimum value. The range of this ratio should be from 0.7, for machines operating at high Reynolds numbers in clean, near-ideal conditions, to 0.8, for machines that must operate over a wide range of incidence angles (flow coefficients) and/or in dirty, corrosive, or erosive fluids. A typical value for average industrial machines is 0.75.

In compressor or pump design the relative-velocity ratio, or the lower of the velocity ratios for rotor and stator, is often treated not as a limit but rather as an independent, or input, variable. The designer chooses a value that is estimated to give a safe margin from stall in the normal working range under the conditions the machine is likely to encounter in practice. Thus a value of 0.75 might well be selected.

In a 50-percent-reaction diagram the outlet-to-inlet velocity ratios are identical (it is a "symmetrical" diagram). In a high-reaction (over 50 percent) diagram the rotor has the lower velocity ratio, and the stator has the lower value for low-reaction diagrams. In figure 5.12 the velocity ratios of the two blade rows is 0.696, which is too low to give efficient, wide-range diffusion. This diagram could be improved either by reducing ψ from, say, 0.3 to 0.25, or by increasing ϕ, from 0.4 to 0.5.

The second variable that is significant in certain compressor and pump stages is the inlet relative velocity to either stator or rotor. The first stage of a multistage air compressor is usually designed to a maximum relative Mach number, which is found at the tip section of the rotor blades or possibly at the root or inner-diameter section of the stator blades.

For pumps operating under small suction heads, this same relative velocity governs the cavitation performance (see chapter 10).

In neither of these two cases (compressor or pump) is steady blade stress a limiting factor in design (except for compressors of low-molecular-weight gases such as hydrogen or helium). The blade-row velocity ratio and the maximum inlet relative velocity often become the determinants of both the blade speed and the diagram shape. The remaining degree of freedom is expressed alternatively as reaction, or flow coefficient, or blade setting angle or "stagger" (figure 5.13).

Effect of design flow coefficient
In some cases the axial-compressor or pump designer has considerable freedom to choose the design value of flow coefficient, particularly in those cases where inlet Mach number or cavitation are not problems to be avoided. With the same de Haller ratio but different flow coefficients and work coefficients, blade rows of different stagger or setting angles have not only differing design points but different ψ-ϕ characteristics in the off-design regions, figure 5.13. The high-work, high-flow, low-stagger blade rows tend to have a sharp stall, with a large fall in pressure rise and possibly a hysteresis effect that makes recovery from stall more difficult (Wilson 1960). On the other hand, the low-work, low-flow, high-stagger blade rows tend to have an almost imperceptible stall, and a stalled pressure rise higher than at design, with no drop off in pressure differential at the stall itself. These characteristics have recently been confirmed and quantified by Koch (1981).

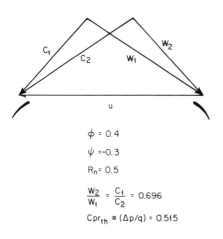

$\phi = 0.4$

$\psi = -0.3$

$R_n = 0.5$

$\dfrac{W_2}{W_1} = \dfrac{C_1}{C_2} = 0.696$

$Cpr_{th} \equiv (\Delta p/q) = 0.515$

Figure 5.12 Axial-compressor and pump velocity diagrams.

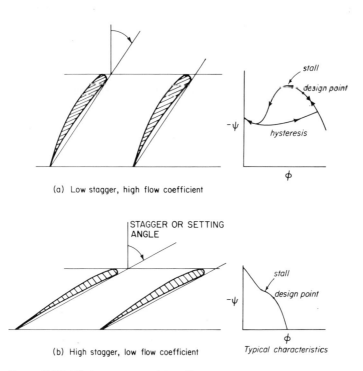

(a) Low stagger, high flow coefficient

(b) High stagger, low flow coefficient

Typical characteristics

Figure 5.13 Blade stagger and its effect on flow characteristics: (a) low stagger, high flow coefficient; (b) high stagger, low flow coefficient.

Energy Transfer in Turbomachines

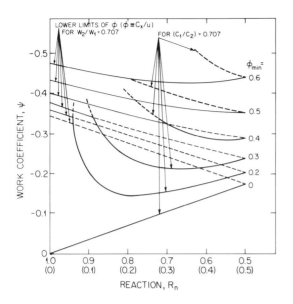

Figure 5.14 Effect of reaction and flow coefficient on stage work.

Choice of flow coefficient

The need to keep the diffusion velocity ratio above some specified limit in both the blade rows of an axial compressor or pump leads to nonintuitive requirements for the minimum flow coefficient for a desired work coefficient in a simple diagram. The results of some calculations for a limiting diffusion ratio of 0.707, approximately the de Haller limit, are shown in figure 5.14. Curves are shown for the lower limit of the flow coefficient for each blade row. The lower of the two curves is then the controlling limit; the higher portion of each pair of curves is shown as a broken line. For example, if one wanted a work coefficient of 0.25 in a diagram of reaction 0.75, the flow coefficient would need to be above 0.33, approximately, for the diffusion limit in the stator to be acceptable. At this value of the flow coefficient the rotor diffusion would be well away from the limiting value.

The attractions of high-reaction and 50-percent-reaction designs become clearer from figure 5.14. For practicable flow coefficients, which would be those over 0.2, more work can be given by diagrams with reactions near 1.0 than by diagrams with intermediate reactions. For the lower flow coefficients (0.2 to 0.3, for high-stagger blades) the work obtainable is higher than the minimum (at reactions between 0.7 and 0.8) but lower than that at $R_n = 1.0$.

In practice, compressors and pumps are not designed with constant reaction along the length of the blades, so that the choice of reaction must be influenced by arguments presented in chapter 6.

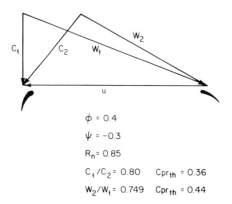

$\phi = 0.4$

$\psi = -0.3$

$R_n = 0.85$

$C_1/C_2 = 0.80$ $\quad Cpr_{th} = 0.36$

$W_2/W_1 = 0.749$ $\quad Cpr_{th} = 0.44$

Figure 5.15 High-reaction axial-inlet diagram.

Design freedom and the radius ratio

The amount of freedom there is to choose the shape of the velocity diagrams is a strong function of the radius ratio, or the ratio of the inner to the outer diameter of the blade row. When there is a strong need to keep outside diameter as small as possible, the radius ratio of the first stage of an aircraft compressor or fan may be made as small as 0.3. The ratio of the outer-to-inner blade-speed-squared will then be 11.11. The enthalpy rise is a function of the blade-speed-squared (and the work coefficient). The desirability of delivering a similar enthalpy rise along the length of the blades from velocity diagrams having the blade-speed-squared varying more than tenfold puts such constraints on the design choices that there can be little variation from one design to another. The axial and tangential velocities have in addition to satisfy the equations of radial equilibrium, discussed in chapter 6.

In industrial design the optimum configuration will normally be one having a higher inlet radius ratio, perhaps of 0.7. Here there can be considerable freedom to choose velocity diagrams. Fashions have developed. In continental Europe, axial-compressor diagrams generally have reactions 30 to 50 percent higher than those used in the United States and Britain. The principal advantages of a high-reaction velocity diagram are the higher (more acceptable) values of the blade-row outlet-to-inlet velocity ratios for the same flow and work coefficients and the lower blade speed that results if the relative Mach number to the rotor blades is to be similar to that for a 50-percent-reaction machine.

A high-reaction machine may be designed with axial inlet and outlet velocity vectors, C_1, which is a major advantage for single-stage units (figure 5.15). The high velocity ratio of the stators makes them less susceptible to flow breakdown when the profiles become roughened by dirt buildup. Normally the stators are affected by dirt more than the rotors.

Energy Transfer in Turbomachines

When severe design constraints are imposed, such as a first-stage "hub-tip" diameter ratio of less than 0.5, "simple" velocity diagrams become, for most practical flow distributions, too approximate to be valuable. General velocity diagrams as shown in figure 5.16 should be used. Some guidance is given in chapter 6. In addition streamline curvature in the axial-transverse plane is often used to improve flow distributions and to modify the vector diagrams. This topic is beyond the scope of this book.

5.4 Radial-flow velocity diagrams

Radial-flow pumps, compressors, and turbines cannot have so-called "simple" velocity diagrams because the rotor peripheral velocity is very different at the point the flow leaves the rotor compared with that at which it enters. Often the flow enters centrifugal pumps and compressors without swirl, and the design-point flow of radial-inflow turbines is usually nominally without swirl at rotor exit. The rotor-tip velocity triangle is therefore the dominant one for radial-inflow turbines and radial-outflow compressors and pumps.

Radial blades in radial-flow rotors
For stress-reduction reasons, high-pressure-ratio centrifugal compressors usually have rotor blades that are radial at outlet. The blades have high aerodynamic loading at outlet. Circulation, plus apparently inevitable flow separation within the rotor channels, combine to produce a flow deviation, or "slip", with the blade alignment. The combined inlet and outlet velocity diagrams of a radial-blade centrifugal compressor will then approximate figure 5.17. Two hypothetical compressor or pump inlet triangles are shown, one for the hub and one for the shroud. Limiting design conditions are often the shroud relative velocity at rotor inlet, W_{s1}, because of relative-Mach-number or cavitation considerations, and the relative-velocity ratio, W_2/W_{s1}. Although, as mentioned earlier, it seems to be impossible to prevent separation within the rotor passages of high-pressure-ratio compressors entirely, the proportion of the flow that is separated, and the relative losses thereby caused, seem to be a strong function of this relative-velocity ratio. It is therefore desirable to keep this ratio above, say, 0.75.

The angle of absolute flow into the diffuser, α_2, has a strong effect on the stability of a radial diffuser, if used (see chapter 4). In general, α_2 should not exceed 70 degrees and should be lower if a large amount of diffusion is to be attempted.

Backward-swept blades for centrifugal machines
The rotor relative-velocity ratio is beneficially increased by using rotor blades that sweep backward, relative to the flow direction figure 5.18.

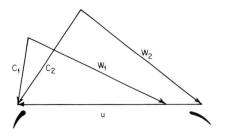

Figure 5.16 General velocity diagram with varying axial velocity and blade peripheral speed.

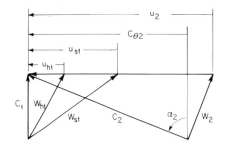

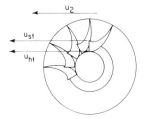

Velocity vectors are shown for compressor or pump
operation, and would be reversed for turbine flow.

Figure 5.17 Radial-flow velocity diagram, with diagram defining blade peripheral velocities.

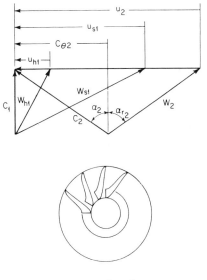

Velocity vectors are shown for compressor or pump
operation, and would be reversed for turbine flow.

Figure 5.18 Velocity diagram for backward-swept radial-flow machine.

In addition, a smaller proportion of the total-enthalpy rise is in the form of velocity. For both reasons the polytropic efficiencies of centrifugal machines with swept-back blades are normally considerably higher than those with straight radial blades.

These benefits are won at the cost of a considerable reduction in maximum possible work per stage, for two reasons. First, the work coefficient is reduced (from about 0.9 for radial-blade machines). Second, the maximum, stress-limited, peripheral velocity must be lower for swept-back blades than for radial blades. In many applications there is no call for maximum work per stage, and blade sweepback angles of up to 50 degrees (and more for pumps) can be advantageously used.

5.5 Correlations of peak stage efficiency with radius ratio and "specific speed"

The hub-tip ratio and the equivalent for centrifugal machines, the blade-height-to-diameter ratio, have obvious effects on the peak efficiency that can be expected from any turbomachine stage. At hub-tip ratios near unity the blade annular passage is so small that it may be all taken up with boundary layers, and it will be strongly influenced by tip-leakage flows and other secondary flows. In the limit the efficiency will fall to zero at a hub-tip ratio of unity.

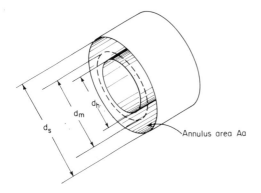

Figure 5.19 Annulus-area notation.

When low hub-tip ratios are used, the choice of velocity diagrams for the whole radial extent of the stage will be compromised by the need to avoid undue separation or high relative Mach numbers at the blade ends. In the limit the efficiency will not go to zero at zero hub-tip ratio, but for compressors and pumps it will be low. Turbines do not experience so sharp a drop-off in efficiency when areas of separation begin to occur, because the overall pressure gradient is favorable.

The hub-tip ratio can be derived in terms of quantities that appear in the machine specifications, as follows (see figure 5.19):

$$1 - \left(\frac{d_h}{d_s}\right)^2 = \frac{d_s^2 - d_h^2}{d_s^2} = \frac{Aa}{\pi d_s^2/4}$$

$$= \frac{\dot{V}}{C_x \pi d_s^2/4}, \quad \text{because } \dot{V} = AaC_x,$$

$$= \frac{\dot{V}N^2}{\phi u_2 (\pi d_s^2/4)(\omega 60/2\pi)^2}, \quad \begin{array}{l} \text{because } \phi \equiv C_x/u_2, \\ N \text{ rpm} = 60\omega \text{ rads/s}/2\pi, \\ \omega/2 = u_2/d_{s2}, \\ \psi \equiv g_c \Delta h_0/u_2^2, \end{array}$$

$$= \frac{\pi}{900} \frac{N^2 \dot{V}}{\phi u_2^3}$$

$$= \frac{\pi}{900} \frac{\psi^{3/2} N^2 \dot{V}}{\phi (g_c \Delta h_0)^{3/2}}$$

$$\left[1 - \left(\frac{d_h}{d_s}\right)^2\right]^{1/2} = \frac{\sqrt{\pi}}{30} \frac{|\psi|^{3/4}}{\phi^{1/2}} \left[\frac{N\sqrt{\dot{V}}}{(g_c \Delta h_0)^{3/4}}\right] = \frac{2\sqrt{\pi}|\psi|^{3/4}}{\phi^{1/2}} N_{s1} \qquad (5.10)$$

where $N_{s1} \equiv \dfrac{N\sqrt{\dot{V}}}{60(|g_c \Delta h_0|)^{3/4}} \equiv$ nondimensional specific speed.

Specific speed is therefore principally a function of the relative flow-passage size and of the design-point values of the work coefficient, ψ, and of the flow coefficient, ϕ.

There are many definitions of specific speed, usually dimensional, but they all have the same relative significance. One used where the old imperial system of units is still in force is N_{s3}:

$$N_{s3} \equiv \frac{(N \text{ rpm})(\dot{V} \text{ gal/min})}{(\Delta H_{01} \text{ ft})^{3/4}}. \tag{5.11}$$

In using N_{s3}, the "gallon" must be specified as U.S. or "imperial." If the mean diameter d_m is chosen so that

$$d_s^2 - d_m^2 = d_m^2 - d_h^2,$$

that is,

$$d_m^2 \equiv \frac{d_s^2 + d_h^2}{2}.$$

Then

$$\frac{d_m}{d_s} = \sqrt{\frac{1 + \lambda^2}{2}}$$

and $\dfrac{d_m}{d_h} = \sqrt{\dfrac{1 + \lambda^{-2}}{2}},$

where $\lambda \equiv d_h/d_s$.

$$\therefore N_{s1} = \frac{\phi_m^{1/2}}{\psi_m^{3/4}} \frac{1}{\sqrt{2\pi}} \sqrt{\frac{1 - \lambda^2}{1 + \lambda^2}}, \tag{5.12}$$

where $\phi_m \equiv \left(\dfrac{C_x}{u_m}\right)$,

$$\psi_m \equiv \frac{-g_c \Delta h_0}{u_m^2}.$$

The derivation of N_{s1} obtainable from equation (5.10) is

$$N_{s1} = \frac{\phi^{1/2}}{\psi^{3/4}} \frac{\sqrt{1 - \lambda^2}}{2\sqrt{\pi}},$$

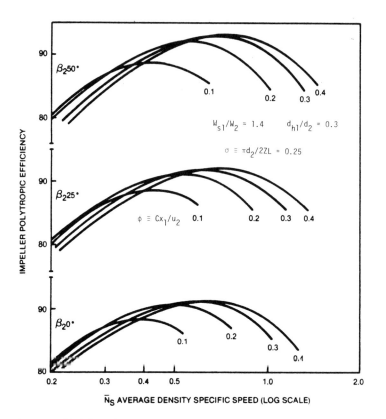

Figure 5.20 (a)

where ϕ and ψ are based on the peripheral speed, u, at the shroud.

The specific speed of hydraulic machines is usually given with the actual total-head rather than the total-enthalpy change:

$$N_{s2} \equiv \frac{N\sqrt{\dot{V}}}{60(g\Delta H_0)^{3/4}}. \tag{5.13}$$

Also a head coefficient, ψ^x, is often used rather than a work coefficient:

$$\psi^x \equiv \frac{g\Delta H_0}{u^2}.$$

In radial-flow machines the rotor velocity, u, is that of the periphery. If b is the axial length of the blades at the outer diameter, d, it can be easily shown that

Energy Transfer in Turbomachines

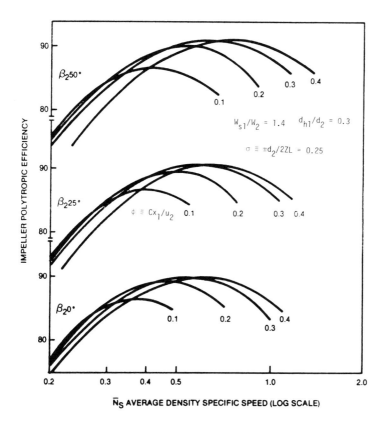

Figure 5.20 (b)

Figure 5.20 Centrifugal-compressor impeller efficiency versus specific speed, flow coefficient and blade angle. From Rodgers 1980. (a) Impeller Mach number 0.77; (b) impeller Mach number 1.54.

$$N_{s2} = \frac{|\phi|^{1/2}(b/d)^{1/2}}{|\psi^x|^{3/4}} \frac{1}{(\pi)^{1/2}}.$$ (5.14)

Here $\phi \equiv (Cr/u)$ at the outer diameter.

The relation between N_{s3}, defined by equation (5.11), using U.S. gallons and the nondimensional N_{s2}, equation (5.13), is

$$N_{s3} = 17{,}163 N_2.$$ (5.15)

Some charts (from Rodgers 1980 and Rohlik 1975) showing the influence of specific speed on the maximum efficiencies for compressors and pumps, and the effects of impeller blade angle and machine size, are shown in figures 5.20 through 5.22 and for radial-inflow turbines in figure 5.23.

Energy Transfer in Turbomachines   215

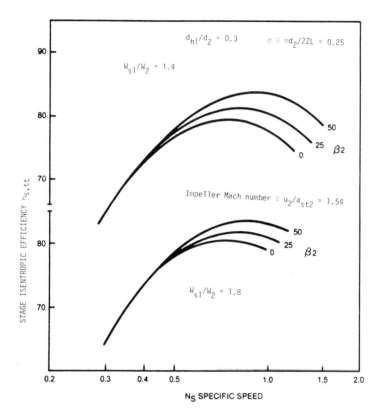

Figure 5.21 Estimated centrifugal-compressor stage efficiency. From Rodgers 1980.

5.6 Choice of number of stages

For any type of compressor or turbine there is an approximate maximum value of pressure ratio that can be handled. This maximum is related to the choice of velocity diagram and usually to a limiting relative Mach number for compressors, and to a limiting blade speed for acceptable blade-root or disk stresses in the case of turbines. In any application these maximum-work stages can be used to find the minimum number of stages that can be used.

The designer can then use this minimum number of stages, or a larger number. Many factors are involved in the choice of the optimum number of stages for any application, but the principal factors affecting design-point efficiency are these.

1. By having more than the minimum number of stages, the designer can use velocity diagrams that give higher total-to-total stage efficiencies,

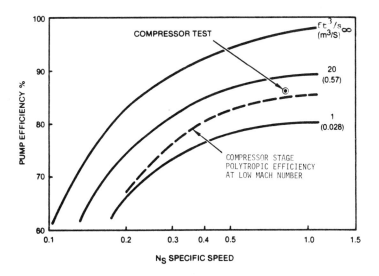

Figure 5.22 Centrifual-pump efficiency versus specific speed and size. From Rodgers 1980.

because of lower values of the loading coefficient, ψ, or higher values of the relative-velocity ratio, W_2/W_1, for compressors and pumps, and/or lower deflections.

2. These more-efficient velocity diagrams usually incorporate lower outlet velocities, so that the total-to-static efficiencies show a further increase.

3. Whether a highly loaded or a lightly loaded velocity diagram is used, increasing the number of stages will entail reducing the blade speed, and with it the axial velocity. The leaving kinetic-energy losses are thereby reduced, again increasing the total-to-static efficiencies.

With regard just to this third factor, the governing efficiency across the total machine, including the diffuser, is the total-to-static value for virtually all process compressors and turbines, all compressors in gas-turbine engines, and· all turbines in shaft-power engines. Jet engines are almost the only application where the kinetic energy of the leaving flow is useful. If perfect diffusers were available, the total-to-static efficiency would equal the total-to-total value, and the number of stages used would have no influence on the machine efficiency. In practice diffusers used after compressors and turbines have pressure-rise coefficients between 0.1 and 0.7. The effect on total-to-static polytropic efficiency·of using from one to four stages of an axial-flow turbine is shown in figure 5.24. In all cases the turbine is represented by the mean-diameter velocity diagram that has 50-percent reaction, a nozzle-outlet angle of $60°$, and a work coefficient of 1.8. The total-to-total polytropic efficiency across the blading (excluding

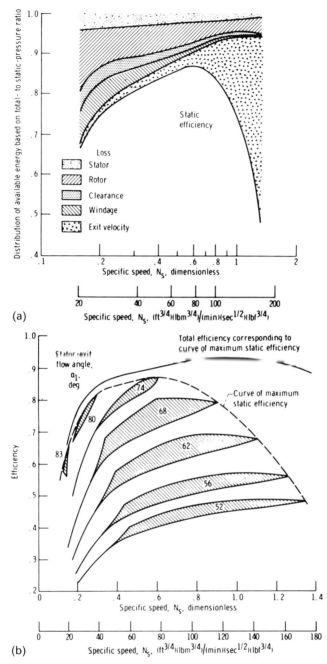

Figure 5.23 Radial-inflow-turbine efficiency versus specific speed. From Rohlik 1975. (a) Loss distribution along curve of maximum efficiency, (b) design point efficiency versus specific speed.

Energy Transfer in Turbomachines

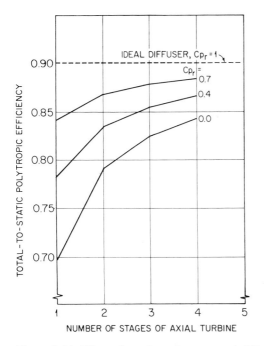

Figure 5.24 Effect of number of stages and diffuser pressure-rise coefficient on total-to-static efficiency of an axial-flow turbine.

the diffuser) is 0.9. The effects on the total-to-static efficiency of using diffusers with pressure-rise coefficients of 0, 0.4, and 0.7 are shown. With an ideal diffuser, the total-to-static efficiency would be constant at 0.9.

References

Koch, C. C. (1981). *Stalling pressure-rise capability of axial-flow compressor stages.* Paper 81-GT-3. ASME, New York.

Rodgers, C. 1980. Specific speed and efficiency of centrifugal impellers. In *Performance Prediction of Centrifugal Pumps and Compressors.* Ed. by S. Gopalakrishnan et al. ASME, New York, pp. 592–603.

Rohlik, Harold E. 1975. Radial-inflow turbines. In *Turbine Design and Application.* Vol. 3. Ed. by Arthur J. Glassman. Special publication SP-290. NASA, Washington, D.C., pp. 31–58.

Wilson, David. 1960. *Patterning stage characteristics for wide-range axial compressors.* Paper 60-WA-113. ASME, New York.

Problems

5.1

Choose a mean-diameter velocity diagram and a blade speed for the single-stage fan delivering cooling air to the condenser of a Rankine-cycle hybrid automotive engine (figure P5.1). Use the following specifications:

Air flow ($\dot{m}$), 2 kg/s.

Inlet total pressure (p_{01}), 99.285 kN/m^2.

Inlet total temperature (T_{01}), 300 K.

Drop in static pressure across condenser ($p_{st5} - p_{st6}$), 1.724 kN/m^2.

Mean air velocity entering condenser (C_5), 4.57 m/s.

Diffuser pressure-rise coefficient (Cpr_{th}), 0.75 [use figure 4.8c to find an average Cpr_{ac} at $(2\delta^*/Y) = 0.03$].

Fan hub-tip ratio (λ), 0.75.

Maximum pressure-rise coefficient in fan at mean diameter (Cpr_{th}), 0.4.

Fan flow coefficient (C_x/u_m), 0.4.

Axial inlet and outlet flow.

Fan total-to-total polytropic efficiency, rotor inlet to stator outlet ($\eta_{pttl\,3}$) 0.9.

Mean specific heat ($\overline{Cp}$), 1,005 J/kg-°K.

Other specified conditions are that ($p_{st6} = p_{01}$) and ($p_{st4} = p_{05}$).

5.2

Calculate and draw the velocity diagrams suitable for a fire pump to pass 1,250 gpm at a delivery (total) pressure of 250 psi. Find the peripheral velocity u, the degree of reaction R_n, and the rotor and stator diffusion ratio, W_2/W_1, for the axial pumps. Comment on the suitability of each design as a single-stage pump.

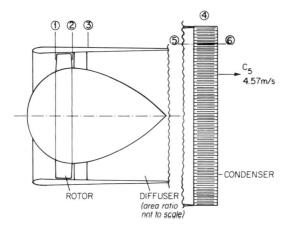

Figure P5.1

Energy Transfer in Turbomachines

(*a*) Axial pump with work coefficient 0.5, flow coefficient 0.4, and axial outlet flow from the rotor (a stator-rotor combination).

(*b*) Axial pump with work coefficient 0.4, flow coefficient 0.3, and axial inlet into the rotor (a rotor-stator combination with axial outlet flow from the stator).

(*c*) A radial-flow (outward) pump with radial (zero-angle) blades giving a relative flow angle at impeller outlet of 12 degrees to the radial direction ("slip"), and an absolute flow angle of 65 degrees at the same point. The diffuser passages have an area ratio of two to one.

(*d*) A radial-flow (outward) pump with 50-degree swept-back rotor blades giving a relative flow angle at impeller outlet of 55 degrees and an absolute flow angle of 60 degrees, all to the radial direction. The diffuser passages have a two-to-one area ratio.

Assume that the pressure rise required is total-to-total, and make an initial guess at 85 percent for the efficiency (defined as theoretical impeller work over actual blading work) for all machines across the stage.

5.3
Identify apparently desirable and undesirable features and sketch the approximate rotor and stator configurations of the velocity diagrams in figure P5.3. The following sketch is given as an example of what is wanted:

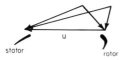

stator   u   rotor

Desirable features are W_2/W_1 acceptable rotor and stator.

Undesirable features are high swirl velocity after both rotor and stator; high relative velocity into stator might lead to high-Mach-number problems.

(a) Axial turbine

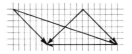

(b) Axial compressor

(c) Axial turbine

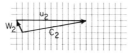

(d) Radial compressor

Figure P5.3

5.4

Calculate the total-to-static polytropic efficiencies from turbine inlet to diffuser outlet of several alternative turbines designed for the same duty. The duty is to expand 12 kg air per second from 4 atmospheres, 650 C total conditions, to atmospheric pressure. (This will be the static pressure at diffuser outlet).

(a) Turbine 1 is a single-stage impulse turbine with a gas angle leaving the nozzle of 70 degrees and axial flow at outlet. Diffuser $Cpr_{th} = 0.65$.

(b) Turbine 2 is a two-stage 50-percent-reaction turbine with axial flow at outlet and a flow coefficient of 0.4. The diffuser $Cpr_{th} = 0.75$, $\psi = 1.0$.

(c) Turbine 3 is similar to turbine 2 except that it has 4 stages. The diffuser $Cpr_{th} = 0.75$.

(d) Turbine 4 is a radial-inflow turbine with an air angle entering the rotor (relative to the radial) of 70 degrees, and a relative flow angle against the direction of rotation of the rotor of 10 degrees. The mean-diameter flow can be assumed to leave the rotor without a swirling component, with a relative flow velocity 50 percent greater than the relative velocity at inlet, and at a radius of one-half the inlet radius. The diffuser $Cpr_{th} = 0.3$. The drop in total pressure in the diffuser is 25 percent of the difference between inlet total and static pressures.

In all cases, the total-to-total polytropic efficiency across the blading (to turbine outlet) can be taken to be 90 percent. Perfect-gas relations or tables may be used, whichever is more convenient. The calculations will involve some iteration. Discuss the reasons for the differences in overall efficiencies.

5.5

Optimize the design (to the first iteration) of a single-stage axial-flow water pump which has the duty of lifting water from a discharge tank to a supply tank, through a vertical lift of 6 m. There is a frictional head loss of 0.25 m in the pipe additional to this lift and a kinetic head equal to that leaving the pump diffuser. The electrical supply available is enough to power a motor of 25 kW output, and it can be assumed that an ungeared motor of any of the usual standard speeds can be used (approximately 875, 1,150, 1,725, or 3,500 rpm).

(a) As a first approximation, assume that a total-to-total efficiency (ideal power over actual power) across the pump of 0.90 can be used, and that the diffuser Cpr is 0.5 (actual static-pressure rise/inlet dynamic head).

(b) Choose a rotor-stator combination and hub-tip ratio, and calculate the conditions for the mean-diameter velocity diagram. Use a flow coefficient of 0.3, axial inlet and outlet flow direction, and a maximum value of Cpr_{th} of 0.30 at the mean diameter.

(c) Assume that the power dissipated in bearings, couplings and disk friction amounts to one kilowatt, so that the power available at the rotor blading is 24 kW.

5.6

Suppose that you have a laboratory "work-horse" centrifugal compressor driven by a constant-speed motor, and you use it at different times for compressing hydrogen and for air. For which gas does the compressor produce more work per unit mass, and why?

5.7

Sketch the velocity diagrams for the constant-axial-velocity, constant-peripheral-velocity turbine and compressor (axial) stages listed in table P5.7. Also sketch in the approximate shapes of the stator and rotor blades. Fill in the missing numbers in table P5.7.

5.8

Calculate the total-to-total efficiency across the blading of an axial water-turbine stage having the following mean-diameter velocity-diagram characteristics:

R_n (reaction) $= 0$,

ψ (work coefficient) $= 2.0$,

ϕ (flow coefficient) $= 0.8$.

All the total-pressure losses to be considered can be ascribed to the nozzles and to the rotor blade row as follows:
for nozzle row, total-pressure losses

$$\Delta p_0 = 0.03 \frac{\rho C_1^2}{2g_c},$$

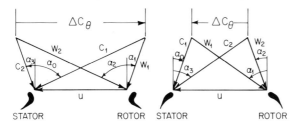

Figure P5.7

Table P5.7

Turbine (T) or compressor (C) velocity diagram	ϕ (C_x/u)	ψ $[\Delta(uC_\theta)/u^2]$	R_n ($\Delta h_{st}/\Delta h_0$)	α_0	α_1	α_2	α_3	(W_2/W_1)
T			0	70°		0		
T				60°	0	0		
T		2	50%	70°				
C	0.3		50%	20°				
C		−0.5		0°				0.75
C		−0.5			70°		0°	

for rotor row, total-pressure losses

$$\Delta p_0 = 0.08 \frac{\rho W_2^2}{2g_c},$$

where $C_1 \equiv$ nozzle exit velocity,

$W_2 \equiv$ rotor exit velocity.

5.9
Why is it unattractive to use a low reaction (having most of the static-pressure rise in the stator) for a single-stage axial-flow cooling fan for electronic equipment? Give two or three reasons if you can.

5.10
Find the flow coefficient that would be required if an axial-compressor stage of 50-percent reaction must have a theoretical pressure-rise coefficient no greater than 0.5. The work coefficient, $g_c \Delta h_0 / u^2$, is to be 1.0.

5.11
Figure P5.11 shows two alternative diagrams for a high-pressure-ratio, high-hub-tip-ratio, single-stage axial air compressor. Give two advantages of each diagram relative to the other. (Both have equal blade velocity and equal enthalpy rise.)

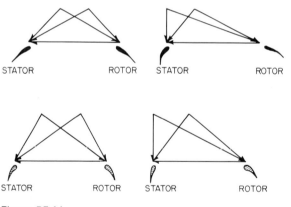

Figure P5.11

Energy Transfer in Turbomachines

6

The analysis and design of three-dimensional free-stream flow for axial turbomachines

In the last chapter several degrees of freedom were shown to be available for the choice of velocity diagrams for axial-flow turbomachines. Some of these degrees of freedom are appropriated by the necessary preliminary selection of change in total enthalpy per stage, and tip blade speed. Both of these quantities are usually decided upon, perhaps by comparison with previous machines, in the early stage of making preliminary specifications.

There usually remain at least three degrees of freedom. This chapter is concerned with describing how one of these freedoms must be given up, at least in machines with low hub-tip diameter ratios (relatively long blades and small hubs), to satisfy, or partially satisfy, the requirement that the flow approach radial equilibrium. That is, if radial flow accelerations are to be avoided, as they usually are to a large degree, the radial distribution of pressure must balance the flow "swirl."

That being so, the choice of velocity diagrams reduces usually to two degrees of freedom. The shape of the velocity diagram may be selected, often within fairly narrow limits, at one radius. And the variation of either the axial velocity or the tangential velocity, which are coupled by the equations of radial equilibrium for an interstage plane, may be chosen, again within sometimes narrow limits. Whether these limits are indeed narrow or broad depends primarily on the value of the hub-tip ratio (the ratio of the inner to the outer diameter of the flow annulus).

A large hub-tip ratio, perhaps defined as between 0.9 and 1.0, can involve only small changes of the velocity diagram with radius. For preliminary design it is usually sufficient to characterize the machine by the velocity diagram at mean diameter. Thus we can usefully refer to a "50-percent-reaction turbine," or to a "low-stagger compressor stage," even though there will be some radial variation of reaction and stagger.

When the hub-tip ratio is less than, say, 0.75, the necessary variation in velocity diagram from hub to tip is large enough for the characterization of the machine by the mean-diameter velocity diagram to be no longer useful. This is not obvious at this point. Having at least partial freedom to choose one velocity diagram and the variation of the diagram with radius, it is in fact possible to choose to have, for instance, constant reaction from hub to tip radius. But it turns out that such a choice produces an unacceptable design for various reasons.

6.1 The constant-work stage

It is normally desirable to design for a constant change in total enthalpy, Δh_0, for the stage at every radius. (The occasions when one would not follow this rule are those where there are strong radial pressure gradients due to flow curvature in the axial-meridional plane. The treatment of this flow curvature is beyond the scope of preliminary design. We shall therefore restrict ourselves to the case of constant Δh_0.)

The variation of work coefficient, ψ, with radius is therefore prescribed:

$$\psi \equiv \frac{-g_c \Delta_1^2 h_0}{u^2}.$$

Now

$$u = \omega r;$$

$$\therefore \psi = \frac{-g_c \Delta_1^2 h_0}{\omega^2 r^2}. \tag{6.1}$$

One design limit stems from this rapid increase of work coefficient with reduction in radius. The maximum permissible work coefficient is chosen at the hub radius (usually coupled with other limits such as of Mach number or reaction). Long blades are necessarily highly "loaded" (that is, ψ is high) at the hub and lightly loaded at the tip.

6.2 Conditions for radial equilibrium

When the fluid flowing through a duct or annulus has a tangential component of velocity, a pressure gradient with radius is naturally set up. The combination of the radial variation of static pressure, the variation of tangential velocity, and the variation of axial velocity, produces the radial variation of total enthalpy and total pressure. In preliminary design we shall deal exclusively with designs having constant-enthalpy conditions at all transverse planes. This requirement then sets the coupling relation between axial-and tangential-velocity variations.

In figure 6.1 the pressure gradient at the annulus δr due to the tangential

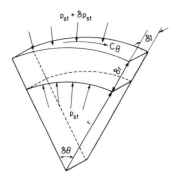

Figure 6.1 Stability of a fluid element.

velocity C_θ alone is

$$\delta p_{st} r \delta\theta \delta l = \frac{r\delta\theta \delta r \delta l \rho_{st} C_\theta^2}{rg_c};$$ (6.2)

$$\therefore \left(\frac{\partial p_{st}}{\partial r}\right)_{C_r} = \frac{\rho_{st} C_\theta^2}{rg_c}.$$

Specifying that the radial component of velocity, Cr, be constant means that we shall not include the effects of flow curvature in the axial-meridional plane on the radial pressure gradient. We want to find how this radial pressure gradient influences the axial-velocity component and the other fluid properties.

By definition,

$$h_0 \equiv h_{st} + \frac{C^2}{2g_c} \quad \text{(equation 2.7)},$$

$$C^2 = C_x^2 + C_\theta^2 + C_r^2,$$

$$h_{st} \equiv u_{st} + p_{st} v_{st} \quad \text{(equation 2.3)};$$

$$\therefore dh_{st} = du_{st} + p_{st} dv_{st} + v_{st} dp_{st}.$$

Gibbs' equation $T_{st} ds_{st} = du_{st} + p_{st} dv_{st}$ (equation 2.28);

$$\therefore dh_{st} = T_{st} ds_{st} + v_{st} dp_{st}$$

and

$$\frac{dh_{st}}{dr} = T_{st} \frac{ds_{st}}{dr} + v_{st} \frac{dp_{st}}{dr};$$

$$\therefore g_c \frac{dh_0}{dr} - g_c T_{st} \frac{ds_{st}}{dr} = g_c \frac{1}{\rho_{st}} \frac{dp_{st}}{dr} + \frac{1}{2} \frac{d}{dr} C_x^2 + \frac{1}{2} \frac{d}{dr} C_\theta^2 + \frac{1}{2} \frac{d}{dr} C_r^2. \qquad (6.3)$$

We can substitute for the radial pressure gradient from equation (6.2), and for the tangential-velocity variation we can use the following construction:

$$\frac{d}{dr}(r^2 C_\theta^2) = r^2 \frac{dC_\theta^2}{dr} + 2r C_\theta^2;$$

$$\therefore \frac{1}{2} \frac{dC_\theta^2}{dr} = \frac{1}{2r^2} \frac{d}{dr}(r^2 C_\theta^2) - \frac{C_\theta^2}{r}.$$

Equation (6.3) becomes

$$g_c \frac{dh_0}{dr} - g_c T_{st} \frac{ds_{st}}{dr} = \frac{C_\theta^2}{r} + \frac{1}{2} \frac{d}{dr} C_x^2 + \frac{1}{2r^2} \frac{d}{dr}(r^2 C_\theta^2) - \frac{C_\theta^2}{r} + \frac{1}{2} \frac{dC_r^2}{dr},$$

or

$$g_c \frac{dh_0}{dr} - g_c T_{st} \frac{ds_{st}}{dr} = \frac{1}{2} \frac{dC_x^2}{dr} + \frac{1}{2r^2} \frac{d}{dr}(r^2 C_\theta^2) + \frac{1}{2} \frac{dC_r^2}{dr}. \qquad (6.4)$$

For $\frac{dC_r^2}{dr} = 0$ and $\frac{dh_0}{dr} = 0$ and $\frac{ds_{st}}{dr} = 0$, we obtain the equation of simple radial equilibrium (SRE):

$$\frac{1}{r^2} \frac{d}{dr}(r^2 C_\theta^2) + \frac{d}{dr} C_x^2 = 0. \qquad (6.5)$$

6.3 Use of the SRE equation for different types of velocity distributions

The designer has considerable freedom to choose either a variation of axial velocity with radius or a variation of tangential velocity with radius, at any one inter-row plane in an axial-flow turbomachine. This choice, and the choice of a velocity diagram at any one radius—for instance, the mean or the hub radius—then determines the axial and tangential velocities in that plane and the neighboring plane in the following way.

In the first plane the choice of either the axial- or the tangential-velocity variation will allow the other to be found from the simple-radial-equilibrium equation. The tangential-velocity variation in the neighboring plane will then be determined because the radial distribution of work (usually constant) is given from the Euler equation as $u_1 C_{\theta 1} - u_2 C_{\theta 2}$. From this tangential velocity variation in the second plane, together with the speci-

Free-Stream Flow for Axial Turbomachines

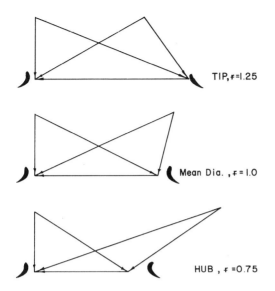

Figure 6.2 Free-vortex turbine diagrams.

fied axial velocity at the reference radius, the distribution of axial velocity can be found from the SRE equation.

A particular case of interest is when

$$rC_\theta = r_m C_{\theta m} = \text{constant},\tag{6.6}$$

where the subscript m signifies the mean-diameter condition. This is the relation for a free vortex. The SRE equation (equation 6.5) and this relation lead directly to the axial-velocity distribution, which is $C_x = \text{constant}$.

Free-vortex designs were formerly almost universal in the turbines of gas-turbine engines and are still widely used although others are supplanting them. Figure 6.2 shows velocity diagrams for the hub, mean-diameter, and tip sections of a turbine with zero-reaction (impulse) conditions at the hub. Even though the hub-tip ratio is not extreme—0.6—the reaction at the tip is 64 percent. This large radial variation in reaction is acceptable in turbines, so long as the value at the hub does not fall below zero. However, it is unacceptable for compressor diagrams for various reasons. Two of these are the consequent large value of relative velocity on to the rotor blades at the tip section, which would limit the relative Mach number unduly, and the large variation in diffusion ratio or de Haller number in both stator and rotor blade rows. Other choices for the radial variation of the tangential component of velocity are shown in figure 6.3.

Obviously, the variation of velocity should be smooth and reasonably

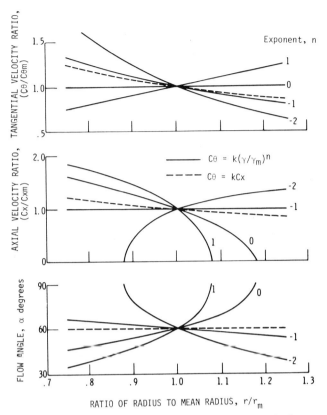

TANGENTIAL VELOCITY RATIO, $(C\theta/C\theta m)$

AXIAL VELOCITY RATIO, (Cx/Cxm)

FLOW ANGLE, α degrees

Exponent, n

$C\theta = k(\gamma/\gamma_m)^n$

$C\theta = kCx$

RATIO OF RADIUS TO MEAN RADIUS, r/r_m

Figure 6.3 Radial variations of velocity and flow angle for a mean-diameter flow angle of 60°. From Whitney and Stewart 1972.

gradual, and it is usual to employ some form of power law. The variation with radius of C_θ, C_x, and flow angle are shown for members of the family of flows in which $C_\theta = ar^n$, where a and n are constants, all for a flow angle at mean diameter of 60°. Also shown is a dashed line in the constant-flow-angle case, $C_\theta = aC_x$, which is similar to free vortex but has the advantage that the stator blades can be designed with no twist. However, it has the large variation in reaction of the free-vortex design.

A general form studied by Carmichael and Lewis (Horlock 1958) is the following:

$$C_{\theta 1} = ar^n - \frac{b}{r},\qquad(6.7a)$$

$$C_{\theta 2} = ar^n + \frac{b}{r},\qquad(6.7b)$$

where a, b, and n are constants for any one type of swirl distribution.

Free-Stream Flow for Axial Turbomachines

Rotor of double-flow low-pressure steam turbine. Courtesy Brown Boveri Co.

6.4 Prescribed reaction variation

There are too many criteria for judging a "good" type of radial-equilibrium solution for us to arrive easily at a simple rule. In axial-compressor
design the most important criterion is to have the relative-velocity ratio,
W_2/W_1 (the de Haller number), stay above, say, 0.7 for both rotor and
stator from hub to tip. It is often also important to keep the relative Mach
number below either, say, 0.9 or (if a transsonic design is acceptable) 1.2
for the rotor-tip conditions and below 0.9 for the stator hub radius.

In compressors and turbines we like to keep the reaction at the hub
above zero and the reaction at the tip below unity, if possible. Since we
know that the distortion of the velocity diagram resulting from extreme
reactions produces high relative Mach numbers, and also gives diffusing
conditions in turbines, a reasonable approach to the choice of an accept-

able radial-equilibrium solution is to start with the requirement that the variation in reaction along the blade length shall be such that the hub shall have some positive value and the tip shall be below 1.0. Then the results can be examined for variations in relative Mach number, in velocity ratio, and in blade angle for the particular mean-diameter velocity diagram chosen for the design being examined.

Let us define new variables $C'_\theta \equiv C_\theta/u_m$, $C'_x \equiv C_x/u_m$, and $r' \equiv r/r_m$, so that the equation of simple radial equilibrium (equation 6.5) becomes

$$\frac{1}{(r')^2}\frac{d}{dr'}[(r')^2(C'_\theta)^2] + \frac{d}{dr'}[(C'_x)^2] = 0. \tag{6.8}$$

The reaction variable R'_n is defined as

$$R'_n \equiv 1 - \frac{1}{2r'}(C'_{\theta 1} + C'_{\theta 2}). \tag{6.9}$$

This is equal to the actual reaction R_n only where the velocity diagram is "simple"—having constant C_x and u from rotor inlet to rotor outlet for any one streamline.

The class of variations of tangential velocity suggested by Carmichael and Lewis (equation 6.7) is defined by

$$C'_{\theta 1} = a'(r')^n - \frac{b'}{r'}, \tag{6.10a}$$

$$C'_{\theta 2} = a'(r')^n + \frac{b'}{r'}. \tag{6.10b}$$

For this class of velocity distributions

$$R'_n = 1 - a'(r')^{n-1}, \tag{6.11}$$

$$\psi' \equiv \frac{C_{\theta 1} - C_{\theta 2}}{u} = \frac{C'_{\theta 1} - C'_{\theta 2}}{r'} = \frac{-2b'}{(r')^2}, \quad \text{for } u_1 \approx u_2; \tag{6.12}$$

$$\therefore a' = 1 - R'_{nm},$$

$$b' = -\frac{\psi_m}{2}.$$

These can be inserted into the SRE equation (6.8) and integrated to give

$$\left[\frac{C'_{x1}}{\phi_m}\right]^2 = \left(\frac{C_{x1}}{C_{xm}}\right)^2 = 1 + \left(\frac{n+1}{n}\right)\left[\frac{1 - R'_{nm}}{\phi_m}\right]^2$$

$$\times \left\{[1 - (r')^{2n}] + \left(\frac{n}{n-1}\right)\left[\frac{\psi_m}{(1 - R'_{nm})}\right][1 - (r')^{n-1}]\right\}, \tag{6.13a}$$

Free-Stream Flow for Axial Turbomachines

$$\left[\frac{C'_{x2}}{\phi_m}\right]^2 = \left(\frac{C_{x2}}{C_{xm}}\right)^2 = 1 + \left(\frac{n+1}{n}\right)\left[\frac{1-R'_{nm}}{\phi_m}\right]^2$$

$$\times \left\{[1-(r')^{2n}] - \left(\frac{n}{n-1}\right)\frac{\psi_m}{1-R'_{nm}}[1-(r')^{n-1}]\right\}. \qquad (6.13b)$$

This form for the prescribed radial variations in axial velocity is more useful than the simpler versions because the constants are velocity-diagram parameters (R'_{nm}, ϕ_m, and ψ_m) that must be chosen in the preliminary-design stage, not arbitrary constants. However, equations (6.13a) and (6.13b) cannot be used for $n = 1.0$, for which value equation (6.8) should be integrated directly.

The actual reaction R_n, is found from its definition, equation (5.8):

$$R_n \equiv \frac{\Delta h_{st}}{\Delta h_0} = 1 - \frac{(C_2^2 - C_1^2)/2}{(u_2 C_{\theta 2} - u_1 C_{\theta 1})} = 1 - \frac{(C_2^2 - C_1^2)/2}{-\psi_m u_m^2}$$

$$= 1 + \frac{1}{2\psi_m}\left[\left(\frac{C_{x2}}{u_m}\right)^2 - \left(\frac{C_{x1}}{u_m}\right)^2 + \left(\frac{C_{\theta 2}}{u_m}\right)^2 - \left(\frac{C_{\theta 1}}{u_m}\right)^2\right]$$

$$= 1 + \frac{1}{2\psi_m}\left[-2\left(\frac{n+1}{n}\right)\frac{(1-R'_{nm})^2}{\phi_m^2}\frac{n}{(n-1)}\frac{\psi_m}{(1-R'_{nm})}\phi_m^2[1-(r')^{n-1}]\right.$$

$$\left. - 2(1-R'_{nm})\psi_m(r')^{n-1}\right]$$

$$= 1 + (1-R'_{nm})\left\{-(r')^{n-1} - \frac{n+1}{n-1}[1-(r')^{n-1}]\right\}$$

$$R_n = 1 + (1-R'_{nm})\left[\frac{2(r')^{n-1} - n - 1}{n-1}\right]. \qquad (6.14)$$

There are too many variables for recommendations to be made that would be useful in all cases. Rather, the use of equations (6.13a) and (6.13b) is best illustrated by an example.

Example (calculation of flow variation across an annulus)

An axial-flow-compressor stage has the following mean-diameter velocity-diagram parameters: $\phi_m = 0.4$, $\psi_m = -0.25$, $R'_{nm} = R_{nm} = 0.50$.

It is desired to use the Carmichael and Lewis tangential-velocity distributions, and to set $R'_n = 0.25$ at $r' = 0.75$. Find the required variations of axial velocity, the reaction variable R'_n, and the true reaction R_n at radial stations where $r' = 0.75$, 0.80, 0.90, 1.0, 1.10, 1.20, and 1.25.

Solution With the substitution of the given values into equations (6.11) and (6.12), we can find a', b', and n.

At mean diameter, $r'_m = 1.0$,

$$R'_{nm} = R_n = 1 - a' = 0.5;$$

$$\therefore a' = 0.5,$$

and

$$b' = -\frac{\psi_m}{2} = 0.125.$$

At the hub, $r'_h = 0.75$,

$$R'_{nh} = 0.25 = 1 - 0.5(0.75)^{n-1};$$

$$\therefore (n-1) = \frac{\ln 1.5}{\ln 0.75} = -1.40942,$$

$$n = -0.40942.$$

Then equations (6.13a) and (6.13b) become

$$\left(\frac{C'_{x2}}{C_{xm}}\right)^2 = 1 - 2.25387\{[1 - (r')^{-0.8188}\} \pm 0.14524\{1 - (r')^{-1.40942}]\}.$$

And equation (6.14) becomes

$$R_n = 1 - 0.5\left[\frac{2(r')^{-1.40942} - 0.59058}{1.40942}\right].$$

The results can be tabulated and plotted as in table 6.1 and figure 6.4.

Table 6.1

	Hub			Mean			Tip
$r' \equiv \dfrac{r}{r_m}$	0.75	0.80	0.90	1.0	1.1	1.2	1.25
$C'_{x1} \equiv \dfrac{C_{x1}}{u_m}$	0.4792	0.4614	0.4291	0.400	0.3735	0.3491	0.3375
$C'_{x2} \equiv \dfrac{C_{x2}}{u_m}$	0.5310	0.5016	0.4482	0.400	0.3555	0.3132	0.2927
$C'_{\theta1} \equiv \dfrac{C_{\theta1}}{u_m}$	0.3958	0.3916	0.3832	0.3750	0.3672	0.3599	0.3563
$C'_{\theta2} \equiv \dfrac{C_{\theta2}}{u_m}$	0.7292	0.7041	0.6609	0.6250	0.5945	0.5682	0.5563
R'_n	0.250	0.315	0.420	0.500	0.563	0.613	0.635
R_n	0.145	0.238	0.386	0.500	0.589	0.661	0.691

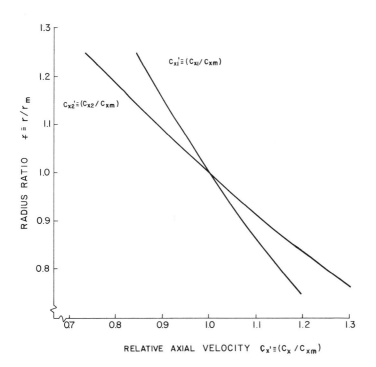

Figure 6.4 Results of calculation of flow variation across an axial-compressor annulus.

Comment The actual reaction variation exceeds that of the reaction variable. The reaction at the hub is still positive, however. The table gives all the information necessary to draw the velocity diagrams and to measure or calculate the relative-velocity ratio, W_2/W_1 (the de Haller number). If a blade speed or a required pressure rise and estimated efficiency were given, the relative Mach numbers could also be calculated.

It would be simple to produce an optimizing computer program if explicit trade-offs for relative Mach numbers at hub and tip, for minimum relative-velocity ratio and for minimum reaction, could be given. All the trade-off curves would be highly nonlinear, however, and I feel that at present the integration of the optimization of such trade-offs is best done in the designer's mind.

References

Horlock, J. H. 1958. *Axial-flow Compressors*. Butterworths, London.

Whitney, Warren J., and Warner Stewart. 1972. Velocity diagrams. In *Turbine Design and Application*. Ed. by Arthur J. Glassman. Special publication SP-290, vol. 1. NASA, Washington, D.C., pp. 69–98.

Problems

6.1

Using figure P6.1, calculate the flow incidence angles on to the rotor blades relative to design incidence at the hub ($r' = 0.75$) and the tip ($r' = 1.25$) of an axial-flow fan having untwisted constant-section inlet guide vanes and rotor blades. There are no downstream stator blades. The mean-diameter ($r' = 1.0$) velocity diagram is 50-percent reaction, with flow coefficient 0.4 and work coefficient 0.3. The inlet guide vanes produce the same flow angle, α_0, at all radii. The solution of the simple-radial-equilibrium equation for this case is $C'_x = (r')^{-\sin^2 \alpha_0}$.

6.2

Find the fluid angles, de Haller velocity ratios, and velocity vectors at hub, mean and tip diameters of a compressor stage with a mean-diameter "simple" velocity diagram ($C_x = \text{constant}$) characterized by $R_{nm} = 0.5$, $\phi_m = 0.4$, and $\psi_m = -0.33$. The fluid compressed is carbon dioxide, and the maximum relative Mach number based on the inlet relative velocity at the mean diameter is to be 0.8. Find the relative Mach numbers at the hub and tip sections for three cases:

free vortex (varying R'_n),

constant R'_n,

$R'_{nh} = 0.2$.

The diameter ratio of the hub is 80 percent. The enthalpy rise is to be constant along the radius. The inlet total temperature is 145 C. Comment on your findings for the constant-R'_n case.

6.3

Find the velocity diagrams at hub and tip diameters for two alternative designs of water pump. One has a free-vortex type of blade "twist," and the other has $C_{\theta 1} = a - b/r$, $C_{\theta 2} = a + b/r$. The hub, mean, and tip diameters are in the ratio 0.6, 0.8, and 1.0. The mean-diameter velocity diagram has axial inlet and outlet, rotor followed by stator, with a flow coefficient of 0.3 and a work coefficient of 0.208.

6.4

Comment on the apparent advantages and disadvantages of a forced-vortex type of flow distribution in which the tangential component of velocity at rotor entrance, $C_{\theta 1}$, is proportional to the square root of the radius. The mean-diameter diagram

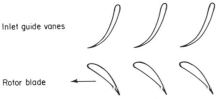

Inlet guide vanes

Rotor blade

Figure P6.1

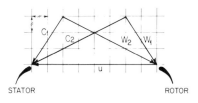

STATOR   ROTOR

Figure P6.4

$(r' = 1.0)$ is shown in figure P6.4; estimate qualitatively the shapes of the diagram at hub $(r' = 0.75)$ and tip $(r' = 1.25)$.

6.5

Calculate the angles and velocities for the diagrams of an axial-compressor stage at mean diameter, and at diameters 20 percent greater and 25 percent smaller. The mean-diameter velocity diagram has 50 percent reaction, a flow coefficient of 0.362 and a work coefficient of 0.25. The relative Mach number of the flow entering the rotor blades is 0.85 at mean diameter. Calculate the relative Mach numbers and the degrees of reaction at the other diameters if the total enthalpy rise is constant across the rotor and if the variation of tangential velocity is given by $C_\theta = a \pm b/r$. The air inlet temperature is 300 K.

6.6

(a) Sketch the tip and hub velocity diagrams for an axial-flow compressor, the mean-diameter velocity diagram of which is defined by $R_{nm} = 0.5$, $\phi_m = 0.4$, $\psi_m = 0.3$, $C_x = $ constant. The tangential component of velocity at entry to the stator is to be constant with radius, and the enthalpy rise must also be constant. The hub-tip diameter ratio is 0.6, and the mean diameter can be taken to be at 0.8 of the tip diameter. (b) By inspecting the diagrams, state where you would look for design limitations, such as high relative Mach number, high deflections, high static-pressure-rise coefficients, and so forth. (c) Comment on the desirability of this type of tangential-velocity distribution.

6.7

It is often desirable in axial-compressor design to have higher axial velocities at the hub than at the shroud, so that the diffusion ratios there will be lessened. By using the equation of simple radial equilibrium, show that a condition for the axial velocity to increase from shroud to hub at the rotor-entry plane is for the distribution of tangential velocity at the same plane to be

$$(C'_{\theta 1})^2 = ar' + (C'_{\theta m1})^2 - \frac{a}{r^2},$$

where a is a constant and is positive; r' is the ratio of the radius with the mean radius; C'_θ is the ratio of the tangential velocity with the mean-radius blade speed; and $C'_{\theta m}$ is the value of that quantity at mean radius.

7

The design and performance prediction of axial-flow turbines

We shall be concerned in this chapter first with methods for choosing the shape, size, and number of turbine blades to fulfill given initial specifications. Second, we shall look at ways of finding the efficiency and the enthalpy drop which will be given by turbine blade rows in different conditions of operation.

7.1 The sequence of preliminary design

This chapter fits in the overall progress of preliminary design in the following way.

1. The starting point will be given turbine specifications. These should include at least

the fluid to be used,
the fluid total temperature and pressure at inlet,
the fluid total or static pressure at outlet,
either the fluid mass flow or the turbine power output required,
perhaps the shaft speed.

These specifications will (presumably) be for the so-called "design point." The specifications should also contain some information on the degree to which the turbine duty will involve operation at design-point conditions, and to what extent the machine will operate at other conditions. The specifications must also give "trade-offs," or an objective function, which enable(s) the designer to aim for maximum efficiency, or minimum first cost, minimum weight, minimum rotating inertia, maximum life, or some combination of these and other measures. (The designer is often left to guess at the trade-offs.)

2. The velocity diagrams at the mean diameter (or some other significant radial station such as at the hub) must be chosen by the approach de-

scribed in chapter 5. This choice will include a selection of the number of stages to be used. This in turn implies some foreknowledge of permissible blade speeds, or an initial choice can be made and changed in later iterations.

3. The velocity diagrams at other radial locations are next chosen by the methods described in chapter 6.

4. The number of blades for each blade row, and their shape and size for each radial station for which velocity diagrams are available, are obtained by the methods given in this chapter.

5. The turbine performance at design-point and off-design conditions is predicted by methods described in this chapter.

6. When iterations involved in these steps have resulted in an apparently satisfactory blade design, it is drawn in detail. Initially, the various blade sections must be "stacked" in such a way that the centers of area are in a straight radial line to prevent the centrifugal forces from bending the blade. The leading and trailing edges must also be "faired" along continuous, regular, straight lines or curves. We do not deal with this stage of the design process in this book.

7. Centrifugal and fluid-bending stresses are calculated for the turbine blades so produced. Approximate preliminary methods are described in chapter 13.

8. For high- or low-temperature turbines, heat-transfer calculations are made to give the transient and steady-state temperature and stress distributions. Some fundamentals needed for heat-transfer analysis are given in chapter 10.

9. The mechanical and thermal stresses are combined. The acceptability of the combined stresses is checked against various criteria of failure. For high-temperature turbines, the most important criterion is usually the maximum permissible creep rate or the maximum rate of crack growth. This aspect of materials selection is discussed in chapter 13. Full mechanical-plus-thermal stress analysis is beyond the scope of this book, however.

10. The natural frequencies of the individual blades are calculated for rotational speeds up to overspeed. The aerodynamic excitations, principally the wakes of upstream and downstream blade rows and struts, or asymmetric inlets or exhausts, are also found. Coincidence of excitations with first- or second-order natural frequencies must be avoided by redesign or by changing the natural frequencies (for example, connected shrouds, lacing wires). The design process to avoid vibration failures is described in chapter 13. A full vibrational analysis is beyond the scope of this book.

7.2 Blade shape, spacing, and number

This is the fourth phase of axial-turbine design, as detailed in section 7.1. It can be broken down into these steps.

1. Calculation of approximate optimum solidity for design-point operation.

2. Interpolation of "induced incidence" from correlations. (However, this quantity may not be used in the design.)

3. Interpolation of outlet-flow deviation from correlations.

4. Choice of leading- and trailing-edge radii.

5. Layout of a range of blade shapes with varying "stagger" or setting angle, and choice of shape giving a passage shape of desirable characteristics.

6. Selection of number of blades from consideration of performance, vibration, and heat-transfer analyses.

Optimum solidity

Suppose that we have a row of blades, whose shape we can change at will. For instance, they may be made of sheet metal and may therefore be bent. The inlet flow approaches the blade row at a predetermined angle. We have to bend the blades so that a desired outlet angle is produced. We try with widely spaced blades and find that we have to bend the blades excessively to compensate for the inevitable separation of the flow from the suction surface.

When we space the blades very close together, the blades do not have to be as greatly curved, and there is no separation. However, it is obvious that there are unnecessary friction losses.

It is clear that there must be a range of spacings for which the losses are near a minimum (see figure 7.1). This optimum solidity must be related somehow to the aerodynamic loading of the blade row. Zweifel (1945) found that minimum-loss solidities could be quite well correlated by setting the tangential lift coefficient, C_L, at a constant value of 0.8, where

$$C_L \equiv \frac{\text{tangential aerodynamic force}}{\text{tangential blade area} \times \text{outlet dynamic head}}.$$

However, current design practice uses somewhat higher values of C_L, between 0.9 and 1.0 (Stewart and Glassman 1973).

The tangential loading is calculated by applying Newton's law to the fluid deflection through a blade row, figure 7.2:

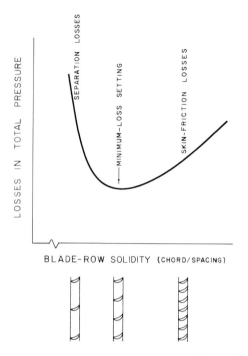

Figure 7.1 Diagrammatic variation of loss with blade setting solidity.

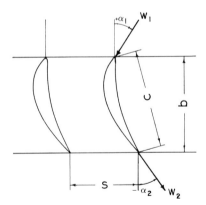

Figure 7.2 Diagram for calculating the tangential lift on a cascade of blades.

Tangential aerodynamic force $F_\theta = \dfrac{\dot{m}(W_1 \sin \alpha_1 - W_2 \sin \alpha_2)}{g_c}$

$$= \frac{sh \cos \alpha_2 \, \rho_{st2} W_2 (W_1 \sin \alpha_1 - W_2 \sin \alpha_2)}{g_c},$$

(7.1)

where $h \equiv$ blade length, subscript $1 \equiv$ blade-row inlet, and subscript $2 \equiv$ blade-row outlet. The mean pressure on the blade to give this force is

$$p_{st} = \frac{sh \, \rho_{st2} W_2 \cos \alpha_2}{g_c bh}(W_1 \sin \alpha_1 - W_2 \sin \alpha_2),$$

(7.2)

where b is the blade axial chord, figure 7.2. This pressure can be converted to a lift coefficient by dividing by the quantity $\rho_{02} W_2^2/2g_c$, which in low-speed or incompressible flow is the outlet dynamic head.

Tangential lift coefficient $C_L = \dfrac{\rho_{02} s W_2^2 \cos \alpha_2 [(W_1/W_2)\sin \alpha_1 - \sin \alpha_2]}{g_c b \rho_{02} W_2^2/2g_c}$

$$= 2\left(\frac{s}{b}\right)\cos^2 \alpha_2 \left[\left(\frac{W_1 \sin \alpha_1}{W_2 \cos \alpha_2}\right) - \tan \alpha_2\right].$$

(7.3)

If the axial velocity, C_x, is constant through the blade row (a usual design condition),

$$C_x = W_1 \cos \alpha_1 = W_2 \cos \alpha_2;$$

$$\therefore C_L = 2\left(\frac{s}{b}\right)\cos^2 \alpha_2 [\tan \alpha_1 - \tan \alpha_2].$$

(7.4)

[The use of the "incompressible" dynamic head in the derivation of this lift coefficient does not preclude it from being used for compressible flow. The correlation based on this defined lift coefficient has been found to be a useful guide for compressible and incompressible flow. See chapter 4, equations (4.4) through (4.8) for a similar argument for an incompressible diffusion coefficient.]

If the optimum lift coefficient is 0.8, the optimum "axial solidity," $(b/s)_{opt}$ is

$$\left(\frac{b}{s}\right)_{opt} = |\, 2.5 \cos^2 \alpha_2 [\tan \alpha_1 - \tan \alpha_2]\,|.$$

(7.5)

The vertical bounding lines indicate that the magnitude of this quantity should be used. The coefficient 2.5 would become 2.0 if the desired value of lift coefficient, C_L, were set at 1.0.

Angle convention

The angles should be measured consistently. One rule is to specify the angle as positive if the flow has a tangential component in the same direction as the blade speed. In figure 7.2, α_1 would be positive and α_2 negative if this were a moving blade row. The magnitudes would therefore be additive.

The blade tangential lift force may be in the direction of rotor-blade movement (for turbine rotors and compressor stators) or in the reverse direction. Under the suggested sign convention, this blade lift force will be negative, and the optimum axial solidity will also seem to be negative, for turbine stators and compressor rotors. Therefore the magnitude of the quantity on the right-hand side of equation (7.5) should be used, or a negative sign should be inserted for these two cases.

Subscript convention

It is not possible of course to have unique definitions for all symbols. Some must serve double or triple duty. When we are considering blade rows by themselves, the subscript 1 pertains to the blade or blade-row inlet and the subscript 2 to the blade outlet. When we are dealing with the complete stage, we try to restrict the subscript 1 to "rotor inlet" and the subscript 2 to "rotor outlet." For the thermodynamic conditions in the engine cycle the subscript 1 means "compressor-inlet conditions." We shall try to make it clear which of these definitions is being used.

Number of blades

Up to this point the number of blades is open to the designer to choose. Axial turbines have been built with

1 to 24 blades for wind turbines,

3 to 30 blades for water turbines,

11 to 110 blades for gas turbines.

The number of blades in any row does not have to be chosen at this stage, although the designer usually has a desirable range in mind. Lightweight aero engines employ a large number of blades to keep the engine length short (axial chord, b, decreases as the number of blades is increased, so decreasing the spacing, s). Stationary industrial machines will often have a small number of large blades, because manufacturing costs are reduced, relative profile accuracies will be increased, blade boundary layers will be relatively smaller (higher Reynolds numbers), and, if used, internal blade cooling may be easier to incorporate. The normal procedure is to make an initial choice of number of blades at this point and to change it later, if desired, from consideration of calculated aspect-ratio losses or of vibration frequencies.

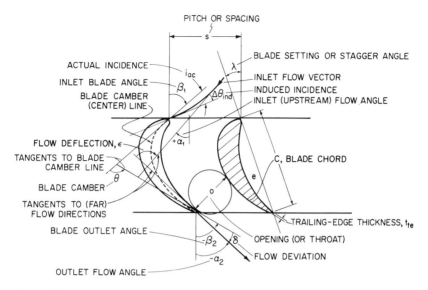

Figure 7.3 Cascade notation for turbine-blade rows.

Induced incidence

The circulation around the blades which enables them to give lift to the flow also causes the incoming streamlines to turn as they approach the blade row, figure 7.3. Dunavant and Erwin (1956) have correlated the induced incidence over a range of conditions, as shown in figure 7.4a. A reasonable correlation of these data is given by

$$\Delta\theta_{\text{ind}} = 14\left(1 - \frac{\alpha_1}{70°}\right) + 9\left(1.8 - \frac{c}{s}\right). \tag{7.6}$$

The validity of this expression outside the range of α_1 between $0°$ and $70°$ is uncertain.

This uncertainty is not of great significance. Turbine blades are often designed without the induced incidence being incorporated. This results in the actual incidence, i_{ac}, being somewhat greater than that calculated, and in the blade camber, θ, being lower than it would be were the induced incidence taken into account. This neglect of the induced incidence is sometimes not so much optional as necessary, particularly for low-reaction rotor blading with little overall acceleration through the blade row. In this case the addition of camber in the leading-edge region can produce an upstream throat, which is obviously undesirable and certainly worse than some positive incidence.

The nominal blade inlet angle, β_1, from which the actual incidence, i_{ac}, is measured in figure 7.3 is the tangent to the blade centerline 5 percent

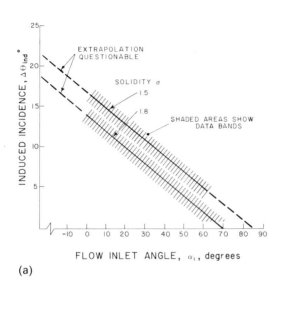

(a)

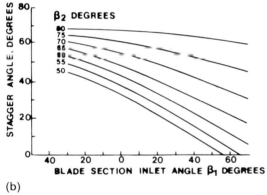

(b)

Figure 7.4 Data for low-speed turbine-blade rows: (a) Induced incidence angle. From Dunavant and Erwin 1956. (b) Stagger angle versus inlet and outlet angles. From Kacker and Okapuu 1981.

of the chordal distance from the leading edge. Similarly, the blade outlet angle, β_2, is the tangent to the blade centerline at 95 percent of the chordal distance from the leading edge. These definitions are used for all NACA/NASA correlations.

The upstream and actual fluid and blade inlet angles, α_1, α_{1ac}, and β_1, and the upstream, actual and induced incidences, i, i_{ac}, and $\Delta\theta_{ind}$, are related by

$$i_{ac} = i + \Delta\theta_{ind},$$

$$\alpha_{1ac} = \alpha_1 + \Delta\theta_{ind},$$

$$\beta_1 + i = \alpha_1 ;$$

$$\therefore \beta_1 + i_{ac} = \alpha_1 + \Delta\theta_{ind}. \tag{7.7}$$

Deviation

The same circulation around an aerodynamically loaded blade causes the leaving-flow direction to deviate from that of the blade trailing edge, as shown in figure 7.3. However, while the induced incidence can be considered or ignored without significant effects on blade performance, the deviation, designated by δ, is of critical importance to the turbine designer. If ignored, the turbine blade will produce a lower change of tangential velocity, and therefore lower torque and work output, than the designer would predict based on velocity diagrams using the blade-outlet angle to indicate flow direction. Conversely, an assumption of a large deviation would result in a turbine which, in practice, would give more work output than it is designed for.

In view of the importance of being able to predict the deviation accurately, it is disappointing to have to report that there is a strong lack of agreement among the several correlations that have been published for turbine-blade deviation. The uncertainty is probably as responsible as any other factor for the mismatching between compressor and turbine, and this has been a disappointing feature of many gas-turbine engines. Particularly for industrial engines the cost of reblading one component when mismatching has been found has seemed too large. The engine efficiency has suffered, often to a considerable extent, because one or both of the major components, even though intrinsically capable of high-efficiency operation, has been forced to work in a low-efficiency area of its characteristic.

Among the alternative correlations for deviation that have been published, we have chosen that due to Ainley and Mathieson (1951) for two reasons. First, in the necessarily limited experience of the author and of those of his associates with turbine design and test information, it has seemed to provide the most accurate and reasonable of the alternatives. Second, the correlation is based on the blade opening or throat, whereas some of the alternatives use other measures of the cascade configuration. The blade throat controls the mass flow passed by a turbine and is a very important specification in the manufacturing drawings. It is therefore advantageous to bring out this quantity directly in the design process, rather than to infer it indirectly subsequently.

The Ainley and Mathieson correlation is normally given graphically, with corrections for high subsonic Mach numbers and for the degree of curvature on the turbine blade downstream of the throat (the radius e in figure 7.3). We have translated these into three analytical expressions, for different Mach-number conditions. The correlations are for flow outlet angle rather than deviation.

1. Predicted outlet angle for hydraulic turbines, and for compressible-fluid turbines with throat Mach numbers up to 0.5:

$$|\alpha_2|_{0<M_t<0.5} = \left[\frac{7}{6}\left(\left|\cos^{-1}\left(\frac{o}{s}\right)\right| - 10°\right) + 4°\left(\frac{s}{e}\right)\right], \tag{7.8}$$

where $o \equiv$ diameter of throat opening (figure 7.3),
   $s \equiv$ blade pitch,
   $e \equiv$ radius of curvature of blade convex surface downstream of throat,
   $M_t \equiv$ Mach number at throat.

2. Predicted outlet flow angle for turbines with sonic flow at the throat ($M_t = 1.0$):

$$|\alpha_2|_{M_t=1.0} = \left|\cos^{-1}\left(\frac{o}{s}\right)\right| - \left(\frac{s}{e}\right)^{(1.786+4.128(s/e))}\left|\sin^{-1}\left(\frac{o}{s}\right)\right|. \tag{7.9}$$

3. Predicted outlet flow angle for throat Mach number between 0.5 and 1.0:

$$|\alpha_2|_{0.5<M_t<1.0} = |\alpha_2|_{0<M_t<0.5} - (2M_t - 1)(|\alpha_2|_{0<M_t<0.5} - |\alpha_2|_{M_t=1.0}). \tag{7.10}$$

Selection of surface curvature

Early turbine blades were usually designed to be plane downstream of the throat, partly because of the limitations of the milling methods normally used for blade manufacture, and partly because designers felt that the flow would not be turned after the throat. Now, with improved casting methods and numerically controlled machining, these limitations do not exist. Some degree of surface curvature on the blade convex (suction) surface downstream of the throat is desirable in subsonic turbine cascades because, if the flow is to remain attached to the blade surface

HYDRAULIC TURBINE GENERATOR UNIT

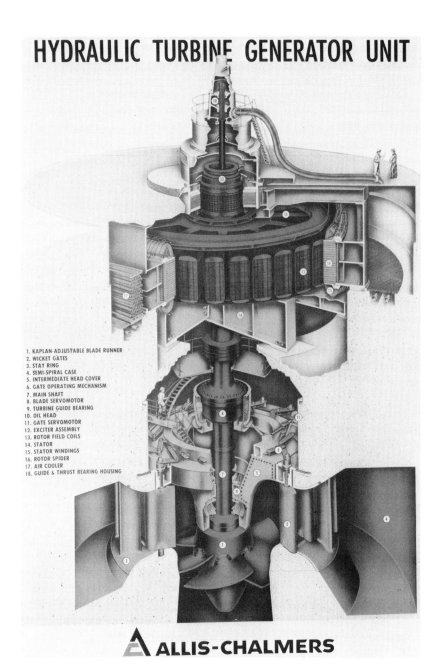

1. KAPLAN-ADJUSTABLE BLADE RUNNER
2. WICKET GATES
3. STAY RING
4. SEMI-SPIRAL CASE
5. INTERMEDIATE HEAD COVER
6. GATE OPERATING MECHANISM
7. MAIN SHAFT
8. BLADE SERVOMOTOR
9. TURBINE GUIDE BEARING
10. OIL HEAD
11. GATE SERVOMOTOR
12. EXCITER ASSEMBLY
13. ROTOR FIELD COILS
14. STATOR
15. STATOR WINDINGS
16. ROTOR SPIDER
17. AIR COOLER
18. GUIDE & THRUST BEARING HOUSING

⟁ ALLIS-CHALMERS

Axial-flow Kaplan hydraulic turbine with adjustable rotor and stator blades.
Courtesy Allis Chalmers Co.

rather than separate, it must continue to accelerate downstream of the throat. This acceleration will favor the preservation of a laminar boundary layer, which has lower losses and a much lower heat-transfer coefficient.

For the preliminary design, ratios of the blade spacing, s, to the surface curvature, e, of up to 0.75 are permissible. A range that appears to give favorably shaped blades is $0.25 < (s/e) < 0.625$.

Blade leading and trailing edges

The turbine-blade leading-edge radius is often specified to be a given proportion of either the blade chord, c, or the spacing, s. Typical values usually fall between $0.05s$ and $0.10s$. In addition specifying the angle between the tangents made by the two blade surfaces (upper and lower, or convex and concave) to the leading-edge circle (say, 20 degrees) can help to define an acceptable blade or passage shape.

The trailing edge is sometimes specified as having a radius, but it is more usual to specify a trailing-edge thickness, t_{te}, along the blade-row-exit plane with the blades being cut off sharply. Typical values are between $0.015c$ and $0.05c$.

Choice of setting angle

One variable remains to define the "framework" of the blade-row configuration at any one radial station after the axial solidity, (b/s), the induced incidence, $\Delta\theta_{ind}$, and the blade curvature, e, and outlet angle β_2, have been found. This variable is the setting angle, λ (figure 7.3). It is related to the axial chord, b, and the blade chord, c, by

$$\cos \lambda = \frac{b}{c}.$$

The setting angle determines the overall blade shape to a large extent. More important, for turbines it determines the passage shape. One does not have the freedom to choose the setting angle arbitrarily. We are restricting our attention to subsonic turbines, and in these the throat or narrowest part of the passage must occur at the trailing edge. Nonetheless, it is possible to arrange that the throat occur at the trailing edge over only a small range of setting angles. Finding which part of the range will be best still appears to be more of an art than a science, even when sophisticated computer programs are used to judge the "quality" of a blade passage. In other words, even computer programs work on a trial-and-error basis, with each new trial being checked for quality by approximation, rather than an optimum setting angle being found by precise computation. A guide for obtaining the setting angles has been published by Kacker and Okapuu (1981), as given in figure 7.4b.

In a computer program to design turbine blades, the criterion of quality would be some function of the calculated stability, and the total losses, of the boundary layers. The type of preliminary design with which we are concerned here has to be an approximation of this criterion, in qualitative rather than quantitative terms. We look for at least a throat at the trailing edge and for a steady reduction in area along the passage and, if possible, gradual curvatures of the passage walls. And then there is the undefinable perogative of the designer: to ask the question, "Does it look right?" An example, given after the following formal restatement of the quantitative component of turbine-blade design, should make the procedure clearer.

Blade-row design procedure

The necessary preliminary work for a blade-row design will include the selection of the velocity diagram for the radius in question, from which the fluid angles will be obtained. The throat Mach number, which for subsonic blade rows can be approximated to be the blade-exit Mach number, can also be found for compressible-fluid turbines. Zweifel's criterion can be used to find the optimum axial solidity (equation 7.5).

The procedure is, then, as follows.

1. Choose one or more values of the spacing/curvature ratio, (s/e), and of the leading-edge radius, r_{le} and trailing-edge thickness, t_{te}.

2. Calculate the throat/spacing ratio, (o/s), from equations 7.8, 7.9, or 7.10, for the appropriate Mach number.

3. Draw the throat-to-trailing-edge regions for the values found. Choose one or more for further exploration.

4. Choose several values of the setting angle, λ, guided by the values suggested by figure 7.4b. Draw in the leading edges using the previously chosen leading-edge radius and a blade inlet angle chosen in the range between α_1 and $(\alpha_1 + \Delta\theta_{ind})$, where α_1 is the fluid inlet angle and $\Delta\theta_{ind}$ is the induced incidence (equation 7.6 and figure 7.4a).

5. Connect the leading-edges with the throat sections with smooth curves. Draw circles in the passages so formed to touch the convex and concave blade surfaces. Acceptable blade passages are those giving continuous reductions of area (or inscribed-circle diameter) up to the throat.

In preliminary design (which for most turbines built in the past will be the final design) the selection of the "best" blade-row design from those produced by this procedure will be largely aesthetic. There may, however, be some constraints favoring one configuration rather than another when the blade-row designs chosen for other radial sections are assembled into a complete blade. For turbines where the need to attain the ultimate

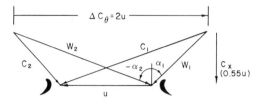

Figure 7.5 Velocity diagram in turbine-blade preliminary design.

in performance is critical, these procedures can produce the input for a computer program that will adjust the blade shape until a favorable, or even a prescribed, velocity distribution around the blade surface is produced.

Example (blade-section preliminary design)

Design and draw at least two neighboring blade sections for a turbine rotor-blade row at a radial station where the velocity diagram is 50-percent reaction, with a work coefficient of 2.0 and a flow coefficient of 0.55. The throat Mach number is 0.75.

Use a leading-edge diameter of one-tenth of the spacing. The trailing-edge thickness should be 0.02c. The ratio of the spacing to the surface curvature downstream of the throat may be chosen at 0.333.

Solution We solve the velocity diagram to find the flow angles (figure 7.5). These will allow the optimum solidity to be obtained. Then we make a table in which we try various values of the setting angle. We draw blade and passage shapes for some of the setting angles and choose that set giving the best passage shape.

Velocity diagram

$$\tan \alpha_1 = +\frac{0.5 \, u}{0.55 \, u},$$

$$\alpha_1 = +42.27° ;$$

$$\tan \alpha_2 = -\frac{1.5 \, u}{0.55 \, u},$$

$$\alpha_2 = -69.86°.$$

Optimum solidity We will use equation (7.5) with the optimum lift coefficient set at 1.0:

$$\left(\frac{b}{s}\right)_{opt} = |2.0 \cos^2 \alpha_2 (\tan \alpha_1 - \tan \alpha_2)| = 0.862.$$

Throat opening With the throat Mach number of 0.75 the outlet angle will be the mean of the values given by equations (7.8) and (7.9). We interpolate to find the value of o/s that will give an outlet angle of $-69.86°$:

$\cos^{-1}\left(\dfrac{o}{s}\right)$	$69°$	$70°$	
$\alpha_{2M_t=0.5}$	$-70.167°$	$-71.333°$	(equation 7.8)
$\alpha_{2M_t=1.0}$	$-68.971°$	$-69.971°$	(equation 7.9)
$\alpha_{2M_t=0.75}$	$-69.569°$	$-70.652°$	(by interpolation)

Interpolating, we find we need $\cos^{-1}(o/s) = 69.71°$ to obtain the outlet angle desired, $-69.86°$. Therefore

$$\frac{o}{s} = 0.3468 \quad \text{or} \quad o = 0.3468s.$$

It is convenient to give all the geometrical parameters in terms of the spacing, s, which will be known from the specifications of the turbine (mass flow, fluid properties, velocity diagram, hub-tip diameter ratio, and number of blades). If we use the value of $(b/s)_{opt}$ derived earlier,

$$b = 0.862s,$$

$$c = \left(\frac{0.862}{\cos \lambda}\right)s.$$

Before the trailing-edge section downstream of the throat can be drawn, an estimate must be made of the intercept, Δs_{te}, of the trailing edge on the outlet plane. We can start by assuming that the trailing-edge slope is $\cos^{-1}(o/s)$, so that in the present case

$$\frac{\Delta s_{te}}{s} = \frac{0.02(b/s)}{(o/s)\cos \lambda} = \frac{0.05}{\cos \lambda},$$

where λ is the setting angle. A first guess at the setting or stagger angle can be made from figure 7.4b to be $40°$.

We can then begin to lay out the blade section on a drawing board, perhaps ten times full size for accuracy. Along the trailing-edge plane we mark a blade pitch, s, and the trailing-edge intercept. Then, as shown in figure 7.6, we find the center of curvature of the convex surface downstream of the throat by striking radii of e and $e + o$ appropriately. We can draw in this part of the blade surface, and it is useful also to draw in the throat circle.

This first iteration of the blade surface downstream of the throat will be common to all subsequent trials of various setting angles. We have chosen a setting angle of 40 degrees to illustrate the remainder of the procedure. We therefore draw the chord line from the trailing-edge point on the convex surface (the point that defines one side of the throat), and the leading-edge line using the axial chord, b. The leading-edge circle is drawn tangent to the chord and leading-edge lines. The initial directions of the concave and convex surfaces were specified to have an included angle of 30

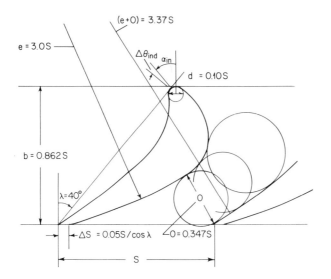

Figure 7.6 Construction of turbine-blade profile.

degrees, and these will be equally disposed around the blade angle. For zero incidence this will be given by the flow angle plus the induced incidence, $\Delta\theta_{ind}$, which is found using equation (7.6). A list can be made up such as the following:

1. Setting angle, λ, deg.   40
2. $b/c = \cos\lambda$ 0.766
3. Solidity, $c/s = 0.862/(b/c)$ 1.125
4. $\Delta\theta_{ind} = [5.546 + 9(1.8 - c/s)]$ deg.   11.62
5. Flow inlet angle, α_{in} deg.   +42.27
6. Flow outlet angle, α_{out} deg.   −69.86

The remainder of the design procedure is nonquantitative and iterative. The leading-edge tangents are joined to the throat surfaces by smooth curves, perhaps using French curves. Initially, we draw just the surfaces bounding a passage, rather than the two surfaces of a single blade. Then a series of circles of increasing size touching both sides of the flow passage is drawn, starting at the throat circle and proceding upstream, to check that the passage does indeed converge steadily to the throat. Vectors showing the range of probable design-point flow angles at inlet, α_{in} to $(\alpha_{in} + \Delta\theta_{ind})$, are drawn to check, by eye, the leading-edge shape. Finally, one blade surface is redrawn to produce a complete blade profile. The trailing-edge thickness is checked and the blade and passages are redrawn if necessary.

The whole procedure may be repeated for other blade setting angles, as was done in the present case for $\lambda = 30°, 45°$, and $50°$. The choice of the best passage and blade shapes from among these alternatives is made purely by the designer's judgment, in this preliminary-design stage. A later stage of refinement would be to examine the velocity distributions in a flow-analysis computer program and, if necessary, to modify the passage shape to produce minimum-loss flow.

7.3 Performance (efficiency) prediction of axial turbine stages

For the purposes of calculating the efficiency, a turbine can be thought of as a perfect machine to which losses have been added. We can categorize the losses as belonging in three groups (although some losses refuse to behave in so orderly a fashion and straddle two or more groups).

1. Losses that appear as pressure losses in the fluid, for instance, boundary-layer friction.

2. Losses that appear as enthalpy increases in the fluid, for instance, disk friction which extracts power from the turbine output and in effect adds this energy as heat to the turbine outlet flow.

3. Losses that reduce the shaft-power output of the turbine through friction, the energy of which is dissipated elsewhere and not to the working fluid, for instance, friction losses in external bearings. Group 3 losses were referred to as "external energy losses" in section 2.8. We need not concern ourselves further here with group 3 losses.

The "perfect machine" of the first paragraph is the turbine expanding from inlet total conditions 1 in figure 7.7 isentropically through the design turbine enthalpy drop, Δh_0, to 2_{th}. The Euler equation, (4.1), and the first law, equation (2.8), show that a turbine designed to produce the flow angles called for in a velocity diagram will in fact produce the required work and enthalpy drop regardless of losses, so long as a sufficient pressure ratio is provided for the flow to attain the designed velocities.

In the treatment of losses outlined here, the various losses are compartmentalized as if all the effects of a given loss, for instance, the tip-clearance loss, could be treated as a pressure drop. In practice, losses have a wider influence. In particular, losses introduce nonuniformities into the flow. These nonuniformities may then induce secondary losses in downstream components, and these might be higher than the original primary losses. For instance, a large turbine tip-clearance loss could produce so large a boundary layer at entrance to the exit diffuser that the flow could stall, producing a loss much greater than the original tip-clearance loss. The diffuser designer may insist that the loss be allocated to the turbine-designer's ledger. The analyst simply calculates the losses in each component, without assigning blame. The overall engine or system designer (and of course all three could be the same person) tries to ensure that upstream components either have lower losses or produce more uniform velocity distributions or uses instead downstream components that are less sensitive to poor velocity profiles.

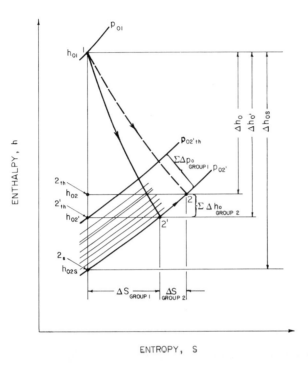

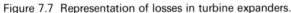

ENTROPY, S

Figure 7.7 Representation of losses in turbine expanders.

Group 1 losses

Group 1 pressure losses simply add together to require a reduction in turbine outlet pressure from $p'_{02\text{th}}$ to p'_{02}. The various loss mechanisms can be considered to be separable and their effects additive. With the pressure loss for each mechanism (for instance, profile loss) expressed as a ratio with the outlet (theoretical incompressible) dynamic head, the isentropic efficiency can be found as follows.

The relative outlet theoretical incompressible dynamic pressure, q_2 from the rotor is $(\rho_{02} W_2^2/2g_c)$, where W_2 is the relative rotor outlet velocity taken from the velocity diagram. If the rotor profile (boundary-layer) losses produce an incremental total-pressure drop of $\Delta p_0 \text{ pro}$, the loss coefficient X_{pro} is defined as

$$(X_{\text{pro}})_{\text{ro}} \equiv \frac{(\Delta p_{0,\text{pro}})_{\text{ro}}}{q_{\text{out,ro}}}, \tag{7.11}$$

where $q_{\text{out,ro}}$ is q_2. Similarly, the stator profile-loss coefficient is defined as

$$(X_{\text{pro}})_{\text{sta}} \equiv \frac{(\Delta p_{0,\text{pro}})_{\text{sta}}}{q_{\text{out,sta}}}, \tag{7.12}$$

Axial-Flow Turbines

where $q_{out,sta} \equiv \rho_{01} C_1^2/2g_c$; the subscript 1 refers here to stator-outlet conditions.

Group 1 losses can be divided among profile losses, represented by the loss coefficients $(X_{pro})_{ro}$ and $(X_{pro})_{sta}$; secondary losses, represented by the loss coefficients $(X_{sec})_{ro}$ and $(X_{sec})_{sta}$; annulus losses, represented by the loss coefficients $(X_{an})_{ro}$ and $(X_{an})_{sta}$; diffuser losses (X_{diff}); and outlet dynamic-pressure losses (X_q). The diffuser and dynamic-pressure losses are not always "charged" to the turbine and may therefore be omitted for some definitions of turbine efficiency. Then the total group 1 pressure loss, $\sum (\Delta p_0)_{g1}$, is given by

$$\sum (\Delta p_0)_{g1} = [X_{pro} + X_{sec} + X_{an}]_{ro} \times q_{out,ro}$$

$$+ [X_{pro} + X_{sec} + X_{an}]_{sta} \times q_{out,sta}$$

$$+ X_{diff} \times q_{out\text{-}diff} + X_q \times q'_{out}, \tag{7.13}$$

where q'_{out} is assessed at a plane defined by the considerations discussed below. In this form the loss coefficients are assessed for each blade row of a single- or multistage turbine.

Short blades may be represented by their mean-diameter velocity-diagram and cascade (such as blade-passage) parameters. For long blades, perhaps defined as those in an annulus with a hub-tip diameter ratio of less than 0.75, the mean (on an area or mass-flow basis) of the profile pressure losses at two or three radial stations could be used. (The use of an average of several parallel-flow pressure drops is thermodynamically unsound, but, because the inaccuracies come from second-order effects, it is pragmatically acceptable.)

In a multistage machine the losses in total pressure must be found for each stage and added together. If the diffuser and dynamic-pressure losses are not included in the group 1 losses, the calculated actual outlet pressure will be the total pressure after the (last-stage) rotor-blade row. The efficiency calculated from this pressure would be the total-to-total efficiency from turbine inlet to rotor outlet. A more usual efficiency is the total-to-static taken to the diffuser outlet. The pressure losses then include the total-pressure loss in the diffuser plus the diffuser-outlet kinetic pressure. If $\Delta p_{0,diff}$ is the diffuser total-pressure loss, the diffuser loss coefficient will then be

$$X_{diff} \equiv \frac{\Delta p_{0,diff}}{q_{out,diff}} = (1 - Cpr)\frac{q_{in,diff}}{q_{out,diff}} - 1, \tag{7.14}$$

where Cpr is the actual pressure-rise coefficient, shown graphically in several charts in chapter 4. In low-Mach-number diffusers, (q_{out}/q_{in}) will be approximately equal to, but larger than, $(A_{in}/A_{out})^2$. The outlet dynamic-pressure loss is that at the diffuser-outlet plane, q_{out}. The loss coefficient, $X_q \equiv \Delta p_0 / q_{out}$, has the value 1.0 if the total-to-static efficiency is to be found, and the value zero if the total-to-total efficiency is desired. If the total-to-static efficiency at the outlet plane of the turbine blading is required, the dynamic pressure at this plane (presumably $q_{out\text{-}ro}$) is used, and X_{diff} is set to zero.

The turbine efficiency can be calculated as follows.

The inlet total conditions, p_{01} and h_{01}, will be known or specified. The useful outlet total pressure, p'_{02}, will be known or defined. As explained in chapter 2, this may be the actual total pressure at the rotor outlet or at the diffuser outlet, if the total-to-total efficiency is to be obtained or p'_{02} may be defined as the static pressure at either of these locations, for the total-to-static efficiency. For a shaft-power turbine exhausting to the atmosphere, the diffuser-outlet static pressure would be atmospheric and would be defined as p'_{02}, and the total-to-static efficiency to diffuser outlet would be used.

Then the theoretical outlet total pressure, p'_{02th}, is obtained from

$$p'_{02th} = p'_{02} + \sum \Delta p_{0,g1} \tag{7.16}$$

The isentropic efficiency based just on the group 1 losses will then be

$$\eta_s \equiv \frac{\Delta h'_0}{\Delta h_{0s}},$$

as indicated in figure 7.7.

For a compressible-fluid turbine expanding an ideal gas, the isentropic efficiency can be obtained analytically:

$$\frac{T_{01}}{T_{02s}} = \left(\frac{p_{01}}{p'_{02}} \right)^{R/\overline{C}p_e},$$

where now the subscripts apply to the inlet and outlet of the whole turbine expander, figure 7.7. T_{02s} is the isentropic outlet temperature for the defined outlet total pressure, p'_{02}:

$$\Delta h_{0s} = \overline{Cp_e}(T_{01} - T_{02s}) = \overline{Cp_e} T_{01} \left[1 - \left(\frac{p'_{02}}{p_{01}} \right)^{R/\overline{C}p_e} \right].$$

The actual enthalpy drop, $\Delta h'_0$, (excluding the effect of group 2 losses) is given by

$$\Delta h'_0 = \overline{Cp_e}(T_{01} - T'_{02th}) = \overline{Cp_e} T_{01} \left[1 - \left(\frac{p'_{02th}}{p_{01}} \right)^{R/\overline{C}p_e} \right]$$

$$= \overline{Cp}_e T_{01} \left[1 - \left(\frac{p'_{02} + \sum (\Delta p_0)}{p_{01}} \right)^{R/\overline{Cp}_e} \right]. \tag{7.17}$$

Then the isentropic efficiency based on group 1 losses is

$$\eta_{se} \equiv \frac{\Delta h'_0}{\Delta h_{0s}} = \frac{[1 - (p'_{02}/p_{01} + \sum (\Delta p_0)/p_{01})^{R/\overline{Cp}_e}]}{[1 - (p'_{02}/p_{01})^{R/\overline{Cp}_e}]}. \tag{7.18a}$$

The related polytropic efficiency is

$$\eta_{pe} = \frac{\overline{Cp}_e}{R} \frac{\ln (T_{01}/T'_{02th})}{\ln (p_{01}/p'_{02})}. \tag{7.18b}$$

Group 2 losses

Bearing, seal, and disk friction take useful torque developed by the blading, and, when the turbine is turning, the work involved is converted into heat. That part of this heat flow that is transferred into the turbine working fluid represents the sum of what are termed the "group 2" losses. The remainder is transferred as heat directly to the atmosphere and is represented simply as an external torque on the shaft. The lost work is the sum of the "group 3" losses.

If the sum of the frictional torques contributing to the group 2 losses is $\sum T_{q2}$, the enthalpy rise produced in the fluid, Δh_{0g2}, is

$$\Delta h_{0g2} = \frac{(\sum T_{q2})\omega}{\dot{m}_e}, \tag{7.19}$$

where ω is the shaft speed, rads/s, and $\dot{m}_e$ is the turbine mass flow. The isentropic efficiency including the group 1 and group 2 losses is

$$\eta_{se} = \frac{(h_{01} - h'_{02}) - \Delta h_{0g2}}{(h_{01} - h_{02s})} = \frac{h_{01} - h_{02}}{h_{01} - h_{02s}}. \tag{7.20}$$

In a compressible-fluid turbine, working on an ideal gas, the isentropic efficiency will be

$$\eta_{se} = \frac{[1 - [p'_{02}/p_{01} + \sum (\Delta p_0)/p_{01}]^{R/\overline{Cp}_e}] - \Delta h_{0g2}/\overline{Cp}_e T_{01}}{[1 - (p'_{02}/p_{01})^{R/\overline{Cp}_e}]}, \tag{7.21a}$$

and the polytropic efficiency

$$\eta_{pe} = \frac{\overline{Cp}_e}{R} \frac{\ln (T_{01}/T_{02th})}{\ln (p_{01}/p'_{02})}. \tag{7.21b}$$

Double-flow low pressure steam turbine unit during assembly. Courtesy Westinghouse Electric Corp.

Group 3 losses

If the sum of the "external" friction torques, defined in the section on group 2 losses, is $\sum T_{q3}$, the shaft work lost per unit mass of working fluid is

$$-\frac{\dot{W}}{\dot{m}} = \frac{(\sum T_{q3})\omega}{\dot{m}_e}. \tag{7.22}$$

The "external" isentropic efficiency, which includes the external losses, is found from

$$\eta_{se} = \frac{(h_{01} - h'_{02}) - \Delta h_{0g2} - [(\sum T_{q3})\omega/\dot{m}_e]}{(h_{01} - h_{02s})} \tag{7.23}$$

For an ideal gas

$$\eta_{se} = \frac{[1 - [p'_{02}/p_{01} + \sum(\Delta p_0)/p_{01}]^{R/\overline{C}p_e} - (\Delta h_{0g2}/\overline{C}p_e T_{01}) - [(\sum T_{q3})\omega/\dot{m}_e \overline{C}p_e T_{01}]}{[1 - (p'_{02}/p_{01})^{R/Cp_e}]} \tag{7.24}$$

Axial-Flow Turbines

Polytropic efficiencies

The corresponding polytropic efficiency for each definition of "internal" isentropic efficiency can also be obtained from equation (2.77):

$$\eta_{pe} = \frac{\ln\left[1 - \eta_{se}(1 - (p'_{02}/p_{01})^{R/\overline{C}p_e})\right]}{\ln\left[(p'_{02}/p_{01})^{R/\overline{C}p_e}\right]}. \tag{7.25}$$

Polytropic efficiencies cannot strictly be found for corresponding "external" isentropic efficiencies, because the end points of the hypothetical polytropic process are not identical to the actual fluid states.

7.4 Loss-coefficient data for axial-flow turbomachinery

In practice, loss data, which are generally obtained from correlations of systematic wind-tunnel or turbine tests, do not fall as neatly into categories as the theoretical breakdown in section 7.3 seems to indicate. Some losses appear to be best correlated by adding factors to other factors, while in other cases the effects of some design variables can best be included by a factor that multiplies other losses.

Whichever method is used, it is obviously necessary to be consistent. Accordingly, we shall give the correlations published by Craig and Cox (1971). These methods were developed for the design of steam turbines but were acclaimed by workers in the gas-turbine as well as the steam-turbine industry. We shall give in addition some of the data confirming and extending the method, published by the Institution of Mechanical Engineers in extensive discussions of the Craig-Cox approach. The data appear to agree well with a later method by Kacker and Okapuu (1981).

Overall profile efficiency

It is sometimes difficult to interpret loss data, particularly when they are plotted as functions of blade-row variables in which the useful output of the blade row (the change in relative tangential flow momentum) is itself varying. In other words, if the profile loss—the friction losses due to the pattern of the flow around the blade profiles—is falling, but the blade-row output is falling faster, then the choice of a minimum-loss condition as an operating point will in fact be disadvantageous. Several of the contributors to the Craig-Cox discussion presented charts of the form shown in figure 7.8, which has been drawn as an approximate average of the published charts. This figure shows very clearly that, other things being equal, optimum profile efficiency for 50-percent-reaction stages is given by a rectangular diagram characterized by $\psi = 1.0$ and $\phi = 0.6$. There may be further small gains by going to even lower loadings. Other losses, particularly kinetic-energy losses at the stage exit and tip-clearance losses, would reduce the values of efficiency given by this chart. The tip-clearance

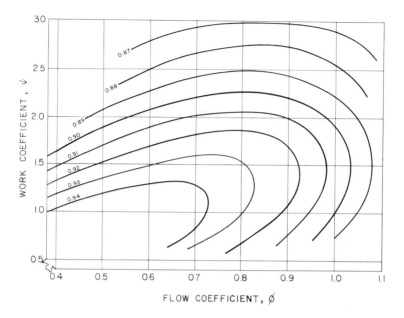

Figure 7.8 Efficiency contours for 50-percent-reaction axial-turbine stages. From Craig and Cox 1971.

loss is an independent one, but the kinetic-energy loss is also minimized in a low-work rectangular diagram, so that one would expect the total-to-static efficiency to be an optimum along with the profile efficiency. In fact figure 7.8 agrees very closely with a predictive curve given by Kacker and Okapuu (1981) for turbine stage efficiency at zero tip leakage, calculated with their modifications of the Ainley-Mathieson method, and including design-point data from a large number of turbines.

Figure 7.8 also shows the relation between higher values of ψ than that for optimum-efficiency and the flow coefficient (for 50-percent-reaction diagrams). The relation is given approximately by the expression

$$\phi_{opt} = 0.32 + 0.3\psi.$$

The detailed correlations of Craig and Cox for the design conditions of operation of blade rows follow. For estimation of performance in off-design conditions, readers are referred to the full Craig and Cox paper.

Profile losses

The basic profile-loss parameter, X_{pro}, is given in figure 7.9 as a function of the modified lift coefficient for the blade row, obtained from figure 7.10, and the blade-row contraction ratio, figure 7.11.

Axial-Flow Turbines

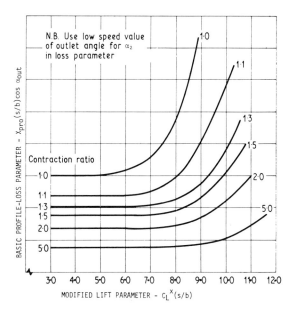

Figure 7.9 Basic profile loss. From Craig and Cox 1971.

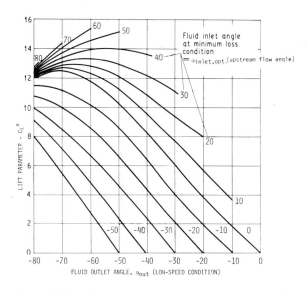

Figure 7.10 Lift parameter versus flow angles. From Craig and Cox 1971.

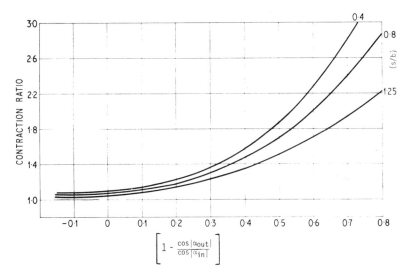

Figure 7.11 Contraction ratio for average profiles. From Craig and Cox 1971.

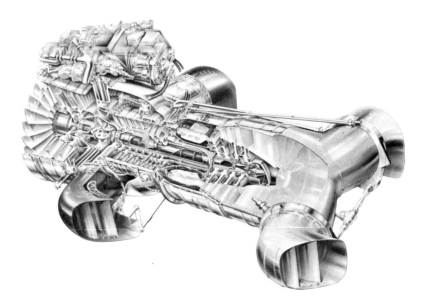

RR Pegasus engine with "vectored thrust" from front fan and turbine for vertical-takeoff aircraft. Courtesy Rolls-Royce Limited.

The profile-loss parameter so obtained would be valid for a throat Reynolds number, Re_0, of 10^5, for a low Mach number (below 0.75), for zero trailing-edge thickness, and for optimum blade-row incidence.

Trailing-edge-thickness correction
Figure 7.12 shows two corrections, a loss-factor increment, ΔX_{pro}, and a profile-loss ratio, given as functions of the ratio of the trailing-edge thickness to the blade pitch, (t_{te}/s), and the outlet flow angle, α_2.

Reynolds-number correction
For throat Reynolds numbers, based on blade-row opening, o, outlet relative flow velocity, and local (static) flow properties at the blade outlet, other than 10^5, figure 7.13 can be used to find a loss-factor ratio. This curve is an approximate mean of several curves given in the discussion of the Craig-Cox paper.

Secondary-flow and annulus losses
Losses resulting from "secondary" (transverse boundary-layer) flows and from friction on the annulus walls are largely a function of the blade-row aspect ratio, (h/b). Figure 7.14 is a combination of mean data presented by V. T. Forster in his discussion of the Craig-Cox paper. One should select the blade row closest to that being studied, or interpolate or extrapolate, to obtain a secondary-loss estimate.

Tip-clearance losses
The simplest of the many loss correlations available is to multiply the "zero-clearance" efficiency by the ratio of the blading annulus area to the casing annulus area for each blade row.

This approach is appropriate for unshrouded blade rows in smooth-wall casings. Blades with rotating shrouds will incur losses from leakage flows through the shroud labyrinths and disturbance of the wall boundary layer(s). Craig and Cox (1971) give some correlations for the clearance losses around blade shrouds and for the losses found in nonsmooth annulus walls. Roelke (1973) gives figure 7.15 showing the effect of reaction on tip-clearance losses.

Example (estimation of axial-turbine efficiencies)
Estimate the design-point polytropic and isentropic total-to-total efficiencies of a single-stage axial turbine of the following specifications. (As is appropriate for preliminary design, only the mean-diameter blading will be given and calculated.)

Fluid: air.

Mass flow, $\dot{m}$: 1.54 kg/s.

Inlet pressure, p_{04}: 2.0 at.

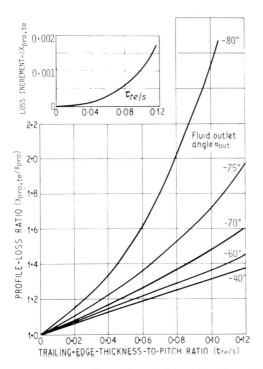

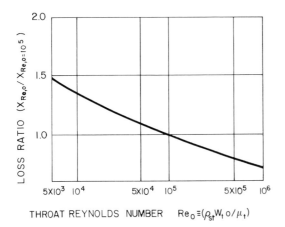

Figure 7.12 Trailing-edge-thickness loss. From Craig and Cox 1971.

Figure 7.13 Effect of Reynolds number on profile losses. From Craig and Cox 1971.

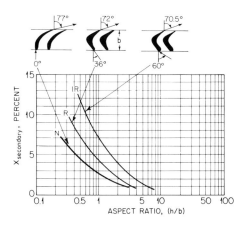

Figure 7.14 Secondary-flow and annulus losses. From Forster's discussion of Craig and Cox 1971.

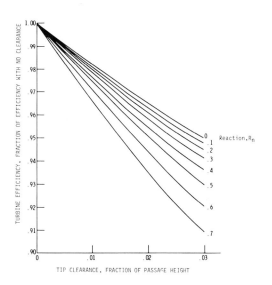

Figure 7.15 Tip-clearance correlation for unshrounded blades. From Roelke 1976.

Inlet temperature, T_{04}: 1,200 K.

Outlet pressure, p'_{05}: 1.0 at (exhaust static pressure).

Mean-diameter diagram: $\psi = 1.0$, $\phi = 0.6$, $R_n = 0.5$.

Number of nozzle blades: 16.

Number of rotor blades: 17.

Tip clearance (on radius): 0.50 mm.

Blades pitch-to-back-radius ratio: $s/e = 0.50$.

Hub-tip ratio: $\lambda = 0.6$.

Optimum tangential lift coefficient: $C_L = 1.0$.

Solidity, $\sigma \equiv c/s$: $\sigma = 1.6$.

Trailing-edge thickness ratio: $t_{te}/c = 0.025$.

The blading design is carried out by the methods suggested earlier in this chapter. Only the principal results will be given here.

An initial estimate of the efficiency must be made to enable the velocity diagram and outlet conditions to be calculated. A polytropic total-to-total efficiency of 90 percent across the blading was chosen. This turned out to be close to the final calculated value. If it had not been, a second iteration would have been desirable.

Some of the results of the calculations follow.

Outer (shroud) diameter, rotor entrance: $d_s = 198$ mm.

Inner (hub) diameter, rotor entrance: $d_h = 118.8$ mm.

Circular pitch at area mean diameter (163.3 mm): $s = 30.18$ mm.

Optimum axial solidity (equation 7.5): $(b/s)_{opt} = 0.882$.

Rotor blade axial chord: $b = 26.6$ mm.

Rotor blade actual chord: $c = 48.29$ mm.

Rotor blade opening/spacing: $o/s = 0.50$.

Relative outlet velocity: C_1 and $W_2 = 439.7$ m/s.

Mach number, rotor exit relative: $M = 0.692$.

Reynolds number, rotor exit: $Re_2 \equiv W_2 \rho_{st2}/\mu_{st2} = 0.58 \times 10^5$.

Flow angle (symmetrical diagram): $\alpha_{in} = 0°$, $\alpha_{out} = 59.04°$.

Reynolds number loss ratio (figure 7.12): $X_{Re,0}/X_{Re,0=10^5} = 1.1$.

Lift parameter for α_{in} and α_{out} (figure 7.10): $C_L x = 11.6$, $C_L x(s/b) = 13.15$ ("modified lift parameter").

Contraction ratio: $[1 - (\cos|\alpha_2|/\cos|\alpha_1|)] = 0.485$ (figure 7.11) $= 1.53$.

Basic profile-loss parameter (figure 7.9): $X_{pro}(s/b)\cos\alpha_2 = 0.033$ (extrapolation), $\therefore X_{pro} = 0.0566$.

Profile-loss ratio (figure 7.12): $X_{pro,te}/X_{pro} = 1.14$.

Profile-loss increment (figure 7.12): $\Delta X_{pro,te} = 0.001$.

Secondary and annulus losses (figure 7.14): $X_{sec} = 0.025$ (for $h/b = 1.49$, interpolating between curves N and R).

Although many of the parameters listed have been calculated for the rotor-blade row, they will also apply exactly or closely to the nozzle-blade row because at mean diameter the flow angles, blade profiles, and relative flow velocities are identical.

Corrected profile-loss parameter $X_{\text{pro, corrected}} = 0.0566 \times 1.1 \times 1.14 + 0.001$
$$= 0.0720.$$

$\sum(X)_{\text{ro, sta(g1)}} = X_{\text{pro, corrected}} + X_{\text{sec}} = 0.0970.$

Total-pressure losses, zero tip clearance,

$$\sum(\Delta p_0) = \sum(X)_{\text{ro}}(\rho_{02} W_2^2/2g_c) + \sum(X)_{\text{sta}}(\rho_{02} C_1^2/2g_c).$$

Outlet total density for $p_{02} = 1.233$ at, $T_{02} = 1{,}076.7$ K, $\rho_{02} = 0.4046$ kg/m^2

$\therefore \sum(\Delta p_0) = 7.588$ kN/m^2.

Isentropic zero-clearance efficiency (equation 7.18)

$$\eta_{\text{se, zero clearance}} = \frac{[1 - (p_{05}'/p_{04} + \sum(\Delta p_0)/p_{04})^{R/\overline{C}p_e}]}{1 - \dfrac{p_{05}'}{p_{04}}^{R/\overline{C}p_e}} = 0.9037$$

for $\overline{C}p_e = 1{,}151$ J/kg-$^\circ$K and $R = 286.96$ J/kg-$^\circ$K.

Tip clearance area ratio, $\dfrac{Aa_{\text{zero tip clearance}}}{Aa_{\text{tip clearance}}} = 0.984.$

Isentropic efficiency, $\eta_{\text{se}} = 0.9037 \times 0.984 = 0.889.$

Polytropic efficiency, η_{pe} $\dfrac{\ln\left[1 - \eta_{\text{se}}\left(1 - \left(\dfrac{p_{05}'}{p_{04}}\right)\right)^{R/\overline{C}p_e}\right]}{\ln\left(\dfrac{p_{05}'}{p_{04}}\right)^{R/\overline{C}p_e}}$ (equation 2.81) $= 0.880.$

This is sufficiently close to the initial estimate of 0.90 for no iteration to be necessary (in preliminary design).

7.5 Turbine performance characteristics

The data and methods given in the previous section make it possible to estimate the design-point performance of a single-stage or multistage axial-flow turbine. These methods can be extended to cover off-design performance (see Craig and Cox 1971).

Another approach, often adequate in preliminary design, is simply to estimate the off-design performance from the characteristics of other, similar, machines. Turbine flow is less complex than is compressor flow, and it leads to less-complex characteristics. Here are some comparisons.

1. In a compressor the rotor blades slice into the fluid, so that the rotor speed largely determines the shape of the flow: pressure-ratio performance characteristic. In a turbine the flow is first "squirted" through nozzles, and the resistance, or area, of these nozzles largely determines the flow char-

acteristics. The speed of the downstream turbine rotor has a rather minor effect on the relationship of flow to pressure drop.

2. Because the fluid in a compressor is flowing generally from low to high pressure, and because turbocompressors have no "positive-displacement" action, sometimes the high-pressure fluid can leak backward through the compressor, and violently. Therefore there is a large region in all turbo-compressor flow:pressure-ratio characteristics where steady-state efficient operation is impossible. In turbines the flow is generally from high pressure to low, and there are no regions of unstable flow in a turbine characteristic "map."

Let us first examine simplified cases to understand turbine characteristics and how they affect the operation of a circuit of which a turbine may be a component.

For the lossless flow of an incompressible fluid through a nozzle, the steady-flow energy equation gives the restricted form of equation (2.10):

$$p_0 - p_{st} = \frac{\rho C^2}{2g_c},$$

yielding

$$\frac{\dot{m}}{A} = 2g_c\rho(p_0 - p_{st}). \tag{7.26}$$

The flow of real fluids through frictional nozzles can be given by multiplying the right-hand side of equation (7.26) by a coefficient of discharge and a compressibility factor. Thus nozzle flow has a flow-pressure-drop relationship generally known as a "square-law" form, figure 7.16a.

The effect of having a rotor downstream of the nozzles is that of adding a small resistance in series with the controlling resistance. The added resistance of the rotor varies with rotor speed, but because the effect is small in any case, the effect of rotor speed on the turbine flow:pressure-drop characteristics is small, figure 7.16b.

The torque, power output, and efficiency of a turbine are of course greatly influenced by rotor speed. Consider a single-stage turbine connected to a test-stand brake and supplied with fluid at constant pressure. To a first approximation, we can deduce from figure 7.16b that the mass flow will be constant regardless of rotor speed (there will in fact be a small variation as rotor speed changes). To predict the general shape of the output characteristics, we will therefore take the axial velocity, C_x, in the velocity diagrams shown below the abscissa of figure 7.17 as constant. The diagram at the center is taken to be the design-point velocity diagram. It is a "simple" diagram, as defined in chapter 5: constant axial velocity

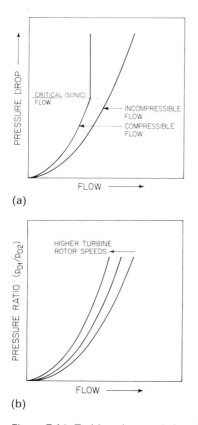

(a)

(b)

Figure 7.16 Turbine characteristics: (a) flow of an incompressible fluid through a nozzle; (b) influence of rotor speed on flow through a turbine.

and constant peripheral velocity. As a further simplification we will consider the off-design velocity diagrams also as if they were "simple," although we know that in compressible-fluid turbines at Mach numbers above, say, 0.5 there would be significant variations in axial velocity through any stage that had been designed for constant axial velocity at best-efficiency point.

We will make one further simplifying assumption: that the fluid outlet angles from rotor and stator blade rows remain at their design-point values even at extreme off-design conditions. For turbines with their intrinsically favorable pressure distribution and generally accelerating relative flow, this assumption is reasonably accurate. The outlet flow vectors are shown as bold lines.

Now suppose that the turbine, running at its design point, is gradually brought to a standstill by the application of the brake. We move left along the abscissa to the origin, where $u = 0$. The effect on the velocity diagram

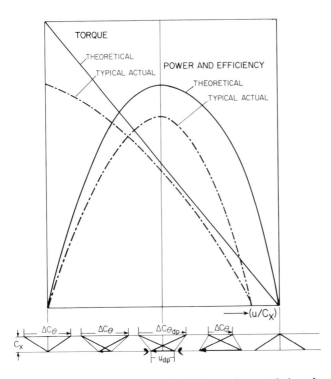

Figure 7.17 Torque, power, and efficiency characteristics of a turbine.

of the peripheral speed, u, reducing to zero while the axial (flow) velocity, C_x and the stator and rotor flow vectors remain constant is that the change of tangential velocity, ΔC_θ, increases to $(\Delta C_{\theta dp} + u_{dp})$. The torque on the rotor is proportional to the product of mass flow (which is constant) and ΔC_θ, and it rises linearly to a maximum value as the rotor is brought to rest. The power output is proportional to the product of torque and shaft speed, and it falls to zero. If, as we suppose, the turbine is supplied with a constant flow of fluid at constant pressure, the input energy is constant, and the turbine efficiency is directly proportional to shaft power output.

The reverse occurs if we gradually reduce the braking torque to zero, the free-running or so-called runaway condition. The peripheral-velocity vector, u, on the velocity diagrams increases in length, and ΔC_θ shrinks, passing through the design-point value and eventually going to zero. At this point the rotor is delivering zero torque, and the work output and efficiency are also zero. The "theoretical" shapes of the characteristics are plotted in figure 7.17 from these arguments, together with typical actual curves.

Test results on actual turbines are shown in figures 7.18 to 7.20.

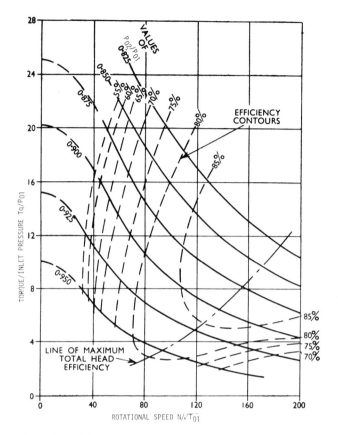

Figure 7.18 (a)

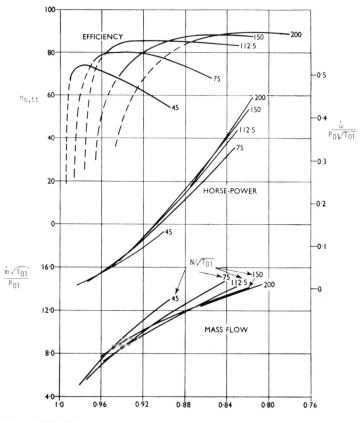

Figure 7.18 (b)

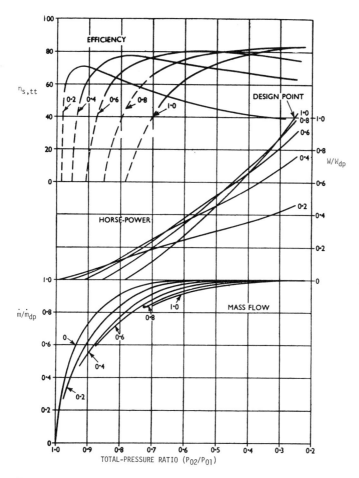

Figure 7.18 (c)

Figure 7.18 Axial-turbine characteristics. From Ainley, 1948. (a) Torque-speed curves of low-pressure-ratio four-stage reaction turbine; (b) pressure-ratio-speed characteristics of four-stage low-pressure-ratio reaction turbine; (c) pressure-ratio-speed characteristics of two-stage low-pressure-ratio impulse turbine.

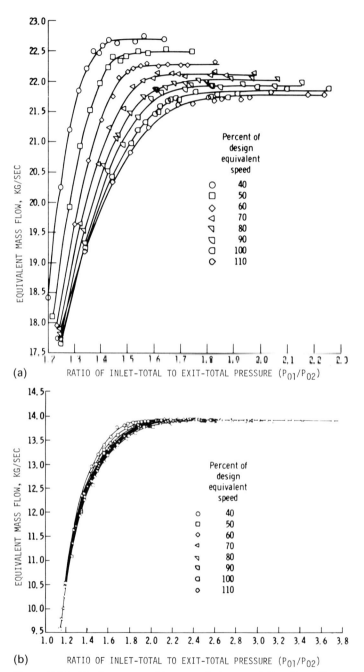

(a)

(b)

Figure 7.19 Single-stage turbine speed characteristics. From Szanca and Schum 1975. (a) Turbine with choked rotor passages; (b) turbine with choked nozzle passages.

Axial-Flow Turbines

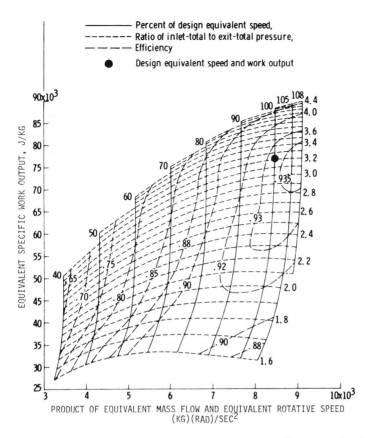

Figure 7.20 Axial-turbine performance "map." From Szanca and Schum 1975.

References

Ainley, D. G. 1948. *Performance of axial-flow turbines.* In Proc. Inst. Mech. Engrs. Vol. 159, War Emergency Proc. 41, pp. 230–244. London.

Ainley, D. G., and G. C. R. Mathieson. 1951. *An examination of the flow and pressure losses in blade rows of axial-flow turbines.* R&M no. 2892 (March). Aeron. Research, Comm, U.K.

Craig, H. R. M., and H. J. A. Cox. 1971. *Performance estimation of axial flow turbines.* Proc. Inst. Mech. Engrs. Vol. 185 32/71, paper and discussion. London.

Dunavant, J. C., and J. R. Erwin. 1956. *Investigation of a related series of turbine-blade profiles in cascade,* Technical Note TN 3802. NACA, Washington, D.C.

Horlock, J. H. 1966. *Axial-Flow Turbines.* Butterworths, London.

Kacker, S. C., and U. Okapuu. 1981. *A mean-line prediction method for axial-flow-turbine efficiency.* Paper no. 81-GT-58. ASME, New York.

Roelke, Richard J. 1973. Miscellaneous losses. In *Turbine Design and Application*. Vol. 2. Ed. by Arthur Glassman. Special publication SP-290. NASA, Washington, D.C., pp. 127–128.

Szanca, Edward M., and Harold J. Schum. 1975. Experimental determination of aerodynamic performance. In *Turbine Design and Application*. Vol. 3. Ed. by Arthur J. Glassman. Special publication SP-290. NASA, Washington, D.C., pp. 103–139.

Stewart, Warner L., and Arthur J. Glassman. 1973. Blade design. In *Turbine Design and Application*. Vol. 2. Ed. by Arthur J. Glassman. Special publication SP-290. NASA, Washington, D.C., pp. 1–25.

Zweifel, O. 1945. The spacing of turbomachine blading, especially with large angular deflection. *Brown Boveri Review*. Vol. 32, no. 12. Baden, Switzerland.

Problems

7.1

Find the apparent optimum solidity and the blade inlet and outlet angles for an axial-flow turbine having an inlet flow angle of $+30°$ and an outlet flow angle of $-60°$. Make the actual incidence zero. Use two stagger angles, $20°$ and $40°$, and comment on their apparent relative advantages. Do your preferences agree with the recommendations of figure 7.4b?

7.2

Explain why a multistage turbomachine would give a higher total-to-static flange-to-flange isentropic efficiency than a single-stage machine having the same velocity diagram and working between the same pressures.

7.3

Calculate the camber angle, and sketch the camber line of an axial-flow turbine blade having inlet and outlet flow angles of $+20°$ and $-50°$. The induced incidence angle is $10°$, the actual incidence angle is $15°$, and the trailing-edge deviation angle is $5°$.

7.4

Choose blade shapes for the mean diameter of a single-stage axial-flow nitrogen turbine. Use a velocity diagram with axial-flow entry and exit: a flow angle out of the nozzle blades of $70°$; and a loading coefficient of 1.5. The nitrogen total conditions are 172 kN/m^2 and 583 K at inlet and 103 kN/m^2 at outlet. The flow is 5 kg/s. The hub-tip ratio at nozzle exit is 0.75. Use an estimated total-to-total polytropic efficiency across the blading of 0.89 to find the blade speed, diameters, and relative nozzle-throat Mach number. Choose other parameters to aim at high efficiency. Specify 19 nozzle vanes, 23 rotor blades, and 0.25 mm tip clearance. Calculate η_{ptt} across the blading based on group 1 losses.

7.5

Find the number of stages required, the mean-diameter blade speed, the mean diameter, the inner and outer diameters at nozzle exit, and the optimum solidity for the first stage of a 10 kW (power from blading) steam turbine for a deep-submergence submarine rescue vessel.

Inlet total conditions: 750 lbf/in^2, 100°R superheat.

Outlet total pressure: 2 lbf/in^2.

Total-to-total isentropic efficiency: 83 percent across blading.

Maximum mean-diameter blade speed: 1,500 ft/s.

Mean-diameter velocity diagrams: $\psi = 1.0$, $\phi = 0.5$, $R_n = 0.5$.

7.6

Design and draw two neighboring blades of the rotor of a turbine having velocity-diagram parameters of $\psi = 2.0$, $\phi = 0.55$, and $R_n = 0.5$. Calculate the spacing using Zweifel's criterion. Use $t_{te}/s = 0.05$ and $r_{1e}/s = 0.1$.

7.7

(a) Choose a suitable blade shape for an axial-flow water turbine of the following specifications.

Mean diameter, d_m: 3.00 in.

Blade speed at d_m: 13.09 ft/s.

Flow axial velocity, C_x: 20.73 ft/s.

Change in tangential velocity, ΔC_θ: 7.40 ft/s.

Reaction at mean diameter, R_n: 0.5.

Axial chord, b: 0.76 in.

Number of rotor blades: 10.

Number of stator blades: 9.

(b) For your final blade shapes estimate the total-total efficiency across the stage blading. The hub diameter is 2.05 in, and the rotor-tip diameter is 3.75 in. There is a tip clearance radially of 0.02 in between the rotor tip and the casing. This turbine was operated with drilling mud in an oil-well drilling string (pipe).

8

The design and performance prediction of axial-flow compressors and pumps

This chapter is the counterpart for axial-flow compressors and pumps of the blade-design methods of chapter 7 for axial-flow turbines. There are, however, major differences in the approaches to blade design. Turbine blades are drawn from first principles for the passages they form between them. Compressor and pump blades are chosen for their own shapes, and at this stage of design are generally selected from a set of precisely shaped blade sections that have been tested in cascade (a straight row of similar blades) in a wind tunnel.

In other respects the phases of design outlined for axial-flow turbines at the beginning of chapter 7 apply also to axial-flow compressors and pumps, except that the heat-transfer calculations of phase 8 are not normally of interest.

8.1 Introduction

It is easy to design a turbine that will work: all that is needed is to squirt fluid out of a nozzle at the blades of a turbine and the rotor will experience a torque.

To design an axial compressor or pump, on the other hand, requires considerable skill. Pressure-increasing machines are a collection of sets of parallel diffusers—the blade rows—in series, with alternate rows moving with respect to the others. All manner of influences can prevent diffusers from operating to increase static pressure, and many early axial compressors worked more as stirring devices. Despite all of these problems many, if not most, axial compressors designed during the decade 1955–1965 and beyond had peak polytropic efficiencies higher than those of axial turbines of that period.

This situation is not as illogical as it sounds. It came about simply because there had been so many failures in axial-compressor design. In several countries and in many private (mainly aircraft-engine) companies an extraordinary effort was made to develop rational compressor-design methods. This many-pronged attack was so successful that the rational design of turbines was left behind.

The reasons for high axial-flow-compressor efficiencies can be summarized as these.

1. Because compressors involve successive diffusing blade rows while, in general, turbines use successive nozzles, compressors are (or were) much more difficult to design. Much more attention has therefore been given to compressor design than to turbine design, and overall a higher standard has been reached.

2. Again because axial-flow compressors use successive diffusers, the permissible stage work is less than in axial-flow turbines, and the stage outlet Mach number (which governs the outlet dynamic head) will be much lower. "Leaving losses" are therefore naturally lower in axial compressors than in axial turbines.

3. "Flow-straightening" diffuser stator blades are used at the outlet of axial-compressor blading (up to three rows being sometimes employed) to produce axial flow into the diffuser. Annular diffusers may suffer flow breakdown unless an axial flow is delivered (see chapter 4). Therefore the annular diffusers downstream from axial flow compressors usually have unseparated flow and achieve a rise in static pressure. Axial turbines seldom have straightening vanes. (Why?) Less diffusion in their downstream annular diffusers is thereby achieved.

The principal published methods of axial-compressor design were developed at the National Gas-Turbine Establishment in Britain and at the National Advisory Committee for Aeronautics (NACA, now NASA) in the United States. We shall give here the NACA method, as modified by Mellor (1956) with other inputs from several sources, such as Horlock (1958). Later NASA-sponsored work produced the pump-cascade correlations summarized here.

8.2 Cascade tests

Compressor and pump blades are normally constructed from a chosen "base profile," such as those listed in tables 8.1 and 8.2 and shown in figures 8.1 and 8.2. The reference, mean, or camber lines can then be given a circular, parabolic, or elliptic shape. However, the camber line of the NACA 65-series is usually defined as in table 8.2. The y coordinates defining the upper and lower (suction and pressure) surfaces can all be multiplied by a factor to increase or decrease the maximum thickness of the blade shape. The base profile is usually specified to give a maximum

Table 8.1
Coordinates of British C4 and C7 compressor-blade base profiles

| x/c | Airfoil form ($\pm Y/c$) | |
(percent)	C4	C7
0	0	0
1.25	1.65	1.51
2.5	2.27	2.04
5.0	3.08	2.72
7.5	3.62	3.18
10	4.02	3.54
15	4.55	4.05
20	4.83	4.42
30	5.00	4.86
40	4.89	5.00
50	4.57	4.86
60	4.05	4.43
70	3.37	3.73
80	2.54	2.78
90	1.60	1.65
95	1.06	1.09
100	0	0

$(t_{max}/c) = 0.10$
$(r_{le}/t) = 0.12$
$(r_{te}/t) = 0.06$

Note: These airfoils are symmetric about their camber lines. For a cambered airfoil, the camber line is usually given a parabolic shape, but it may also be circular.

thickness of 10 percent of the camber-line length. Experimental data on compressor-blade performance are usually taken from a single row of blades arranged in a linear row, called a linear "cascade" (figure 8.3) rather than in an annulus, to simplify the test conditions. Cascade tests are usually made with blades of maximum thickness equal to 10 percent of the chord length, which is a little shorter than a curved camber line.

If we ignore for the moment the possibility of varying the base profile, the camber-line shape, and the maximum thickness, we still have three variables (figure 8.4) to specify a linear cascade:

the blade camber, θ,

the setting or stagger angle, λ, and

the spacing of the blades, s or the solidity, $\sigma \equiv (c/s)$.

To choose a set of values, the resulting linear cascade should be tested over a range of fluid inlet angles from negative to positive stall to obtain

Table 8.2
NACA 65-series compressor-blade base profile

| Station, x/c (percent) | Base profile, $\pm(y/c)_{pro}$ (percent) | Ordinates and slope of mean line (camber line) for $C_{L,th} = 1.0$ | |
		Ordinate, y/c (percent)	Slope, dy/dx
0	0	0	
0.5	0.772	0.250	0.42120
0.75	0.932	0.350	0.38875
1.25	1.169	0.535	0.34770
2.5	1.574	0.930	0.29155
5.0	2.177	1.580	0.23430
7.5	2.647	2.120	0.19995
10.0	3.040	2.585	0.17485
15.0	3.666	3.365	0.13805
20.0	4.143	3.980	0.11030
25.0	4.503	4.475	0.08745
30.0	4.760	4.860	0.06745
35.0	4.924	5.150	0.04925
40.0	4.996	5.355	0.03225
45.0	4.963	5.475	0.01595
50.0	4.812	5.515	0
55.0	4.530	5.475	−0.01595
60.0	4.146	5.355	−0.03225
65.0	3.682	5.150	−0.04925
70.0	3.156	4.860	−0.06745
75.0	2.584	4.475	−0.08745
80.0	1.987	3.980	−0.11030
85.0	1.385	3.365	−0.13805
90.0	0.810	2.585	−0.17485
95.0	0.306	1.580	−0.23430
100.0	0	0	
Leading-edge radius	0.687		

Note: The amount of camber is expressed as design lift coefficient, $C_{L,th}$, for the isolated airfoil, and that system has been retained. Ordinates and slopes are given in the third and fourth columns for the camber corresponding to $C_{L,th} = 1.0$. Both ordinates and slopes are scaled directly to obtain other cambers. Cambered blade sections are obtained by applying the thickness perpendicular to the mean line at stations laid out along the chord line (see figure 8.2).

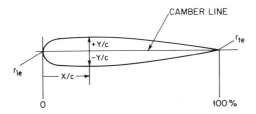

Figure 8.1 Construction of C4 and C7 compressor-blade base profiles (see table 8.1).

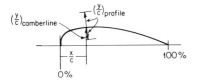

Figure 8.2 Coordinates of NACA 65-series airfoils (see table 8.2).

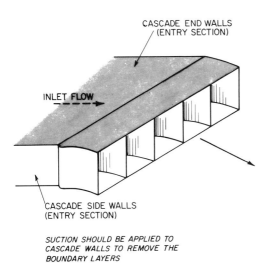

SUCTION SHOULD BE APPLIED TO
CASCADE WALLS TO REMOVE THE
BOUNDARY LAYERS

Figure 8.3 Axial-flow compressor-blade cascade.

Axial-Flow Compressors and Pumps

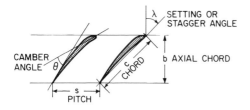

Figure 8.4 Blade-setting variables.

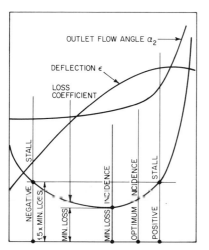

FLUID INLET ANGLE α_1 → AND INCIDENCE

Figure 8.5 Cascade-test results.

the total pressure loss and the change of outlet angle as a function of the flow inlet angle or incidence, figure 8.5. It is desirable also to take some readings over a range of Reynolds and Mach numbers of interest.

The magnitude of undertaking a testing program with so many variables at a time (1938–1957) when there were no automatic readout and data-processing systems makes it understandable that organizations tended to remain attached to the base profile and camber-line shape that they chose initially for testing.

In Britain test results such as those shown in figure 8.5 were correlated for a range of setting angles and cambers into a single curve in which the deflection, for instance, was given as a ratio with the "nominal" deflection arbitrarily set at 80 percent of the stalling deflection. Such correlations were not attempted in the NACA method. Separate curves for each blade setting were grouped by Mellor (1956) to produce plots for each blade camber over a range of useful setting angles and useful incidences (figure 8.6). This useful range was defined as that from negative stall to positive stall, these being the incidences, or inlet flow angles, where the total pressure loss rose to 1.5 times its minimum value (figure 8.5). The definition of incidence (figure 8.7) is that for aircraft wings: the angle between the inlet flow direction and the chord line i^* (whereas the more usual definition of incidence in turbomachinery is the angle between the inlet flow direction and the blade inlet angle i). Some geometrical relations between the NACA airfoils and blades with circular-arc camber lines of the same mean height are shown in figure 8.8.

8.3 The preliminary design of single-stage fans, pumps, and compressors

This section is the counterpart to phase 4 of the design procedure for axial turbines. We suppose that the stage velocity diagram has been chosen for the diameter of concern. If the diffusion ratios, W_2/W_1 and C_1/C_2, have been held at above the de Haller limit of about 0.71 (chapter 5), we are assured that viable blade cascades can be found.

Axial-compressor cascade selection

Fluid inlet and outlet angles for each blade row will be specified in the velocity diagram. For blade sections and settings of axial-flow compressors and fans these angles can be selected easily and rapidly by means of the Mellor-NACA charts, figure 8.6 (Mellor 1956). The procedure for axial-flow pumps is given subsequently.

1. Mark on tracing paper the ordinates and scales of the Mellor charts, as shown in figure 8.9. Show the desired fluid inlet and outlet angles.

2. Place the sheet, correctly aligned, over the 65-series cascade-data charts shown in figure 8.6. Note the cascade designations and setting angles that include the desired fluid angles between positive and negative stall.

3. Choose the most suitable blade and setting. Because the curves give relative rather than absolute losses, it is not possible to select the most efficient blade and setting from these data alone. However, the following is a general design rule. Minimum loss for axial-compressor blade rows having moderate loading (typical of mean-diameter conditions) and of moderate hub-tip ratio (above 0.6, say) is usually given by blade settings at a solidity of about unity and with approximately zero actual incidence. Hub (inner diameter) sections will then have higher solidity and shroud sections will have lower solidity. The actual incidence, i, is obtained from the chordal incidence, i^*, and $(\beta_1 - \lambda)$ given by figure 8.8, through $i = i^* - (\beta_1 - \lambda)$ for the theoretical lift coefficient, $C_{L,th}$.

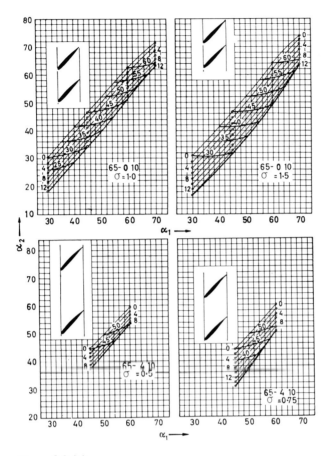

Figure 8.6 (a)

Axial-Flow Compressors and Pumps

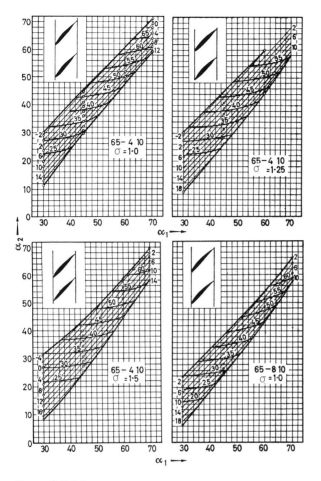

Figure 8.6 (b)

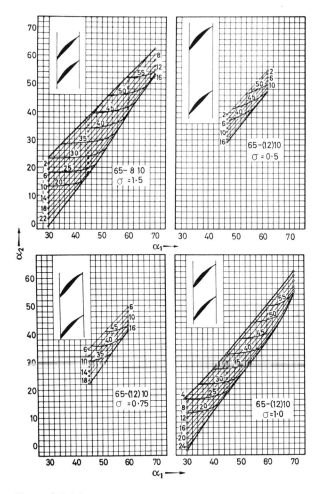

Figure 8.6 (c)

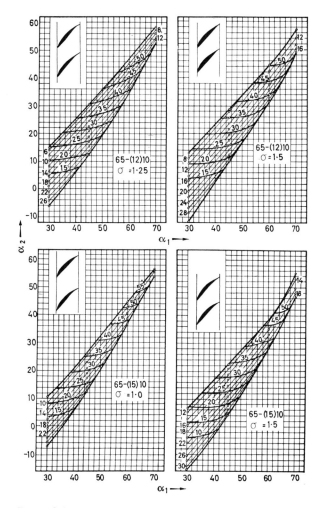

Figure 8.6 (d)

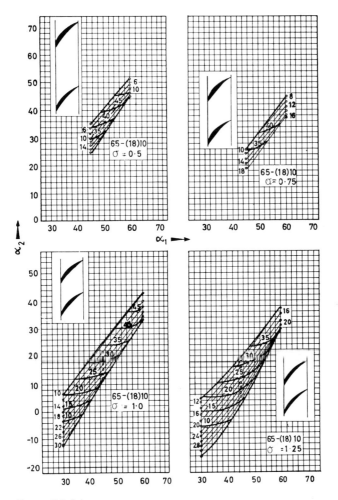

Figure 8.6 (e)

　　　　Axial-Flow Compressors and Pumps

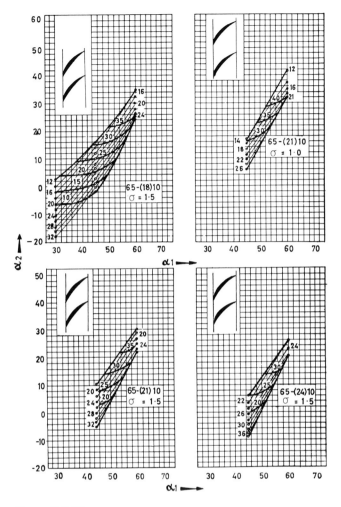

Figure 8.6 (f)

Figure 8.6 NACA 65-series cascade data, Mellor charts. From Horlock 1958 and Mellor 1956. (a) Cambers 010 and 410; (b) cambers 410 and 810; (c) cambers 810 and (12)10; (d) cambers (12)10 and (15)10; (e) camber (18)10; (f) cambers (18)10, (21)10, and (24)10.

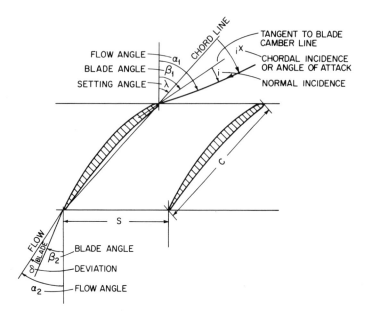

Figure 8.7 Incidence definitions.

The meaning of the NACA cascade designation is as follows. The first two numbers indicate the basic airfoil section (in this case the 65-series). The middle number, one or two digits, is the theoretical lift coefficient. The camber angle corresponding to this lift coefficient may be found from figure 8.8. The third number, usually ten, is the maximum profile thickness as a percentage of the chord length. Then the solidity, c/s, the setting angle, λ, and the chordal incidence, i^*, are given as parameters.

The Mellor-NACA charts are for nearly constant axial velocity across the blade rows. Where the axial velocity changes by more than 10 percent, correction factors for deviation and loss must be applied. A recent method is that by Starke (1980).

For high-pressure-ratio multistage axial-flow compressors, and for fans that must face varying circuit resistances, it is not always desirable that the design-point condition be selected at the peak-efficiency point (which will normally be close to positive stall). Considerations of off-design operation may dictate design-point operation closer to negative stall. Such considerations are discussed below with regard to stage design of multistage axial compressors.

8.4 Performance prediction of axial-flow compressors

This section parallels the treatment of turbines in section 7.3. A real compressor can be thought of as an ideal machine taking in a gas at

Axial-Flow Compressors and Pumps

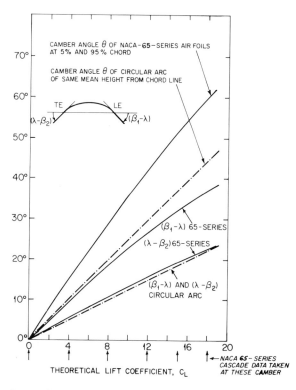

Figure 8.8 Airfoil relationships.

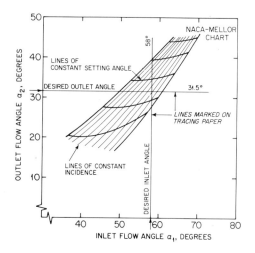

Figure 8.9 Use of NACA-Mellor charts.

Axial-Flow Compressors and Pumps

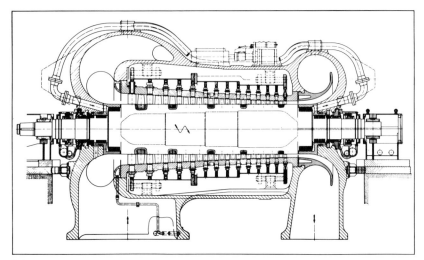

(a)

(b)

Axial-flow process compressor with adjustable stator blades: (a) cross section;
(b) view during assembly. Courtesy Sulzer Brothers Limited.

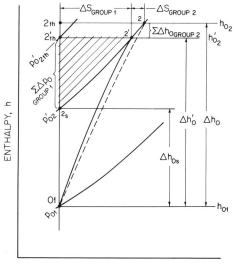

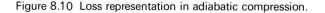

ENTROPY, s

Figure 8.10 Loss representation in adiabatic compression.

p_{01}, h_{01}, and delivering it at p'_{02th}, h'_{02}, (figure 8.10) with added losses which make the actual delivery conditions p_{02}, h_{02}. To make them manageable, we treat these losses as if they were separable. In fact they are to some extent interactive and connected.

Group 1 losses

Group 1 losses are pressure losses resulting from fluid friction. Examples are boundary-layer friction on blade and casing walls and flow-separation losses in areas of stall and blade trailing edges.

The sum of group 1 pressure losses is designated $\sum \Delta p_{0g1}$ in figure 8.8. These pressure losses reduce the outlet pressure from the theoretical p'_{02th} to p'_{02}, which is identical to p_{02}.

Group 2 losses

Group 2 losses take shaft power and degrade it to an enthalpy increase in the gas discharge condition. For instance, the "windage" gas friction on the compressor disks usually appears in the discharge gas state either by heat transfer through the annulus (hub) wall or, more likely, by a mass exchange of the main flow with the gas partly trapped in the cavity around the disks. In either case the result of all such degradations of shaft power is an increase in discharge enthalpy from h'_{02} to h_{02}. The sum of group 2 losses is designated $\sum \Delta h_{0g2}$ in figure 8.8.

Group 3 losses

Again, as for turbines, group 3 losses are losses of shaft power through friction, the energy from which is dissipated away from the working fluid. Examples are the friction losses in external bearings and seals. The power losses appear as increases in compressor power requirements, whereas in turbines group 3 losses decrease power output. In chapter 2 group 3 losses are designated "external energy losses" (section 2.8). Their treatment is obvious, and we do not need to be concerned further with group 3 losses here.

Isentropic efficiency

By definition, $\eta_{sc} \equiv \dfrac{\Delta h_{0s}}{\Delta h_0} = \dfrac{\Delta h_{0s}}{\Delta h_0' + \sum \Delta h_{0g2}}$, (8.1)

$\eta_{sc} = \dfrac{\overline{Cp}_{1-2s}(T_{02s} - T_{01})}{\overline{Cp}_{1-2'}(T_{02}' - T_{01}) + \overline{Cp}_{2'-2} \sum \Delta T_{0g2}}$

$= \dfrac{[(T_{02s}/T_{01}) - 1]}{(\overline{Cp}_{1-2'}/\overline{Cp}_{1-2s})[(T_{02}'/T_{01}) - 1] + (\overline{Cp}_{2'-2}/\overline{Cp}_{1-2s}) \sum (\Delta T_{0g2}/T_{01})}$

$= \dfrac{R/(r^{\overline{Cp}_{1-2s}} - 1)}{(\overline{Cp}_{1-2'}/\overline{Cp}_{1-2s})\{[(p_{02}' + \sum \Delta p_{0g1})/p_{01}]^{R/\overline{Cp}_{1-2'}} - 1\} + \sum \Delta h_{0g2}/\overline{Cp}_{1-2s}T_{01}}$,

where $r \equiv p_{02'}/p_{01}$.

$\eta_{sc} = \dfrac{r^{R/\overline{Cp}_{1-2s}} - 1}{(\overline{Cp}_{1-2'}/\overline{Cp}_{1-2s})\{[r + (\sum \Delta p_{0g1})/p_{01}]^{R/\overline{Cp}_{1-2'}} - 1\} + \sum \Delta h_{0g2}/\overline{Cp}_{1-2s}T_{01}}$. (8.2)

In preliminary design the specific-heat ratio $(\overline{Cp}_{1-2'}/\overline{Cp}_{1-2s})$ is often taken to be unity, and the group 2 losses are often merely guessed at by deducting one or two points from the value of isentropic efficiency calculated without accounting otherwise for group 2 losses.

The final state after compression is p_{02}, h_{02}. This point can be considered to be arrived at directly by the process line shown dashed in figure 8.10. Or two processes could be conceptually involved: points 01 to 2', with just group 1 (pressure) losses; and 2' to 2, with just shaft-power (energy) losses (group 2). Or we could think of an ideal (isentropic) process over the actual enthalpy rise from 01 to 2_{th}, followed by a pure increase of entropy from 2_{th} to 2, a throttling process, being the sum of the entropy increases due to group 1 and group 2 losses.

Calculation of group 1 losses

We give here simplified, approximate methods for estimating losses in the following categories:

1. profile diffusion,
2. trailing-edge thickness,
3. high-Mach-number shocks,
4. end-wall boundary-layer friction, and
5. blade-end clearance.

Of the published loss correlations we have chosen the method of Koch and Smith (1975) as a basis, with that of Lieblein (1959) providing simplifications. We have also made our own approximations, as seems appropriate for preliminary design.

Blade-profile losses

Lieblein showed that the losses around the blade profile appeared as a boundary-layer "momentum" thickness, θ_{te}, at the trailing edge or in the wake, θ_2. (θ_2 will be larger than θ_{te} for highly loaded blades because there will be a mixing loss as the suction-surface and pressure-surface boundary layers join to form the wake.) Lieblein also showed that as the aerodynamic loading on a compressor blade increased, the diffusion on the suction surface increased, but that on the pressure surface stayed approximately constant.

He therefore defined an "equivalent diffusion ratio" (Deq) as the ratio of the suction-surface peak velocity to the outlet velocity:

$$Deq \equiv \left[\frac{W_{max,\,suction\,surface}}{W_{out}} \right], \tag{8.3}$$

and he showed that this could be correlated by

$$Deq = \frac{\cos \alpha_1}{\cos \alpha_2} \left[1.12 + 0.61 \frac{\cos^2 \alpha_1}{\sigma} (\tan \alpha_1 - \tan \alpha_2) \right], \tag{8.4}$$

where α_1 and α_2 are the blade-row inlet and outlet mean flow angles, and σ is the solidity (chord/spacing, c/s). This correlation could be extended to cover operation of the blade row at incidence, i, defined here as the difference between the inlet flow angle with that for minimum loss, $i \equiv (\alpha_1 - \alpha_{1\,opt})$:

$$Deq = \frac{\cos \alpha_1}{\cos \alpha_2} \left[1.12 + 0.0117 i^{1.43} + 0.61 \frac{\cos^2 \alpha_1}{\sigma} (\tan \alpha_1 - \tan \alpha_2) \right]. \tag{8.5}$$

Koch and Smith (1976) introduced factors correlating the airfoil maximum-thickness ratio, t_{max}/c, and the streamtube contraction ratio, Aa_2/Aa_1. They also used cascade data for which the boundary layers were turbulent to a greater extent than were those used by Lieblein, and which therefore were somewhat more representative of the conditions in a compressor:

$$Deq = \frac{W_1}{W_2}\left[1 + 0.7688\left(\frac{t_{max}}{c}\right) + 0.6024\Gamma\right]\left[(\sin\alpha_1 - 0.2445\sigma\Gamma)^2\right.$$

$$\left. + \left(\frac{\cos\alpha_1}{1 - [0.4458(t_{max}/c)/\cos(\alpha_1 + \alpha_2)/2][1 - [1 - (Aa_2/Aa_1)]/3]}\right)^2\right]^{1/2},$$

where the circulation, Γ, is given by

$$\Gamma = \frac{r'_1 W'_{\theta1} - r'_2 W'_{\theta2}}{\sigma W'_1} \tag{8.7}$$

and $r' \equiv r/r_m$,

$r_m \equiv$ mean radius $\equiv (r_h + r_s)/2$,

$W' \equiv W/u_m$,

$u_m \equiv$ blade peripheral speed at r_m.

The subscripts 1 and 2 are for inlet to and outlet from the blade row. The symbol W is used for relative velocity rather than C to indicate that the correlation can be used for moving as well as for fixed blade rows.

The Lieblein and the Koch-Smith correlations, equations (8.5) and (8.6), can be regarded as alternatives, with equation (8.6) being used, presumably, where greater precision is desired.

Leiblein's data can be correlated with blade-outlet momentum thickness, θ_2, by

$$\left(\frac{\theta_2}{c}\right) = 0.00258e^{0.886 Deq}. \tag{8.8}$$

The momentum thickness can be regarded as an intermediate parameter which need not be defined.

Koch and Smith's data are correlated by

$$\left(\frac{\theta_2}{c}\right) = 0.00138e^{1.1127 Deq} + 0.0025 \tag{8.9}$$

and by a trailing-edge boundary-layer shape factor, H_{te}:

$$H_{te} = 1.26 + 0.795(Deq - 1)^{1.681}. \tag{8.10}$$

The shape factor is the ratio of the boundary-layer displacement thickness δ^* to the momentum thickness θ, but again may be treated as an intermediate parameter.

In the Koch-Smith correlations the values of θ_2/c and H_{te} given by equations (8.9) and (8.10) are for nominal conditions of

1. an inlet Mach number below 0.05,

2. no contraction of the streamtube (annulus) height h,

3. a Reynolds number, Re (based on chord and blade-row inlet velocity) of 1×10^6, and

4. hydraulically smooth blades.

Multipliers are given for conditions other than nominal. These are reproduced in figure 8.11. The corrected values of θ_2/c and H_{te} for each blade row can then be used in the following relation, due to Lieblein, to find the blade-profile total-pressure loss:

$$\frac{\Delta p_{0,\text{pro}}}{[\rho_{01} W_1^2/2g_c]} = +2\left(\frac{\theta_2}{c}\right)\frac{\sigma}{\cos\alpha_2}\left(\frac{\cos\alpha_1}{\cos\alpha_2}\right)^2\left[\frac{2}{3-1/H_{te}}\right]$$

$$\left[1-\left(\frac{\theta_2}{c}\right)\frac{\sigma H_{te}}{\cos\alpha_2}\right]^3. \tag{8.11}$$

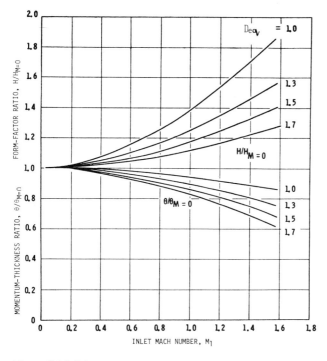

Figure 8.11 (a)

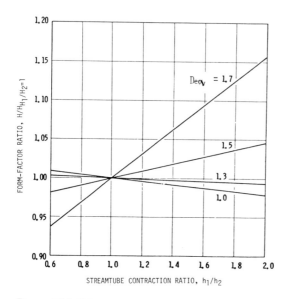

Figure 8.11 (b)

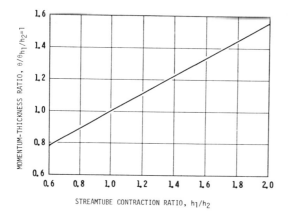

Figure 8.11 (c)

Figure 8.11 Multipliers for blade momentum thickness and form factor. From Koch and Smith 1976. (a) Effect of inlet Mach number on calculated trailing-edge thickness and form factor; (b) effect of streamtube height variation on calculated trailing-edge form factor; (c) effect of streamtube height variation on calculated trailing-edge momentum thickness; (d) effect of Reynolds number and surface finish on calculated trailing-edge momentum thickness.

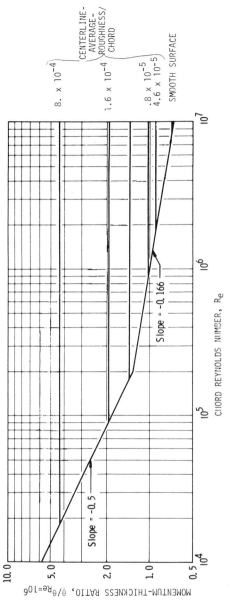

Figure 8.11 (d)

If the more-approximate Lieblein approach is used (equation 8.5) a constant value may be taken for H_{te} of 1.08.

High-Mach-number leading-edge losses

Koch and Smith give the following correlation, ascribed to D. C. Prince:

$$\Delta p_{0,le} = -p_{01} \ln \left\{ 1 - \frac{t_{le}}{b \cos \alpha_1} [1.28(M_1 - 1) + 0.96(M_1 - 1)^2] \right\}. \quad (8.12)$$

Here t_{le} is the blade thickness at the leading edge and b the axial chord. In all these expressions the subscripts 1 (and 2) denote "relative conditions at inlet to (and outlet from) the blade row." Also, to conform with figure 8.10 and equation (8.2), the total-pressure losses are treated as positive, although for consistency pressure losses are by definition negative.

Stage free-stream efficiency

By adding the profile and leading-edge total-pressure losses for a rotor and a succeeding (or preceding) stator-blade row (depending upon whether the stage is defined as a rotor-plus-stator or a stator-plus-rotor), and inserting these in equation (8.2), the stage isentropic compressor free-stream efficiency is obtained, $\eta_{se,fs}$. (Group 2 losses are not accounted for at this point and are set to zero in equation 8.2.)

For greater accuracy, Koch and Smith recommend two further steps and an iteration. The combination of θ_2/c and H_{te} will give a boundary-layer displacement thickness, δ^*, which would in effect thicken the blade profile and change the velocity distribution. For a more exact calculation, beyond the scope of this book, the calculation should be iterated until the calculated value of δ^* conforms with the trailing-edge relative velocity. The second step is to calculate the mixing losses that will occur in the wake downstream of the trailing edge. However, it is pointed out that these are small except at blade loadings higher than are recommended here, and therefore this step also is considered to be beyond the scope of this book.

End-wall losses

The effects on stage efficiency of hub-tip ratio, blade-tip clearance, axial gap between blade rows and, to a lesser extent, blade aspect ratio (blade height over chord, h/c) can be estimated using the Koch-Smith formulation of end-wall losses. We give here a simplified, approximate version.

The test data were correlated by Koch and Smith with the sum of the relative displacement thicknesses of the two end-wall boundary layers plotted against the stage pressure-rise coefficient relative to the maximum pressure-rise coefficient of which the stage is capable (figure 8.12a). Rotor-blade clearance is a parameter. (There is no guidance for the type of disk-

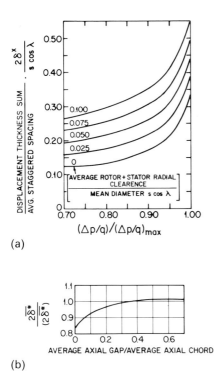

(a)

(b)

Figure 8.12 Correlations of end-wall losses. From Koch and Smith 1976. (a) Radial-clearance multiplier; (b) axial-gap multiplier.

and-drum construction where the stator blades are unshrouded and also have an end clearance). A multiplier for blade-row axial gaps (relative to blade axial chord) different from 0.35 is shown in figure 8.12b.

The stage isentropic total-to-total efficiency is then given by a modification of the stage free-stream efficiency, already obtained:

$$\eta_{sc,stage} = \eta_{sc,fs} \left[\frac{1 - (2\delta^*_{12a}/h)(2\delta^*/2\delta^*_{12a})}{(2\delta^*_{12a}/h)(2\delta^*/2\delta^*_{12a})} \right], \tag{8.13}$$

where $2\delta^*_{12a}$ is the value of end-wall boundary-layer displacement thickness (for the two end walls) given by figure 8.12a and h is the mean blade height.

For preliminary-design purposes it is probably sufficiently accurate to estimate the relative static-pressure-rise coefficient with which to enter figure 8.12a. For somewhat greater accuracy one can use the velocity diagram and the blade-row fluid-turning curves (figure 8.6) to obtain $C_\theta/C_{\theta max}$, assuming that $C_{\theta max}$ occurs at positive stall.

Then with the estimate that the stage efficiency will fall by about 20 percent from the operating point to the stall point, we can assume that:

$$\frac{(\Delta p/q)}{(\Delta p/q)_{\max}} = 0.8 \left(\frac{C_\theta}{C_{\theta \max}} \right).$$

(8.14)

Aspect ratio as an independent variable

As in the case of axial-flow turbines the designer has considerable freedom to choose the number of blades in a blade row. A large number of blades leads to a shorter compressor, and vice versa. As the number of blades in rotors and stators is changed, there will be proportional changes in the excitation frequencies from blade wakes, and nonlinear changes in blade natural frequencies, thus allowing the designer to use the choice of number of blades as a principal method of avoidance of critical excitations.

In aircraft and some ground-transportation engines the requirement for the smallest possible engine often outweighs other considerations, and high-aspect-ratio compressors result from the need to reduce compressor (and turbine) lengths. When there is design freedom to do otherwise, the designer should be aware that the use of low-aspect-ratio compressor blading brings many benefits. In experimental study of such blading Reid and Moore (1980) confirmed earlier studies, not all published, that low aspect ratios produce

1. higher peak pressure ratio, higher stage efficiency, greater stall margin;
2. improved performance over the whole blade span;
3. good operation at higher diffusion factors and higher incidences; and
4. improved high-Mach-number performance.

In addition substantial cost savings result from the use of fewer, larger, blades and vanes.

Example (calculation of stage efficiency)

This example is of the calculation of the efficiency of the hub (inner diameter) section of a low-speed fan. The flow is almost incompressible, and several of the factors that would normally be accounted for in the calculation of a compressor stage can be ignored. The example then demonstrates the method but not the potential complexity.

The problem is to calculate the total-to-total isentropic efficiency, from stage inlet to stage outlet, of the hub section of an axial-flow fan (figure 8.13) of the following specifications.

$p_{01} = 101 \text{ kN/m}^2$.
$T_{01} = 310 \text{ K}$.
Fluid = air.

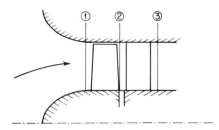

Figure 8.13 Calculation planes for axial-flow-fan example.

Table 8.3
Choice of hub-section blade profiles

	Rotor	Stator
Flow inlet angle	59.04°	35.90°
Flow outlet angle	43.31°	0°
Blade sections	65(12)10	65(18)10
Solidity, σ	1.25	1.25
Stagger, λ	49°	15°
Relative inlet velocity, m/s	16.676 (W_1)	10.592 (C_2)
Work coefficient, ψ_h	0.4343	

$\overline{Cp}_c = 1.005$ kJ/kg-°K.
$R = 286.96$ J/kg-°K.
λ (hub-tip diameter ratio) = 0.5.
ϕ_h (flow coefficient at hub) = 0.6.
u_h (blade speed at hub) = 14.30 m/s.
$(\Delta p/q)_{th,h}$ (pressure-rise coefficient at hub) = 0.5 (rotor)

Axial-flow inlet and outlet flow, rotor-stator stage
Shroud diameter, $d_s = 0.4$ m.
Rotor-blade radial clearance = 1 mm.
Number of rotor blades = 16.
Number of stator blades = 15.
    The hub velocity diagram was calculated and compressor-blade sections were chosen from figure 8.6 as given in table 8.3 and figure 8.14.
    The axial gap between rotor and stator is 0.35 of the blade axial chord (no boundary-layer correction will be needed from figure 8.12b).
    We first calculate the circulation for rotor and stator from equation (8.7). The ratio of the (area) mean diameter to the tip diameter,

$$\frac{d_m}{d_s} = \sqrt{\frac{1}{2}(1 + \lambda^2)} = 0.791.$$

Therefore $r_h' \equiv r_h/r_m = 0.632.$

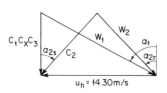

Figure 8.14 Hub velocity diagram of axial-flow fan.

Table 8.4
Loss calculation for compressor stage

	Rotor	Stator
Circulation at hub, Γ_h (equation 8.7)	0.08178	0.2965
Assumed (specified), $(t_{max}/c)_h$ (recognizing that figure 8.4 is for t_{max}/c of 0.10	0.125	0.10
$(W_1/W_2)_h$	1.4142	1.2345
Deq (equation 8.6)	1.6294	1.5249
θ_2/c (equation 8.9)	0.010958	0.01003
H_{te} (equation 8.10)	1.625	1.5291
Circular pitch, s_h m	0.0491	0.0524
Mean density, ρ_{st} kg/m^3	1.412	1.412
Mean viscosity, μ_{st} Ns/m^2	1.845×10^{-5}	1.845×10^{-5}
Reynolds number, $\rho_{st} W_1 c_h/\mu_{st}$	78,300	53,060
$\theta/\theta_{Re=10^6}$ (figure 8.9d)	2.06	2.45
θ_2/c corrected for Re	0.02257	0.02457
$\Delta p_0/q_{in}$ (equation 8.11)	0.0267	0.0273
Inlet dynamic head $\rho_{01} W_{in}^2/2$ N/m^2	196	79.2
Total pressure loss, profile Δp_0 N/m^2	5.252	2.165
$\sum(\Delta p_0)$ N/m^2		7.417

We give principally the results of calculations with comments in table 8.4. The number of significant figures carried along in the calculation is much greater than the accuracy warrants.

To use this total-pressure drop in equation (8.2) to find the "free-stream" isentropic efficiency requires an iteration. We first assume that the efficiency is 0.90 so that we can calculate the isentropic enthalpy rise, Δh_{os}. We use this and find that the calculated value of $\eta_{sc,fs}$ is 0.924; we use a value of 0.925 in a new calculation and arrive at a final value of 0.926.

We now have merely the end-wall losses to account for, using figure 8.12a and equations (8.13) and (8.14). The curves of figure 8.12a are for the average rotor and stator clearance (we assume that the stator blades have zero clearance, and we therefore use half the rotor-blade radial clearance) normalized by the mean-diameter value of $s \cos \lambda$. We know this last quantity for the hub radius, but we do not know precisely how it varies with radius. The spacing will increase proportional to radius,

Axial-Flow Compressors and Pumps

and the stagger or setting angle, λ, will increase slightly with radius for the rotor blades, which is the appropriate blade row because it is the one having clearance. The value of (radial clearance/$s_h \cos \lambda_h$) for the rotor is 0.031; the average rotor-stator value is half this: 0.0155, and an estimate of the mean-diameter value is 0.011.

For the rotor, $C_\theta/C_{\theta\,max}$ is estimated from figure 8.6 to be about 0.25. (Very conservative airfoil cambers were chosen to allow for blade-surface fouling in the projected application.) The value of $2\delta^*/s_m \cos \lambda_m$ in figure 8.12a must then be estimated by extrapolation: 0.13. The hub value is then estimated at about 0.19. Then $(2\delta^*/h) \simeq 0.0816$, the multiplier from equation (8.13) is 0.957, and the stage efficiency is $0.957 \times 0.926 = 0.887$.

This example, and the loss correlations that preceded it, were for single compressor stages and for total-to-total efficiencies. For a full calculation of the efficiency across a multistage axial-flow compressor and diffuser, similar calculations to those given here would be made for the hub, mean, and shroud streamlines for all stages (with the stage efficiencies being the average of the three values) to arrive at a total-to-total isentropic efficiency across the blading. (This may be too complex. The major aircraft-engine companies currently spend about $1 billion to develop a new engine, and, according to one senior compressor designer, "momentous decisions are made on the basis of mean-line analyses.") Then the total-pressure losses in the diffuser (see chapter 4) would be added, and, if a total-to-static efficiency were desired, the dynamic pressure at diffuser outlet would also be treated as a loss.

Some other considerations of multistage-compressor design follow.

8.5 The design and analysis of multistage axial compressors

At first sight it may seem that one can simply add compressor stages together (with, of course, progressive reductions in annulus area or blade height to accommodate the increase in density) to reach any desired pressure ratio. In fact high-pressure-ratio "fixed-geometry" compressors have severe starting and off-design losses, for the reasons explained in this section, so that the practical limit of pressure ratio for such machines working on air is about ten to one. This range can be considerably extended, perhaps up to thirty to one, by arranging for the stator blades, and even the rotor blades, on some of the blade rows to be pivoted to new setting angles for certain conditions of operation. Or interstage "bleeds" may be designed to blow off air to keep stages away from stall during starting.

However, let us suppose that we are given design specifications acceptable for a fixed-geometry compressor—say, a pressure ratio of six to one —to compress air from normal ambient temperature and pressure. We can calculate the isentropic enthalpy rise and, making an estimate of the appropriate efficiency, we can estimate the actual enthalpy rise, figure 8.15.

The first two questions to be answered are related: What type of first-stage velocity diagram and blades should be used? And how many stages should be employed?

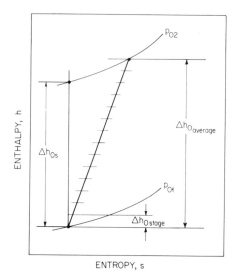

Figure 8.15 Stage enthalpy rises in a multistage compressor.

The choice of the first-stage diagram will entail the determination of the mean-diameter and shroud-diameter peripheral speed. We may then decide to keep the shroud diameter constant for later stages (or the mean or the hub diameters could have been held constant with different advantages and disadvantages). While the first-stage velocities will be chosen largely on the question of the relative Mach number allowable, subsequent stages will have lower Mach numbers, and there will be some freedom to modify somewhat the velocity diagram used for the first stage to yield higher-work downstream stages. The number of stages to give the overall (estimated) enthalpy rise will result from these choices of first-stage and subsequent velocity diagrams.

Normally, the axial velocity at design point is kept approximately constant (at the reference diameter) throughout the machine. In practice, this is accomplished by changing the annulus area between each blade row and "fairing" the annulus shape between these locations.

Off-design performance of a multistage axial compressor

The most important off-design situation for a compressor is when it is started from rest, usually against a fixed circuit resistance. In a gas-turbine engine without interstage air bleed-off to atmosphere, the circuit resistance will in addition increase sharply when heat is added to the compressed air, for instance, by the combustor "lighting off."

The following argument is made in rather specific terms so that the reader may have a distinct picture in mind. However, the argument may obviously be generalized.

Let us take the case of an air compressor designed to produce a pressure ratio of 6 : 1 at design point from atmospheric air at inlet. The first stage is designed to a maximum relative Mach number of 0.9 with respect to the rotor-blade tips. Let us suppose that this stage design is then repeated fourteen times with successively shorter blades to give the overall pressure ratio desired. (The relative Mach number will of course be greatly reduced as the temperature and the speed of sound increase through the machine.) If the polytropic efficiency η_{tt} across the blading at design point is 0.9, the density ratio (static) from inlet to outlet will be about 3.5. Therefore, since the design-point axial velocity is kept constant (by the use of similar stages), and because the mass flow is constant (no air bleed), the annulus area at outlet is smaller than that at inlet by a factor of 3.5. See figure 8.16a.

Now let us consider what will happen at off-design conditions. By "off-design" we mean that we impose on the compressor a different combination of shaft speed and massflow (or, more strictly, inlet Mach number). Let us take the following conditions in turn.

1. Low shaft speed (low u and low Mach number) and low mass flow to give the design-point value of C_x/u ($\equiv \phi$) at the first stage (line 1, figure 8.16b). At low speed there will be little or no increase of density. So in succeeding stages, u is constant, but C_x increases because the cross-sectional annulus area decreases and ϕ goes up beyond ϕ_{DP} into the region of negative stall (line 1). With so many of the later stages in negative stall, the small pressure rise produced by the first stages will be more than dissipated by losses, and the result will be a negative pressure ratio—that is, there would have to be external pressurization to force this amount of flow through the compressor. This is shown as point 1 in the compressor characteristics of figure 8.16c.

2. Low speed, low mass flow such that the first two or three stages are in positive stall (line 2). In this case there are more stages working, and a small pressure ratio should be produced across the compressor. The point is plotted as 2 in the compressor-characteristics chart (figure 8.16c).

3. Low speed, even lower mass flow—see line 3 and point 3 in figures 8.16b and c.

4. High (design-point) speed, higher-than-design-point mass flow, line 4. There will be some increase in C_x because the stages will be working less at higher flow coefficients, but the increase in flow coefficients along the compressor will not be as strong as at low speeds because there will be some increase in density. But when the later stages run into negative stall at full speed, the high local velocities around the blades will sooner or later cause local sonic velocities to be reached, with consequent shock losses. The mass flow will not be susceptible to further increase, and a vertical constant-speed line will result on the compressor characteristics.

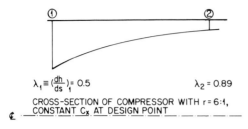

$$\lambda_1 \equiv \left(\frac{dh}{ds}\right)_1 = 0.5 \qquad\qquad \lambda_2 = 0.89$$

CROSS-SECTION OF COMPRESSOR WITH r = 6:1,
¢ CONSTANT C_x AT DESIGN POINT

Figure 8.16 (a)

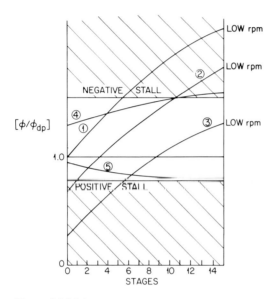

Figure 8.16 (b)

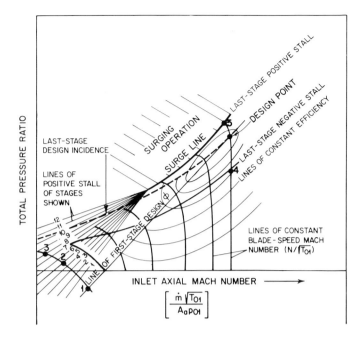

Figure 8.16 (c)

Figure 8.16 Off-design operation of a high-pressure-ratio axial-flow compressor: (a) cross section of compressor; (b) variation of flow coefficient with various combination of rotational speed and mass flow; (c) typical multistage axial-flow-compressor characteristics.

5. High (design-point) speed, lower-than-design-point mass flow. Each stage will give more than design-point work because of the reduced mass flow. The density will be increased to above that of design. This increase of density will further reduce C_x and ϕ. The line 5 has been shown as one where the last stage is on the verge of stalling. If it were to be pushed into stall, by a momentary reduction in mass flow, for instance, the next-to-the-last stage would suddenly experience the full compressor outlet pressure which, being so close to positive stall, it could not deliver but would go into stall itself. This domino stall sequence would generate what is called a "surge" flow breakdown. No known high-pressure-ratio axial compressors could withstand repeated full-speed, full-density surging for long without losing blades, so large are the transient forces imposed.

Intermediate speeds and mass flows would complete the characteristics to show not only lines of positive and negative stall and lines of design-point incidence for various stages but also lines of constant isentropic or polytropic efficiency, which are also drawn in.

8.6 Compressor surge

Compressor surge is therefore essentially a compressible-flow high-Mach-number phenomenon. The boundary between steady-flow and surging operation is called the "surge line" (figure 8.16c), and it exists strictly only where the blade-speed Mach number and axial Mach number are in the compressible region, say, above 0.3. (However, in most representations of compressor characteristics, the surge line is incorrectly continued down to the zero-mass-flow origin. At lower speeds the compressor can operate in a stable manner with one or more low-pressure stages stalled. Whereas low-speed stalled operation depends very little on downstream conditions, surging is a function of the capacity and loss characteristics of the system into which the compressor discharges.

Let us suppose that, when operating the previously discussed compressor in the condition represented by point 5 in figure 8.16, the circuit resistance is increased slightly so that the last stage stalls. The next-to-last stage is exposed to the full outlet pressure when it is working near positive stall and will in turn stall. In this manner the stall propagates backward through the compressor, with reverse flow becoming fully established when the stall reaches the first stage. The downstream circuit will then discharge itself through the compressor, which is still rotating forward at full speed. After a period of time, the length of which will depend on the volume and loss characteristics of the circuit, the pressure will have fallen to the point where most if not all of the compressor stages will again be able to operate unstalled. Forward flow becomes re-established, the discharge pressure gradually rises, and, unless the circuit resistance has been lessened by some control change, the compressor will surge again.

The period of this surge cycle may be several seconds for a process compressor discharging into a large chemical reaction vessel, or may be a fraction of a second for an aircraft gas-turbine compressor. A surge sounds like a succession of pops or explosions, depending on the machine Mach number and on how well the ducts are muffled.

8.7 Axial-compressor stage stacking

As was intimated in section 8.6, most early compressors were designed with all stages at their optimum-efficiency condition, which is near positive stall at the design point. Such compressors frequently had very poor "sagging" surge lines even though the design-point efficiency of the whole machine could be high. A so-called "knee" in the surge line means that the compressor may run into severe stalling or mild surging during the normal run up to full speed, necessitating the installation of devices such as interstage blow-off valves to increase the flow coefficients of the low-pressure stages. Such measures may be necessary for two reasons. First, even half-speed surges can impose severe fluctuating loads on the blading and possibly also on the thrust bearing. Second, there tends to be a hysteresis connected with both stalling and surging. Once a blade row has stalled, it often has to be brought to well below the stalling incidence before it again becomes unstalled. Therefore, although the starting line of a compressor may only just graze the surge line at perhaps 60-percent speed and then pass through a normally unsurged region, the machine might in fact remain surged all the way up to full speed.

The first compressor with which I became involved (just the building and testing phases) exhibited this behavior to a high degree. Although of only four-to-one pressure ratio and thirteen stages (which were therefore, on the average, lightly loaded), it would prevent the gas turbine of which it was a part from being started if there was any unsteadiness of flow (shown by a tuft in a window in the outlet duct) even at 5-percent speed. This compressor had low-stagger blades, which were found later to have a large hysteresis loop when stalled (Wilson 1960) (figure 5.11a) and a very poor inlet, which probably helped to force many stages into stall. Koch (1981) has quantified the stalling characteristics of a wide range of compressor stages.

The starting and surging conditions influence stage "stacking." This term is used to mean the choice of types of stages, and their points of design operation, to be used in different parts of a compressor. The earlier description of surging and of the flow coefficients imposed on different stages showed that the low-pressure stages operate in the region from design-point incidence into full positive stall. Therefore it would seem sensible to choose a blade-row design that has little hysteresis in positive stall—for instance, one having a high stagger angle—and the design-point flow coefficient such that the blade row operates toward negative stall. Likewise, the high-pressure stages normally operate in the range between design point and negative stall and should therefore be chosen to have good negative-stall and high-Mach-number characteristics—for instance, low-stagger blade rows.

This type of compromise involves a trade-off of a slight reduction in peak efficiency for a broader range of high efficiency and a "pushed-out" surge line. This philosophy was applied to a small degree in the author's aerodynamic design of a Ruston & Hornsby compressor of 5 : 1 design-point pressure ratio and a mass flow of 110 lbm/s. The first build and test, with no adjustments or stage tests whatsoever, gave the characteristics shown in figure 8.17. Despite a missing region in the characteristic (in which data could not be taken because of instabilities in the naval-destroyer steam-turbine drive, and not because of any deficiencies in the

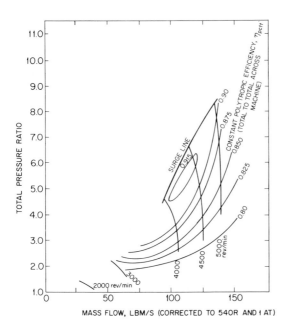

Figure 8.17 Compressor characteristics, Ruston & Hornsby TG.

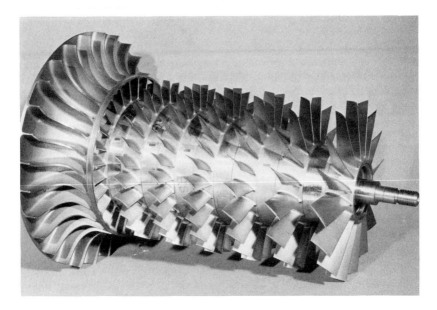

Axial-centrifugal compressor rotor for Allison T63/250 engine. Courtesy Detroit Diesel Allison Division of General Motors Corporation.

Axial-Flow Compressors and Pumps

compressor), it can be seen that there is a good surge line, a broad plateau of high efficiency, and a peak polytropic total-to-total blading efficiency of over 0.91.

A second aspect of stage stacking is the need to adjust the annulus area to allow for boundary-layer growth. Larger boundary layers also mean lower stage efficiencies. To some extent these can be estimated by the Koch-and-Smith method introduced in the previous section. However, there has to be some guesswork in choosing values of wall-boundary-layer displacement thicknesses that will give the incremental amounts by which the casing radius should be increased, or will give annulus "blockage" factors, and the values of reduced efficiency. The following observations are made on the basis of personal and vicarious experiences.

1. In compressors that have blade rows designed to operate away from stall, wall boundary layers grow through the first three or four blade rows and then remain at constant thickness. In compressors in which the blade rows operate near stall, boundary layers have been found to continue growth through the machine. See DeRuyck and Hirsch (1980).

2. If the designer assumes large boundary layers or low efficiencies, these assumptions tend to become self-fulfilling prophecies. The reason is that they produce increased blade-row incidences and greater likelihood of local stalling. Therefore it is actually more conservative to assume high efficiencies, even if they are on the high side of what is actually attained.

3. The use of the "work-done factor" should be avoided. It will not be defined here for fear of contaminating any remaining pure minds reading these words. Suffice it to say that it is a thermodynamically unacceptable "fudge" factor invented to bring cascade data and compressor data into line. The early British cascade data were taken without boundary-layer bleed, and hence with boundary-layer growth through the cascade, leading to accelerating flow. The NACA and NASA data reproduced here were taken with boundary-layer bleed and do not require correcting factors other than those described.

When high-pressure blades seem likely to be very short, with relatively large clearance losses, an axial-centrifugal stage-stacking arrangement is sometimes used. Matching the characteristics of these different types of compressor stages is difficult, however, and long and costly development programs have often been resorted to.

8.8 Axial-flow pump design

Compressor-design methods, such as those given in the earlier part of this chapter, can be used for other fluids, including Newtonian liquids (water, for instance), with very few modifications. (The design of machines to work on non-Newtonian liquids, such as free-molecule flows or fluids having viscosities dependent on shear rates, is well beyond the scope of this book.)

We present this section on pump design because the cascade data given earlier may not wholly cover the axial-pump region for two reasons. First, an overriding design consideration is often the avoidance of cavitation. This consideration puts a premium on the use of a low axial velocity and low relative velocities in the choice of velocity diagrams. Cavitation avoidance is reviewed in chapter 10.

Second, axial-flow pumps are most usually single-stage designs, or even single rotors without stators. This factor again puts a premium on low axial velocities to give low leaving losses.

These two considerations combine to lead to velocity diagrams of low flow coefficient and to blades of high stagger or setting angles. The cascade data given in figure 8.6 do not adequately cover the area of interest for pump designs. Accordingly, we give in figure 8.18 cascade data for double-circular-arc hydrofoils (Taylor et al. 1969).

These data were taken for NASA by United Aircraft Research Laboratories in a series that also included multiple-circular-arc hydrofoils and slotted hydrofoils. We have chosen to give only the double-circular-arc data because this configuration appears to be the best general choice for pumps, and because it is popular also for axial-flow compressors [1] The coordinates of the tested shapes are given in table 8.5 and the cascade nomenclature in figure 8.19.

The information given in Taylor et al. (1969) includes loss coefficients, diffusion coefficients, trailing-edge boundary-layer thicknesses, and cavitation indexes. We are reproducing just the turning-angle data in the form of a carpet plot, the use of which we illustrate by means of an example. Readers are recommended to go to the original reports for the profile-loss and other performance information. Or the performance-analysis methods given earlier in this chapter may be used at least as a comparative guide. We do not have the machine test data to know if absolute accuracy is given when the methods are applied to liquid flows.

---

1. The NACA 65-series are sometimes recommended for relative Mach numbers from 0 to 0.78, and double-circular-arc (DCA) blades for $0.70 < M_{rel} < 1.20$. But the DCA blades and data are obviously also useful at low Mach numbers, for fans, for instance.

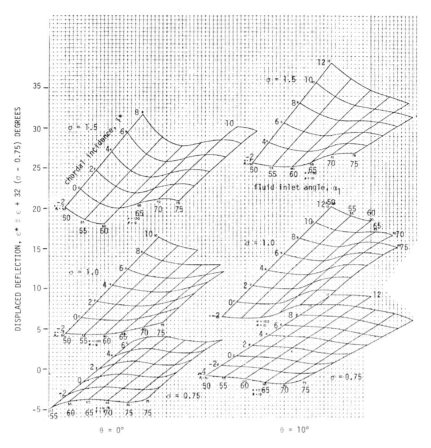

Figure 8.18 (a)

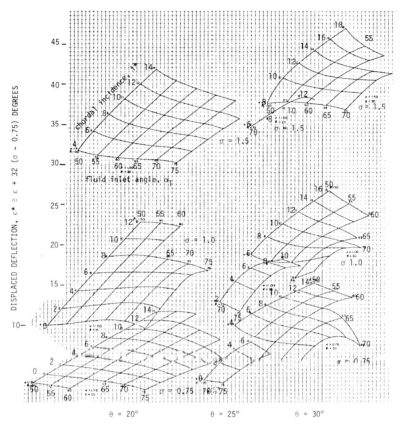

Figure 8.18(b)

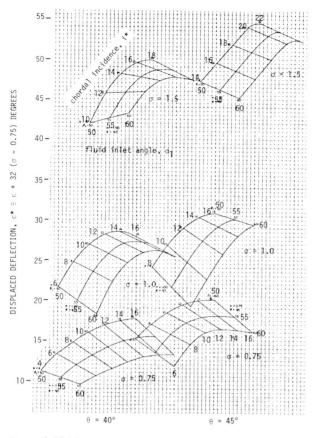

Figure 8.18 (c)

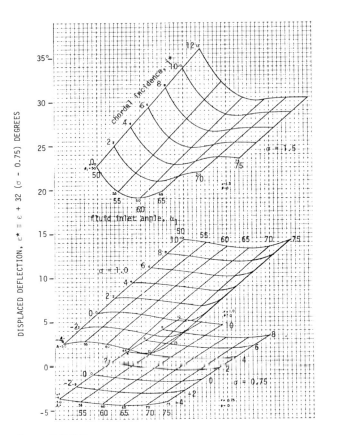

Figure 8.18(d)

Axial-Flow Compressors and Pumps

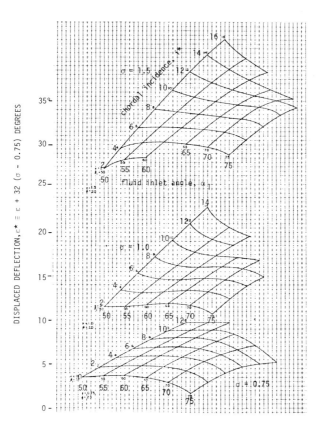

Figure 8.18(e)

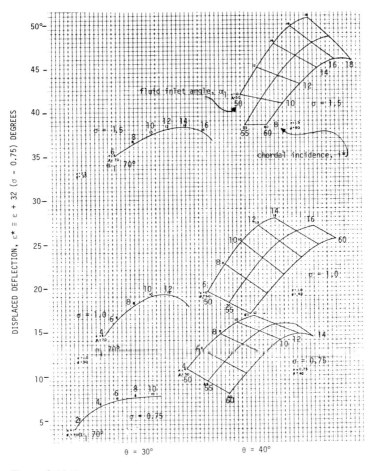

Figure 8.18(f)

Figure 8.18 Turning data for double-circular-arc hydrofoils. From Taylor et al. 1969. (a) Plots for cambers 0° and 10° for $(t_{max}/c) = 0.06$; (b) plots for cambers 20°, 25° and 30° for $(t_{max}/c) = 0.06$; (c) plots for cambers 40° and 45° for $(t_{max}/c) = 0.06$; (d) plots for camber 0° for $(t_{max}/c) = 0.10$; (e) plots for camber 20° for $(t_{max}/c) = 0.10$; (f) plots for cambers 30° and 40° for $(t_{max}/c) = 0.10$.

Table 8.5
Coordinates for double-circular-arc profiles

Camber angle θ (deg)	0°		10°	20°	
Thickness (t_{max}/c) ratio (%)	6%	10%	6%	6%	10%
Chordal station (%) (x/c)	y/c upper surface (%)				
0	0.10	0.10	0.10	0.10	0.10
8.33	0.93	1.53	1.62	2.29	2.94
16.67	1.67	2.80	2.92	4.14	5.29
25.00	2.23	3.77	3.94	5.56	7.09
33.33	2.67	4.43	4.65	6.57	8.37
41.67	2.90	4.87	5.09	7.18	9.12
50.00	3.00	5.00	5.23	7.38	9.38
58.33	2.90	4.87	5.09	7.18	9.12
66.67	2.67	4.43	4.65	6.57	8.37
75.00	2.23	3.77	3.94	5.56	7.09
83.33	1.67	2.80	2.92	4.14	5.29
91.67	0.93	1.53	1.62	2.29	2.94
100.00	0.10	0.10	0.10	0.10	0.10
	y/c lower surface (%)				
0	−0.10	−0.10	−0.10	−0.10	−0.10
8.33	−0.93	−1.53	−0.23	0.42	−0.19
16.67	−1.67	−2.80	−0.42	0.76	−0.35
25.00	−2.23	−3.77	−0.58	1.03	−0.47
33.33	−2.67	−4.43	−0.68	1.22	−0.55
41.67	−2.90	−4.87	−0.76	1.33	−0.61
50.00	−3.00	−5.00	−0.77	1.38	−0.62
58.33	−2.90	−4.87	−0.76	1.33	−0.61
66.67	−2.67	−4.43	−0.68	1.22	−0.55
75.00	−2.23	−3.77	−0.58	1.03	−0.47
83.33	−1.67	−2.80	−0.42	0.76	−0.35
91.67	−0.93	−1.53	−0.23	0.42	−0.19
100.00	−0.10	−0.10	−0.10	−0.10	−0.10

Table 8.5 (continued)

Camber angle θ (deg)	25°	30°		40°		45°
Thickness (t_{max}/c) ratio (%)	6%	6%	10%	6%	10%	6%
Chordal station (%) (x/c)	y/c upper surface (%)					
0	0.10	0.10	0.10	0.10	0.10	0.10
8.33	2.64	3.00	3.67	3.77	4.48	4.15
16.67	4.76	5.41	6.58	6.74	7.95	7.41
25.00	6.40	7.25	8.80	9.01	10.58	9.87
33.33	7.55	8.55	10.35	10.58	12.40	11.59
41.67	8.24	9.33	11.27	11.54	13.49	12.61
50.00	8.47	9.58	11.58	11.84	13.84	12.95
58.33	8.24	9.33	11.27	11.54	13.49	12.61
66.67	7.55	8.55	10.35	10.58	12.40	11.59
75.00	6.40	7.25	8.80	9.01	10.58	9.87
83.33	4.76	5.41	6.58	6.74	7.95	7.41
91.67	2.64	3.00	3.67	3.77	4.48	4.15
100.00	0.10	0.10	0.10	0.10	0.10	0.10
	y/c lower surface (%)					
0	−0.10	−0.10	−0.10	−0.10	−0.10	−0.10
8.33	0.79	1.10	0.48	1.80	1.17	2.15
16.67	1.41	2.00	0.88	3.26	2.12	3.89
25.00	1.89	2.69	1.18	4.40	2.87	5.23
33.33	2.20	3.19	1.40	5.19	3.40	6.19
41.67	2.40	3.48	1.54	5.68	3.72	6.76
50.00	2.47	3.58	1.58	5.84	3.82	6.95
58.33	2.40	3.48	1.54	5.68	3.72	6.76
66.67	2.20	3.19	1.40	5.19	3.40	6.19
75.00	1.89	2.69	1.18	4.40	2.87	5.23
83.33	1.41	2.00	0.88	3.26	2.12	3.89
91.67	0.79	1.10	0.48	1.80	1.17	2.15
100.00	−0.10	−0.10	−0.10	−0.10	−0.10	−0.10

Example (use of hydrofoil carpet plots)

The problem is to recommend blade shapes for an axial-flow, rotor-only pump. The design specifications are the following.

Flow = 8,000 gpm, 0.5047 m³/s.

Static-head rise = 15 ft, 4.57 m.

With an assumed efficiency and outlet-dynamic-head allowance and an approximate boundary-layer allowance, the specifications used for the velocity-triangle calculation were

Flow = 8,034 gpm, 0.5069 m³/s.

Theoretical total-head rise = 19.45 ft, 5.928 m.

After trials of various flow coefficients and hub-tip diameter ratios, it was decided to use a hub-tip diameter ratio of 0.8, a peripheral speed at the hub of 40.0 ft/s (12.19 m/s), and an axial velocity of 25.0 ft/s (7.62 m/s). From these values the velocity-diagram values in the first part of table 8.6 were arrived at. The problem is to recommend blade shapes for those flow angles. We do this by continuing the table. We will select blade forms for two values of solidity: 0.75 and 1.0.

The blades were chosen by calculating the ordinate ε^x for figure 8.17, drawing a horizontal line corresponding to the value of ε^x, and choosing an operating point near positive or negative stall, or in the middle of the range, for plots of the appropriate solidity. Each separate "carpet plot" is for a cascade of blades of fixed camber and set at a given solidity, with setting or stagger angle and angle of attack (incidence) varied (figure 8.17). Usually the designer has a choice of blades of two or more camber angles and therefore can interpolate between them.

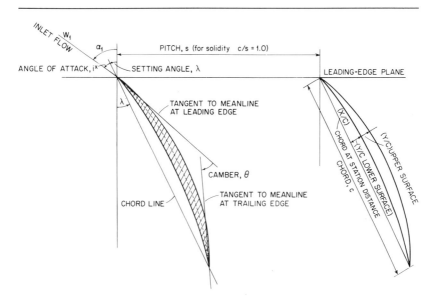

Figure 8.19 Hydrofoil cascade nomenclature.

Table 8.6
Preliminary design of axial-flow pump

	Hub	Mean	Tip
Fluid inlet angle to rotor, deg.	57.99	60.94	63.43
Fluid outlet angle from rotor, deg.	44.26	51.20	56.30
Radius ratio	0.8	0.9	1.0
Deflection, $\varepsilon \equiv (\alpha_{in} - \alpha_{out})$, deg.	13.73	9.74	7.13
Solidity, $\sigma \equiv c/s$ (first choice)	0.75	0.75	0.75
Ordinate $\varepsilon^x \equiv \varepsilon + 32(\sigma - 0.75) = \varepsilon$	13.73	9.74	7.13
Possible blades: camber, θ deg.	45	30	15
Angle of attack (chordal incidence) i^* deg.	7.3	6.15	8.25
Solidity, σ (second choice)	1.0	1.0	1.0
Ordinate $\varepsilon^x \equiv \varepsilon + 32(\sigma - 0.75)$	21.73	17.74	15.13
Possible blades: camber θ deg.	30	20	10
Angle of attack (chordal inc, i^*) deg.	10.6	7.2	7.5

References

DeRuyck, J., and C. Hirsch. 1980. *Investigations of an axial-compressor end-wall boundary-layer prediction method*. Paper 80-GT-53. ASME, New York.

Emery, James C., L. Joseph Herrig, John R. Erwin, and Richard Felix. 1959. *Systematic two-dimensional cascade tests of NACA 65-series compressor blades at low speeds*. Report 1368. NACA, Langley, Va.

Horlock, J. H. 1958. *Axial-Flow Compressors*. Butterworths, London.

Kock, C. C. 1981. *Stalling pressure-rise capability of axial-flow compressor stages*. Paper 81-GT-3. ASME, New York.

Koch, C. C., and J. H. Smith, Jr. 1976. Loss sources and magnitudes in axial compressors. *Trans. ASME J. Eng. Power*, A 98 (July): 411–424.

Lieblein, Seymour. 1959. Loss and stall analysis of compressor cascades. *Trans. ASME J. Eng. Power* (September): 387–400.

Mellor, George L. 1956. The NACA 65-series cascade data. Gas-Turbine Laboratory charts. MIT, Cambridge, Mass.

Reid, L., and R. D. Moore. 1980. *Experimental study of low-aspect-ratio compressor blading*. Paper 80-GT-6. ASME, New York.

Starke, J. 1980. *The effect of the axial-velocity-density ratio on the aerodynamic coefficients of compressor cascades*. Paper 80-GT-134. ASME, New York.

Taylor, W. E., T. A. Murrin, and R. M. Colombo. 1969. *Systemic two-dimensional cascade tests. Double-circular-arc hydrofoils*. Vol. 1. Report CR 72498. NASA, Cleveland, Ohio.

Wilson, David Gordon. 1960. *Patterning stage characteristics for wide-range axial compressors*. Paper 60-WA-113. ASME, New York.

Problems

8.1

Choose the mean-diameter stage arrangement (rotor-stator or stator-rotor), the rotational speed, the hub and tip diameters velocity diagrams at hub and tip diameters, and blades for a human-powered axial-flow irrigation pump of the following specifications:

Vertical lift = 2 m.

Steady power to blading = 150 W.

Delivery-pipe length = 8 m.

Delivery-pipe diameter = 0.1 m.

Pump hub-tip ratio = 0.7.

The delivery pipe has three 45-degree bends, each contributing an estimated pressure drop equal to a length of pipe of 15 diameters. The friction factor Cf of the pipe is 0.007 in the relation

$$\Delta p = 4Cf\left(\frac{L}{d}\right)\left(\frac{\rho C^2}{2g_c}\right).$$

Assume that one pipe dynamic head is lost at outlet (the use of a diffuser would be desirable). Assume a pump efficiency of 0.88, total-to-total across the blading.

8.2

If you have chosen the mean-diameter velocity diagram for the single-stage cooling fan of problem 5.1, choose suitable blades from figure 8.6. Then calculate the total-to-total isentropic efficiency, using the methods of chapter 8. Convert your value to a polytropic efficiency, and compare it with the value given in problem 5.1. You will have to assume that the loss values you obtain for the mean diameter apply over the whole annulus.

8.3

(a) On the axial-compressor characteristics provided in figure P8.3, draw in lines of constant design-point incidence for the first stage and for the last stage. The design point is shown. You will have to make various simplifications because you are not given many data that would be required for a more precise calculation. State your assumptions.

(b) Then select suitable mean-diameter blade cascades from figure 8.6. Find the flow angles for positive stall of the first stage and negative stall of the last stage, and draw lines on the characteristics when these apply. Assume that the velocity diagrams are 50-percent reaction, that the work coefficient is 0.25 for the first stage and 0.4 for the last stage, that the flow coefficients are 0.3 for each stage, and that the solidities are unity.

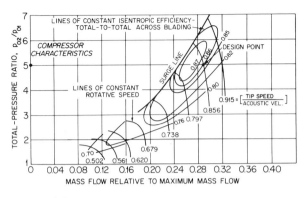

Figure P8.3

(c) From your results discuss the flow conditions in the blading as the compressor is started from rest in a flow circuit with a "square-law" resistance curve going through the design point.

8.4
(a) Choose NACA 65-series compressor blades for both rotor and stator of a helium compressor having the following specifications. Choose blades that give a deflection about 20 percent less than that for positive stall. For this preliminary design choose to have all mean-diameter velocity diagrams similar.

$p_{01} = 981$ KN/m^2.

$T_{01} = 40$ C.

$p_{02} = 2 \times p_{01}$ (at last-stage exit).

Maximum blade speed = 500 m/s.

Reaction = 0.5.

Work coefficient = 0.5 (mean-diameter, first stage).

$W_2/W_1 = 0.7$.

$\eta_{ptt} = 0.90$.

(b) Find the mean-diameter blade speed necessary to give an integral number of stages.

8.5
Find the flow coefficients relative to the design-point flow coefficients for the first-stage inlet and last-stage outlet for the two conditions marked on the axial-flow compressor characteristics shown in figure P8.5. At these two locations the flow is axial and equal in magnitude at the design point. The axial Mach number before the first stage at design point is 0.4.

8.6
Use the charts of figure 8.6 and figure 8.18, as appropriate, to list all the alternative stator and rotor blades (cambers, solidities, setting and incidence angles) for a jet-propulsion pump for a high-speed vessel

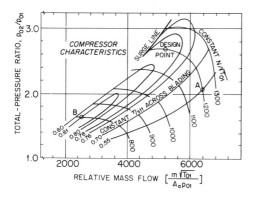

Figure P8.5

The pump is a rotor-stator combination having axial inlet and outlet flow at design speed and flow. The flow coefficient is equal to $\tan 20°$, and the minimum value of the velocity ratio for either stator or rotor, W_2/W_1 or C_x/C_2, is 0.75. The axial velocity should be held constant through the stage.

9

Preliminary design methods for radial-flow turbomachines

The advantages and disadvantages of using radial-flow machines in preference to those having predominantly axial flow are first reviewed. It is shown that a radial-flow pump or compressor can be designed with many times the head or enthalpy rise per stage than is possible with axial-flow machines. This gives radial-flow compressors and pumps an advantage where it is desirable to reduce the number of stages in an application to a minimum. Radial-flow machines also have very large cost, and even possibly efficiency, advantages over multistage axial machines when the size is small enough for the Reynolds number of the flow to become small, and when the blade tip clearances become relatively large. Radial-flow machinery in almost every size will cost less to manufacture than the equivalent multistage axial machinery. In the larger sizes the efficiencies will be, in general, lower than for axial machines.

In contrast, radial-flow turbines are shown to produce generally lower head or enthalpy drops per stage than is possible with axial-flow machines. The areas of application of radial-flow turbines are limited predominantly to those in which the lower manufacturing cost of small, single-stage radial turbines is of overriding importance and to special cases where the configuration is advantageous, such as large Francis water turbines.

Simple design methods are given, and illustrated with examples, for most types of radial-flow machines. The complex flow and blade shapes of Francis-type water turbines are not covered in any detail.

9.1 The difficulties of precise design

Radial-flow machines present a challenge and an enigma. The preliminary design of a radial-flow machine can be quite simple. However, the flow is extremely complex. Attempts over many years to solve the three-dimensional flow analytically and numerically, even with the most advanced

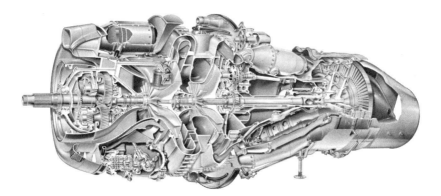

RR Dart turboprop engine with two-stage centrifugal compressor, two-stage turbine driving compressor, and single-stage power turbine. Courtesy Rolls-Royce Limited.

computer programs, have not yielded procedures that guarantee good performance in machines designed with their aid.

The two most challenging categories of radial-flow turbomachines are radial-inflow water turbines and high-pressure-ratio radial-outflow compressors. I know of no instance in either of these categories where a new machine, when run in the as-designed configuration, achieved outstanding performance. Normally a fairly extensive program of development, of models in the case of water turbines, has to be undertaken, in which "cut-and-try" methods, perhaps disguised, play a prominent part. What is meant by a "new" machine has to be defined for this statement to have validity. Some new machines in fact are merely small extrapolations from, or interpolations between, previous designs. Most new machines are designed by a person or by a team with years of experience in designing and developing radial-flow machines. Features of those machines that have given good performance will obviously tend to be carried forward as part of design practice. This is the art that, for radial-flow machines particularly, complements the science.

It should not be inferred that flow analysis is of no utility. It seems at least justifiable to claim that application of advanced methods of flow analysis to, particularly, the flow in the rotor of a radial-flow turbomachine will probably greatly reduce, but equally probably not eliminate, the period of subsequent development testing that will be required.

For the less demanding types of radial-flow turbomachines, for instance, low-head-rise centrifugal pumps and compressors, the preliminary design methods given in this chapter may be wholly sufficient. Before these methods are developed, the areas of potential application of radial-flow machines should be discussed.

9.2 Advantages and disadvantages and areas of application

Radial-flow machines have advantages that enable them to rule unchallenged in some applications. In others, they are used preferentially just over the smaller end of the size range.

Air and gas compressors, and pumps

More work per stage—a larger head rise, or larger pressure ratio—can be given by a radial-flow machine than by an axial-flow stage. Often this fact is attributed to the "centrifugal effect"—the radial pressure gradient that will be set up when a mass of fluid is rotated by an impeller. But the enthalpy rise is still the result of the torque given by the impeller to accelerate the fluid and is evaluated by Euler's equation (5.1). Therefore we can look to the velocity diagrams of radial-flow and axial-flow machines to find out why the former should give more work per stage (figure 9.1).

The work coefficient, ψ, which is the specific work for a given blade speed, is limited in an axial stage by the requirement to keep the relative-velocity ratios above some limit, such as the de Haller limit of 0.72 (figure 4.7). In a radial-flow machine the relative velocity of the flow entering the rotor is low simply because it comes in at a smaller radius than that of the outlet flow. If one uses the same de Haller limit for radial-flow rotors as for axial, the permissible increase in outlet tangential velocity is similar to the increase in rotor peripheral velocity between the entering and leaving radii.

In figure 9.1 velocity diagrams for axial-entry stages are compared. The diagrams for the radial-flow machines (figure 9.1b and c) are for a configuration in which the diameter of the inducer-blade tips, or of the inner surface of the shroud, is one-half of the rotor periphery. The radius of the hub at impeller (or inducer) entry is one-half that of the shroud at the same plane. The axial-machine diagram (figure 9.1a) is for a hub-tip diameter ratio of 0.6, which has been regarded (chapter 6) as the lower limit at which the simple design methods outlined in this book can be used with success. For simplicity, both diagrams have constant axial velocity with radius at the rotor-entry plane.

For the axial stage, the limiting diagram insofar as relative-velocity ratio is concerned is at the hub. The diagram has been drawn with a hub flow coefficient of 0.5 (so that the tip value is 0.33) and with a relative-velocity ratio of 0.71. The angle of flow into the rotor at the hub is then 63.43°.

The velocity diagram for the shroud streamline to have the same enthalpy rise as that at the hub is shown in broken lines.

In the radial machine the limiting diagram for rotor relative-velocity ratio is that involving the shroud streamlines at the impeller inlet. The reason for this is that both the hub and the shroud streamlines (or stream

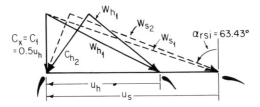

(a) Axial-flow compressor diagram: axial entry and exit; free vortex;
     hub-tip ratio 0.6; relative-velocity ratio 0.71

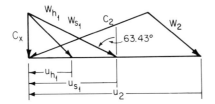

(b) Radial-flow diagram: axial entry; inducer-periphery diameter ratio 0.5;
     inducer hub-tip ratio 0.5; relative-velocity ratio 0.71

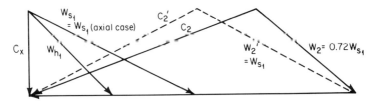

(c) Radial-flow diagram similar to (b) but with identical shroud inlet relative
     velocity to (a)

Figure 9.1 Axial- and radial-flow velocity diagrams.

surfaces) emerge from the rotor at the same radius—the rotor periphery
—and with the same relative flow angle. (That of course is a simplifying
assumption. In fact the flow is extremely nonuniform within a centrifugal-
impeller channel. Even in those few cases where streamlines have been
found experimentally, the patterns do not hold for other machines. Therefore we are really referring to the mixed-flow mean streamline here.) Then
the inlet streamline having the larger relative velocity, which is obviously
the shroud streamline, will undergo the larger flow deceleration.

Therefore we have chosen W_{s1} in the radial diagram to have the same
relative flow angle, 63.43°, as W_{h1} in the axial diagram and W_2 to be
0.72 W_{s1}, just as W_{h2} was 0.72 W_{h1} in the axial case. (It will be shown later
that this flow angle happens to be close to the optimum for the relative
flow at the inlet shroud for most radial flow machines.) When we complete

the diagram, we find that the work coefficient for the radial machine is 0.69, whereas for the axial machine at the same peripheral speed, which is for the shroud radius of the axial diagram, the work coefficient is 0.23. The radial machine has exactly three times the work coefficient of the axial machine using the same relative-velocity ratio.

But is it fair to use the same peripheral speed, and the same relative-velocity ratio, in the two cases? The peripheral speed might be set by stress limits. However, this will be the case only for compressors of low-molecular-weight gases such as hydrogen and helium and for experimental air compressors of very high pressure ratio, such as 10 : 1. We regard these as special cases. In more usual compressors and pumps the highest relative velocity at inlet will normally set the limiting speed because of Mach-number limits in compressors and cavitation limits in pumps.

In both the axial and radial diagrams, figure 9.1a and b, the highest relative velocity at inlet is W_{s1}, the relative velocity of the shroud stream-lines. But as these two diagrams are drawn for the same value of peripheral speed, the relative velocity as drawn for the radial diagram is much lower than that for the axial diagram. In figure 9.1c the radial diagram has been enlarged so that W_{s1} is the same as for the axial diagram, 9.1a. The work that can be given by the radial stage is now 10.56 times that for the axial stage.

Now we return to the second question: Is it fair to use the same relative-velocity ratio for the radial machine as for the axial? We don't know, but a safe answer is that we probably should not. The flow entering a radial-flow impeller has to go through such sharp turns, and has to withstand such continued lateral acceleration, that it would be natural to expect that, for the flow to withstand the tendency to separate from the walls of the channel, the deceleration rate should be lessened. In figure 9.1c a broken line shows an extreme case in which W_2 has the same value as W_{s1}. That is, there is no deceleration of the mean flow. The work coefficient falls from 0.69 to 0.50, and the ratio of radial-machine work to axial work falls to 7.74, which is still considerable.

To produce a relative-velocity vector in either the higher-work direction, W_2, or the lower-work direction, W_2', the rotor blades would have to be slanted backward with respect to the direction of rotation. They are termed "backward-swept" blades. The blade-root bending stress increases with higher degrees of "sweepback" or backslope and with relatively longer blades (in the axial direction). However, in radial-flow compressors it is merely desirable, not necessary, that the flow be unseparated in the rotor channels. Rotors designed for very high pressure ratios must use a high peripheral speed, and radial, or near-radial, blades. The relative-velocity ratio in the rotor becomes too low for efficient flow, and separation within the rotor must be expected. When the relative-velocity-ratio limitation is relaxed, the impeller flow will be highly separated, and it will not be possible to obtain much diffusion in a following diffuser fed with highly nonuniform flow. Despite this, a radial-flow compressor can give fifteen or more times the enthalpy rise of an axial stage with the same inlet relative Mach number. The efficiency will, however, be much lower—for instance, 83 percent instead of 93 percent.

To sum up, radial-flow machines can produce much higher enthalpy rises per stage than can axial machines at the same relative inlet velocity principally because there is no direct coupling between the magnitude of the relative inlet velocity and that of the peripheral speed and of the tangential component of the absolute velocity at outlet. Furthermore flow separation in the radial-machine rotor may actually increase the enthalpy rise, whereas in the axial rotor passage separation acts to reduce the deflection of the flow, and with it the enthalpy rise.

Relative efficiencies. The efficiency of radial pumps and compressors is usually lower than that of alternative axial machines, for four reasons.

1. The flow in the rotor is much more complex in the radial-machine rotor than in the more streamlined passages of an axial compressor or pump. In addition to the sharp turn from the axial to the radial direction, which is required and will introduce losses, the severe aerodynamic loading and the long flattened passage shape encourages the growth of secondary flows in the boundary layers (figure 9.2). These secondary flows can combine to facilitate flow separation within the channels and impose highly nonuniform flow at the entrance to the diffuser. (The "relative eddy" shown in figure 9.2 is in fact hypothetically produced by irrotational potential flow passing through a centrifugal impeller. It can explain the deviation ("slip") of the flow from the blade direction.)

2. The absolute outlet velocity, C_2, is high in a radial-flow compressor or pump, so that much of the enthalpy rise is in the form of velocity. Even an ideal conical diffuser fed with almost uniform inlet flow cannot convert much more than 80 percent of velocity head into pressure head (figure 4.14).

3. Radial diffusers of any type (see figure 4.23) have much lower actual pressure-rise coefficients than do axial-flow diffusers, particularly when the radial flow at inlet is highly nonuniform.

4. Radial-flow machines have relatively large wetted areas, both on the surface of the flow channels and on the backfaces of the disks, and on the rotating shrouds, if used. Friction must therefore be relatively large.

Effect of backslope. The first two of the above-listed causes of reduced efficiency can be made much less penalizing by utilizing large angles of backslope, when stress limits allow (as in hydraulic pumps and low-pressure-ratio fans and blowers). As the blade angle is increased to, perhaps, 50 degrees from the radial direction, the flow angle will be

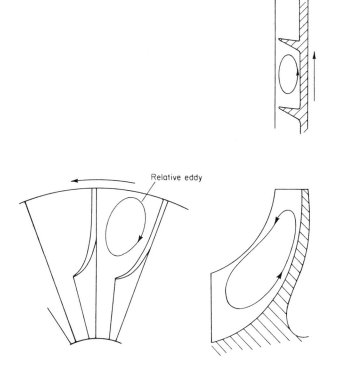

Figure 9.2 Secondary flows in radial-flow pumps and compressors.

greater than this and the magnitude of the absolute outlet velocity (the diffuser-entry velocity) will be greatly reduced. While the work per stage is also reduced, the efficiency increases both because of the reduced loading on to the diffuser and because of the reduced diffusion in the rotor. (The mean relative flow may even accelerate in the rotor.) The uniformity of the flow entering the diffuser will also be improved. Kluge (1953) showed the results of tests on similar impellers with only the blade angle at outlet varying and found a peak efficiency at an angle of about 50° (figure 9.3). The efficiency levels of Kluge's results are low. For high Reynolds numbers and near-optimum specific speeds it should be possible to reach ten points or more higher than Kluge's curve, as indicated by the shaded area. Large centrifugal pumps can give efficiencies along this upper line.

Effect of size. When the tip diameter of the first stage of an axial compressor has to be much below one meter, the blades in the high-pressure stages (assuming that the overall pressure ratio is four or above) become small, fragile, expensive to produce, and severely affected by tip-clearance losses. (Tip clearances have been greatly reduced following the introduction of high-speed preloaded ball bearings, in place of the lightly loaded

Centrifugal compressor with sweepback and splitter blades for moderate pressure ratio. Courtesy Detroit Diesel Allison Division of General Motors Corporation.

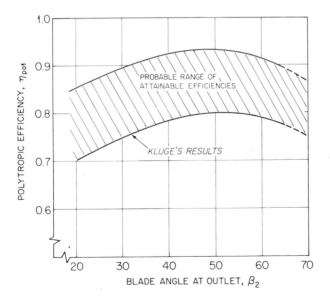

Figure 9.3 Effect of radial-flow-impeller blade angle on polytropic efficiency.

Radial-Flow Turbomachines

plain bearings previously required, and of abradable shroud materials.) Therefore for small, high-pressure-ratio compressors the only choice is a radial machine.

Axial-flow pumps, because they operate on incompressible fluids, retain the same blade sizes in the high-pressure stages and could therefore be made in smaller sizes than would be economic for compressors. However, the high-backslope multistage centrifugal pump is preferred in practice even for the larger sizes, giving very high efficiencies (often well over 90 percent). Axial-flow pumps presently seem to be used only for high-flow applications, when the head rise is small enough for a single stage to be sufficient (high-specific-speed applications).

Specific-speed effects. Radial-flow machines are also to be preferred when rotational speeds are so limited that the blade lengths of axial-flow machines would be very small. (In other words, the hub-tip ratio at machine entry might be above 0.85. As developed in chapter 5, a high hub-tip ratio signifies a low specific speed.) The efficiency of a radial machine may exceed that of an axial-flow machine if the blade lengths are small compared with the tip diameter (which will in turn influence the blade clearance); see figure 5.19.

Cost considerations. Radial-flow compressors and pumps are normally much less expensive to produce, for the same flow and head rise, than are their axial-flow equivalents, because fewer stages are required, and because one-piece rotors, and possibly one-piece stators, can be produced, often by highly accurate but inexpensive investment casting.

Space considerations. Radial-flow compressors and pumps may be shorter than their axial-flow equivalents, but they have a much larger frontal area. Whether or not there is a weight penalty depends upon many factors.

Radial-inflow water and gas turbines
Some of the considerations made with regard to the areas of application of radial-flow pumps and compressors apply equally to radial-flow turbines. In particular, the remarks about the effects of size, specific speed, cost and space requirements all apply almost equally to radial turbines.

Relative work output. There is one very large difference that affects the relative applications of radial-flow turbines, on the one hand, and compressors and pumps, on the other. Whereas radial-flow compressors and pumps can be designed to give ten or twenty times the enthalpy rise per stage of the equivalent axial machines, radial-flow turbines produce, in some cases, less than the enthalpy drop possible with an axial turbine of the same rotor diameter and speed.

Velocity diagrams. The reasons for this rather extraordinary lack of symmetry lie in the velocity diagrams and in the very different nature of accelerating flow in a favorable pressure gradient to decelerating flow in an unfavorable pressure gradient.

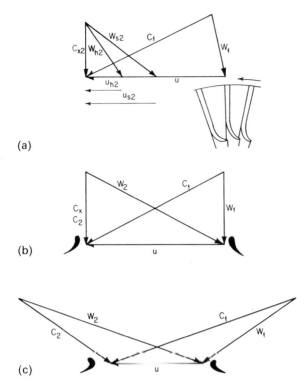

(a)

(b)

(c)

Figure 9.4 Radial-inflow compared with axial-flow turbine velocity diagrams: (a) radial-inflow turbine; (b) high-efficiency axial-flow turbine; (c) high-work axial-flow turbine.

The velocity diagram of a low-specific-speed radial-inflow turbine (one having the tip diameter of the rotor blades at the outlet plane considerably smaller, say under 60 percent, of the rotor peripheral diameter) will be similar to the equivalent radial-outflow compressor, figure 9.4a.

The blades of hot-gas expansion turbines are virtually always radial to minimize blade bending stress. Because there is no velocity-ratio limit in accelerating flow, there is no fluid-mechanic compromise involved in using radial blades. The peak efficiency appears to occur when the relative flow angle is similar to that which would be found to give peak efficiency if the same rotor were used as a compressor. This means that the relative flow should be aimed slightly against the blade rotation, so as to provide a gradual increase of aerodynamic loading. The work coefficient would then be about 0.9.

The diagram for an axial turbine of peak efficiency is shown in figure 9.4b. This is a 50-percent-reaction turbine with a work coefficient of unity.

The diagram of a high-work turbine is shown in figure 9.4c. It is possible

Radial-Flow Turbomachines

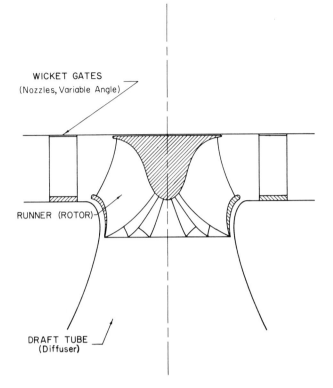

WICKET GATES
(Nozzles, Variable Angle)

RUNNER (ROTOR)

DRAFT TUBE
(Diffuser)

Figure 9.5 Francis hydraulic turbine of high specific speed.

to obtain a work coefficient of 3.0 without a major fall in the efficiency. However, a single-stage machine would not have axial outflow, and the diffuser performance would suffer unless straightening vanes were used at diffuser entrance.

Cryogenic and hydraulic turbines. Cryogenic turbines are speed limited because of the low speed of sound at low temperatures. Accordingly, higher-work nonradial blades can be used. Hydraulic turbines are also not stress limited, and there is considerable freedom to choose the best blade shape. The very-high-specific-speed Francis turbines, which are nominally radial inflow but may have some stream surfaces going radially outward through the blading, have blade shapes that are the result of decades of patient experiments aimed at eliminating flow reversals and cavitation (figure 9.5). The blade shapes that have evolved cannot be fully designed by the methods of this book and must be regarded as a special case.

Relative efficiencies. Radial-inflow turbines, possibly excluding Francis hydraulic turbines, have lower efficiencies than axial turbines which could be substituted. The reasons are similar to those for the lower efficiency of radial compressors and pumps—more turning in the rotor, higher aerodynamic loading, larger wetted area, and in addition the significant one that the conical diffusers which are almost always combined with radial turbines work very poorly with nonuniform or slightly swirling inlet flow. It is very difficult to design the outlet section of the blades of a radial-inflow turbine (the exducer) so that the flow emerges without swirl. Even if the designer is successful in doing so, this condition will occur at only one flow coefficient. At slightly off-design conditions the flow will inevitably have swirl, which will cause the type of flow breakdown illustrated in figure 4.19.

Disincentives to multistaging. A less-general reason for radial-inflow turbines having generally lower efficiencies than equivalent axial turbines is that, despite the lower work coefficient given by radial turbines, they are almost never used in multistage configurations. Any radial-flow machine makes severe demands on ducting. Centrifugal compressors and pumps have sufficiently large advantages over axial-flow compressors and pumps that the heavy additional costs of the ducts and casings are compensated for. Radial-inflow turbines do not have these advantages. The configuration of axial-flow turbines makes the design and manufacture of multi-

Figure 9.6 Hot-gas radial-inflow-turbine rotor

Radial-Flow Turbomachines

stage turbines relatively easy. For the highest-efficiency turbine therefore, the designer can choose a multistage axial-flow turbine with an optimum diagram (figure 9.4b) and a low blade speed. This will probably have several percentage points advantage in efficiency over a single-stage radial-inflow turbine of high blade speed because the multistage axial turbine will have a low exit velocity, which will then be diffused more efficiently because the flow will be more uniform (and because the annular diffusers used with axial turbines can incorporate rudimentary straightening vanes, so that flows with higher degrees of swirl can be efficiently diffused); these factors are additional to the intrinsically higher efficiency of the more direct flow in axial turbines.

Rotating inertia. Radial turbines have higher rotating inertia than do axial turbines designed to meet the same specifications with minimum inertia. This is the result of the necessarily long blades and consequently long hub.

Thermal inertia. The long hub, with a continuous surface scrubbed by hot gases and extending over a large radial range, results in high thermal stresses, particularly in gas-turbine engines during start-up and shutdown. The absolute magnitude of thermal stress is a function of size. Thermal stresses are managed adequately in small hot-gas turbines (up to, say, 250 mm rotor diameter) by incorporating deep cuts ("scallops") between the blades at the periphery (figure 9.6).

Reversible operation. Radial-inflow turbines have the potential advantage that the direction of rotation could be reversed by the incorporation of means to swivel the nozzle vanes (figure 9.7). The efficiency in the reverse direction would be lower than in the ahead direction because the exducer would now add swirl to, instead of removing it from, the leaving flow. However, for transportation vehicles such as ships and tanks which operate for only short periods in reverse, but require rapid response, this concept could offer advantages. It has been incorporated into several proposed propulsion arrangements.

Summary of applications. The radial-inflow turbine emerges from the foregoing as being used in some special applications, such as for Francis hydraulic turbines and for cryogenic turbines, where low speeds permit optimal blade shapes to be used, and for small single-stage hot-gas expanders where low manufacturing cost is of major importance. Two examples in the latter category are small exhaust-gas turbosuperchargers for internal-combustion engines and small short-mission gas-turbine engines. Experimental ceramic turbines have achieved greater success in the radial configuration than in the axial.

9.3 Preliminary design methods for radial-flow turbomachinery

Although the design of radial-flow turbomachinery is based, as is axial machinery, on Euler's equation, the design method differs because the

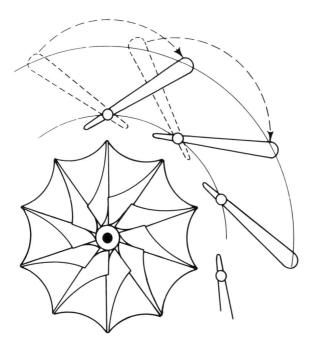

Figure 9.7 Reversible-rotation radial-inflow turbine.

velocity diagram at rotor outlet can be considered separately from that at rotor inlet.

For compressors and pumps with no swirl in the inlet flow, the outlet velocity diagram chosen will completely determine the work, pressure ratio, and flow. Conversely, the rotor-inlet velocity diagram will be controlling for radial-inflow turbines designed for zero outlet swirl. In both cases, compressor-pumps and turbines, the inducer or exducer section can be designed subsequently.

Compressor-pump rotor-outlet velocity diagrams

Two outlet diagrams combined with inlet diagrams are shown in figures 5.5 and 5.17. From the steady-flow energy equation (2.9) and Euler's equation (5.1), the specific work done by the rotor on the flow (in adiabatic conditions) is

$$\frac{\dot{W}}{\dot{m}} = -g_c \Delta_1^2 h_0 = -(u_2 C_{\theta 2} - u_1 C_{\theta 1}). \tag{9.1}$$

(The rotor will also perform frictional work through the frictional torques exerted by the casing, the fluid scrubbing on the back of the disk and on the back of a rotating shroud, if used, and by bearings and seals. When the frictional work appears as heat in the working fluid, there will be an additional enthalpy rise.)

For a compressor working on a fluid that can be approximated to an ideal gas, equation (9.1) can lead to the following:

$$-\frac{\dot{W}}{\dot{m}} = g_c \Delta_1^2 h_0 = g_c \overline{Cp}_c (T_{02} - T_{01}) = u_2 C_{\theta 2} - u_1 C_{\theta 1}$$

$$= g_c \overline{Cp}_c T_{01} \left(\frac{T_{02}}{T_{01}} - 1 \right);$$

$$\therefore u_2 C_{\theta 2} - u_1 C_{\theta 1} = g_c \overline{Cp}_c T_{01} \left[\left(\frac{p'_{02}}{p_{01}} \right)^{(R/Cp_c)/\eta_{pc}} - 1 \right], \qquad (9.2)$$

where p'_{02} is the defined useful total outlet pressure for which η_{pc} is specified.

For a pump working on an incompressible liquid, equation (9.1) may be combined with the appropriate definition of pump efficiency, η_i, equation (2.61), to give

$$\frac{\dot{W}}{\dot{m}} = u_2 C_{\theta 2} - u_1 C_{\theta 1} = \frac{1}{\eta^i} [H'_{02} - H_{01}] \frac{g}{g_c}. \qquad (9.3)$$

Design process

The design process is then as follows.

1. The specifications will give the pressure ratio, $p_{02'}/p_{01}$, or the head rise, $H'_{02} - H_{01}$, to be obtained.

2. The efficiency corresponding to the definition of the useful outlet pressure, p'_{02}, or head, H'_{02}, must be estimated. The experienced designer can usually make a fairly accurate guess based on an interpolation or extrapolation of previous known results. The inexperienced designer will probably be less accurate, but this will merely mean the possibility of two, rather than one, iterations, because the efficiency can be better predicted once preliminary dimensions and velocity diagrams have been chosen. The specific-speed versus efficiency envelopes, figure 5.19, can be used as guides, with estimates of efficiency decrements being made for high tip clearance, low Reynolds number, high Mach number, high pressure ratio, rough surface finishes, and so forth.

3. Initially at least the designer usually specifies that there be no rotation in the inlet flow (by requiring that, if necessary, there be straightening or "de-swirl" vanes) so that $C_{\theta 1} = 0$. Then $C_{\theta 2}$ is obtained from equation (9.2) or (9.3). The outlet velocity diagram can now be selected.

4. The outlet velocity diagram is specified by two angles, α_2 and α_{rel2}. The first of these, α_2, is the absolute angle of flow leaving the rotor. (Under the conventions by which velocity diagrams are made, it would be the mean flow direction if the rotor-blade wakes, the other boundary layers, and the flow nonuniformities within the rotor-blade passages were to be fully mixed to give a uniform flow at the rotor-outlet diameter.) If the rotor flow passes into a diffuser, at least the first part of this will be a vaneless diffuser. The angle α_2 can then be chosen to give the desired margin from diffuser stall, using figure 4.20. In preliminary design a reasonable initial value to use for α_2 if the diffuser configuration has not been selected would be in the range 60° to 70°. If the rotor is to exhaust direct into a plenum-collector, α_2 could be set somewhat higher, say 75°, to reduce the losses in the outlet kinetic energy, so long as the rotor relative-velocity ratio, W_2/W_{s1}, is held above, say, 0.8.

5. The rotor-outlet relative flow angle, α_{rel2}, at the design point can be specified by the designer. The normal range is from $-10°$, for a radial-bladed impeller with a large number of blades, to $-60°$, for an impeller with a few highly backswept blades. A correlation due to Stodola between the angle of the relative flow leaving the rotor, α_{rel2}, and the impeller blade angle at the periphery, β_2, with the number of blades, Z, and the flow coefficient at best-efficiency (design) point, ϕ_{be2} ($\phi_2 \equiv Cr_2/u_2$), is

$$\tan \alpha_{rel2} = \tan \beta_2 + \frac{\pi \cos \beta_2}{Z \phi_{be2}}. \tag{9.4}$$

Another correlation, by Eck, is given later.

6. The outlet velocity triangle is now tentatively fixed. The work coefficient, $\psi = C_{\theta2}/u_2$, can be measured or calculated for impellers with no inlet swirl. The peripheral speed, u_2, can be calculated from equations (9.2) or (9.3), where now

$$u_2 C_{\theta2} - u_1 C_{\theta1} = u_2^2 \psi. \tag{9.5}$$

7. It may be found that the blade speed for a single-stage machine is too high for stress reasons (see chapter 13). Therefore several stages must be used. The number of stages may be tentatively chosen at this point. Usually not all the swirl coming from the upstream diffusers is removed for the second and later stages, and the inlet velocity diagram may therefore need to be calculated to account for $C_{\theta1}$.

There are reasons other than high stresses for choosing to use several stages rather than one. The overall total-to-static efficiency will normally increase with the number of stages, because the outlet kinetic energy will decrease. The relative velocity at rotor inlet will decrease with number of stages, alleviating cavitation and high-Mach-number problems. Last,

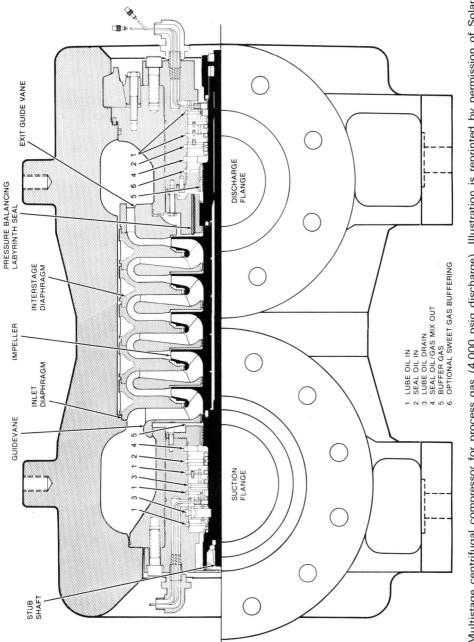

EXIT GUIDE VANE

PRESSURE BALANCING
LABYRINTH SEAL

INTERSTAGE
DIAPHRAGM

IMPELLER

INLET
DIAPHRAGM

GUIDEVANE

STUB
SHAFT

DISCHARGE
FLANGE

SUCTION
FLANGE

1. LUBE OIL IN
2. SEAL OIL IN
3. LUBE OIL DRAIN
4. SEAL OIL/GAS MIX OUT
5. BUFFER GAS
6. OPTIONAL SWEET GAS BUFFERING

Multistage centrifugal compressor for process gas (4,000 psig discharge). Illustration is reprinted by permission of Solar Turbines International, an operating group of International Harvester; all rights reserved.

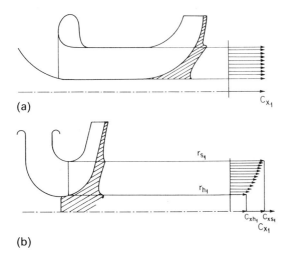

(a)

(b)

Figure 9.8 Axial-velocity variations in machine inlets: (a) gently curved inlet; (b) sharply curved inlet.

using multiple stages will lower the required shaft speed and increase the size of the impellers, which may be desirable for small machines.

8. The rotor inlet can now be designed. Sometimes mechanical requirements dictate a minimum hub diameter, d_{h1}, at the inlet. Alternatively, "overhung" impellers used in many single-stage applications give the designer freedom to specify a desirable hub-tip ratio. In either case the shroud diameter at rotor inlet, d_{s1}, is now chosen to give minimum relative velocity at that point.

If a small shroud diameter is used, the blade peripheral speed, u_{s1}, will be small, but the axial velocity, C_{x1}, will be high because of the reduced area. Accordingly, the relative inlet velocity at the shroud, W_{s1}, will be high. If the area is increased to reduce the axial velocity, the peripheral speed is increased, and again W_{s1} may be high. It is obvious that there must be an inlet shroud diameter giving minimum relative velocity. This can be found analytically or iteratively. An example that follows will illustrate the method.

A value of C_{x1} that is constant with radius can be used for each calculation if a gently curved inlet configuration can be used (figure 9.8a). If the inlet flow is sharply curved, the axial flow velocity (if unseparated) will increase with radius, and a simple function of C_{x1} with radius can be assumed (figure 9.8b). A power-law variation of axial velocity with radius may be assumed:

$$C_x = C_{xh} + (C_{xs} - C_{xh}) \left[\frac{r - r_h}{r_s - r_h} \right]^f. \tag{9.6}$$

Radial-Flow Turbomachines

The mean velocity, C_{xm}, for this velocity distribution is given by

$$\frac{C_{xm}}{C_{xs}} = \frac{C_{xh}}{C_{xs}} + \frac{2[1 - (C_{xh}/C_{xs})]}{(1 + \lambda)} \frac{(1 - \lambda)}{(f + 2)} + \frac{\lambda}{(f + 1)}, \tag{9.7}$$

where $\lambda \equiv r_h/r_s$, the hub-tip ratio.

As an example of the use of this relation, suppose that an inlet of hub-tip ratio 0.6 is being designed. Experimental results from a model of the inlet, or an approximation to a potential-flow solution or a designer's intuition, are used to characterize the axial-velocity distribution by the following values of the two variables involved:

$$\frac{C_{xh}}{C_{xs}} = 0.5, \quad f = 0.8.$$

Then $C_{xm}/C_{xs} = 0.798$.

The velocity diagram can be constructed from equation (9.6). The use of these preliminary-design methods is best illustrated by an example.

Example (radial turbosupercharger preliminary design)

Calculate the principal dimensions of the centrifugal compressor, the rotor diameter of the radial-inflow turbine of a diesel-engine turbosupercharger, and the "back" pressure (total) on the engine exhaust (figure 9.9). Some design information and suggestions are listed in table 9.1.

The vaneless-diffuser radius ratio should be designed to be 80 percent of that found for the stability limit. Both the compressor and turbine operating conditions at their design points should be for velocity diagrams at the rotor peripheries in which the tangential velocity is given by Eck's correlation:

$$\left[\frac{C_{\theta\,ac}}{C_{\theta\,th}} \right] = \left\{ \frac{1}{1 + [2\cos\beta/Z(1 - d_s/d)]} \right\}, \tag{9.8}$$

where the shroud diameter, d_s, is for the inducer inlet for compressors and the exducer outlet for turbines, d is the rotor peripheral diameter, and β is the blade angle at the rotor periphery (figure 9.10).

Use the same ratio of shroud-to-rotor-periphery diameter ratio for the turbine as is found optimum for the compressor. Bearing and windage losses are 2 percent of the compressor fluid power.

It is sufficient to make graphical or other approximate interpolations of the compressible-flow functions and to make only one iteration of, for instance, static conditions. Part of the design process will be to find the compressor-shroud diameter for minimum diffusion of the shroud streamline.

Summary of calculations The principal dimensions of the turbocharger were found to be as follows (together with other data).

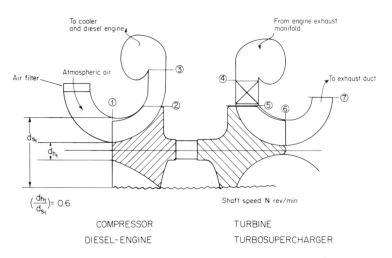

Figure 9.9 Diagram showing calculation planes of a radial-flow turbosupercharger.

Table 9.1
Turbocharger design input data

	Compressor		Turbine
Mass flow, kg/s	1.0		1.04
T_{01} K	300	T_{04}	800
p_{01} N/m²	1×10^5	p_{04}	Find engine back pressure
p_{st3} N/m²	2×10^5	p_{st7}	1.2×10^5
Fluid	Air		Combustion products, 100% theoretical air
$\overline{Cp}$ J/kg-°K	1,010		1,172
Blade angle with radial β_2 (deg)	30		0
Specific speed, N_s (equation 5.10)	0.1		
Polytropic efficiency, $\eta_{p,ts(1-3)}$	0.82	$\eta_{p,ts(4-7)}$	0.82
Flow angle leaving rotor, α_2 (deg)	60		
Flow angle entering rotor (deg)	0		70
Number of rotor blades, Z	17		13
Polytropic efficiency, $\eta_{p,tt(1-2)}$	0.96	$\eta_{p,tt(5-6)}$	0.96

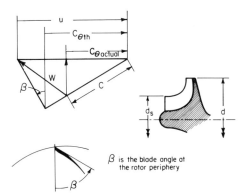

β is the blade angle at
the rotor periphery

Figure 9.10 Diagram defining inputs to Eck's correlation of tangential velocity.

Shaft speed, $N = 30,084$ rev/min.
Compressor rotor, $d_2 = 232.1$ mm.
Compressor rotor, $d_{s1} = 120.0$ mm.
Compressor rotor, $d_{h1} = 72.0$ mm.
Compressor rotor, $b_2 = 6.44$ mm.
Compressor stator, $d_3 = 371.4$ mm.
Turbine rotor, $d_5 = 207.7$ mm (assuming $d_{s6}/d_5 = d_{s1}/d_2$).
Engine back pressure, $p_{04} = 1.61 \times 10^5$ N/m².

Calculations The sequence of the calculations was to find the shaft speed from the specific speed and then to solve the compressor-rotor outlet velocity diagram. A number of iterations on the inlet-shroud diameter was necessary to find the value that gives minimum relative velocity there. The vaneless-diffuser stability limit was found, and hence the diffuser diameter ratio for the specified safety factor. Last the turbine rotor diameter was found, assuming the same diameter ratio as on the compressor rotor and the engine back pressure.

Enthalpy rise

$$\frac{T_{03}}{T_{01}} = \left(\frac{p'_{03}}{p_{01}}\right)^{(R/\overline{Cp}_c)/\eta_{p,ts}},$$

where $p'_{03} \equiv p_{st3}$ when $\eta_{p,ts3}$ is used,
$T_{03}/T_{01} = 2^{0.3484} = 1.2732,$
$\Delta T_{013} = 0.2732 \times 300 = 81.95°K,$
$\overline{Cp}_c = 1,010$ J/kg-°K;
$\therefore\ \Delta h_{013} = 82,772$ J/kg   and $(g_c\Delta h_{013})^{3/4} = 4,879.93.$

Inlet volume flow Guess C_{x1}, axial velocity at inlet 110 m/s (127.3 m/s) (revised, second-iteration, values added in parentheses):

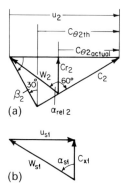

(a)

(b)

Figure 9.11 Turbocharger-compressor velocity diagrams: (a) rotor outlet; (b) rotor-shroud inlet.

$$a_{01} = \sqrt{g_c \gamma R T_{01}} = \sqrt{1.4 \times 286.96 \times 300}$$

$$= 347.16 \text{ m/s},$$

$$\frac{a_0}{a^*} = \sqrt{1.2} = 1.0954,$$

$$a^* = 316.93 \text{ m/s},$$

$$M_1^* = 0.3471 \quad (0.4017).$$

From table 30 of *Gas Tables* (Keenan and Kaye) $M_1 \approx 0.32$, $\rho_{st1}/\rho_{01} \approx 0.95058$,
                                                    (0.372)   (0.90886)

$$\rho_{01} = \frac{p_{01}}{R T_{01}} = \frac{10^5}{286.96 \times 300} = 1.1616 \frac{\text{kg}}{\text{m}^3},$$

$$\rho_{st1} = 1.1042 \text{ kg/m}^3 \quad (1.0557),$$

$$\dot{m} = \rho_{st1} \dot{V}_1,$$

$$\dot{V}_1 = 0.9056 \text{ m}^3/\text{s} \sqrt{\dot{V}_1} = 0.9516 \quad (0.9472 \text{ m}^3/\text{s}, 0.97725),$$

$$N = \frac{60 N_s (g_c \Delta h_0)^{3/4}}{\sqrt{\dot{V}_1}},$$

$$= 30{,}767 \text{ rev/min} \quad (30{,}084).$$

Compressor-outlet velocity diagram Geometrically, $u_2/C_{\theta 2 \text{th}} = 1 + (\tan 30°/\tan 60°) = 1.3333$; see figure 9.10. From Eck's correlation

$$\frac{C_{\theta ac}}{C_{\theta th}} = \frac{1}{1 + (2 \cos \beta_2)/Z(1 - d_{s1}/d_2)},$$

where $\cos \beta_2 = \cos 30°$, $Z = 17$, and d_{s1}/d_2 is not yet known. We need to optimize the inlet velocity diagram in figure 9.11 first.

Radial-Flow Turbomachines

Inlet-velocity diagram The procedure to select minimum W_{s1} is as follows.

1. Choose a value of d_{s1}.
2. Find $u_{s1} = (N\pi/60)d_{s1}$.
3. Calculate $A_a = (\pi d_{s1}^2/4)(1 - 0.6^2)$.
4. Find $\dot{m}\sqrt{T_{01}}/Ap_{01}$.
5. Find $A/A^* = 0.040421/(\dot{m}\sqrt{T_{01}}/Ap_{01})$.
6. Find $\dfrac{\rho_{st1}}{\rho_{01}} \to \rho_{st1} \to C_{x1} \to W_{s1}$.

(In table 9.2 second-iteration values are again given in parentheses.)
The optimum value of d_{s1} is found to be ~ 120 mm:

$$\alpha_{re1,s1} = \cos^{-1}(C_{x1}/W_{s1}) = 56.04°.$$

We still need to iterate on (d_{s1}/d_2). We start by estimating this ratio as 0.6. Two further iterations are required to give a close agreement with the estimated d_2; see table 9.3. Then C_2 and W_2 are calculated. The minimum value of W_{s1} is found to be 227.9 m/s so that the De Haller velocity ratio (diffusion index) $W_2/W_{s1} = 0.838$. This should lead to no diffuser-induced separation.

Impeller-outlet and diffuser-inlet width, b_2

$$C_2 = 261.43 \text{ m/s},$$

$$T_{02} = 381.95 \text{ K},$$

$$a_{02} = \sqrt{g_c \gamma R T_{02}} = 391.72 \text{ m/s},$$

$$a_2^* = 357.61 \text{ m/s},$$

$$M_2^* \equiv C_2/a_2^* = 0.7310,$$

$$M_2 \approx 0.698,$$

$$\rho_{st2}/\rho_{02} \approx 0.793,$$

$$\rho_{02}/\rho_{01} = (T_{02}/T_{01})^{1/(n-1)} = 1.7683,$$

where

$$\frac{1}{n-1} = \left[\frac{Cp}{R}\eta_{p,tt} - 1\right] = 2.360$$

and

$$\eta_{p,tt} = 0.96 \quad \text{across rotor},$$

$$\rho_{02} = 1.7683 \times 1.1616 = 2.0541 \text{ kg/m}^3,$$

$$\rho_{st2} = 1.628 \text{ kg/m}^3$$

$$\dot{m}_2 = \rho_{st2} Cr_2 Aa_2$$

Table 9.2
Calculation of optimum shroud diameter

d_{s1} mm	100	125	120
u_{s1} m/s $= 1610.97 d_{s1}$	161.10	201.37	193.32
	(1575.22)	(157.22)	(196.90)	(189.03)
Aa m$^2 = 0.5027 d_{s1}^2$	0.005026	0.007854	0.0072389
A/A^*	1.1730	1.8329	1.6893
M_1 (approximate)	0.61	0.34	0.3718
ρ_{st1}/ρ_{01}	0.8357	0.9445	0.9341
C_{x1} m/s $= \dfrac{0.86088}{(\rho_{st1}/\rho_{01})Aa}$	204.93	116.06	127.31
$W_{s1} = \sqrt{C_{x1}^2 + u_{s1}^2}$	260.67	232.42	231.47
	(258.47)	(228.56)	(227.90)

Table 9.3
Calculation of rotor outside diameter

	Source			
d_2 mm	Guess	200	236	232
$\dfrac{d_{s1}}{d_2}$	$d_{s1} = 120$ mm	0.6	0.5085	0.517
$\dfrac{C_{\theta 2ac}}{C_{\theta 2th}}$	$\left[1 + \dfrac{2\cos\beta_2}{Z[1-(d_{s1}/d_2)]}\right]^{-1}$	0.7970	0.8283	0.8257
$\dfrac{C_{\theta 2ac}}{u_2}$	$\left(\dfrac{C_{\theta 2ac}}{C_{\theta 2th}}\right)\left(\dfrac{C_{\theta 2th}}{u_2}\right)$	0.5977	0.6212	0.6193
$u_2 \dfrac{m}{s}$	$\left[\dfrac{\Delta h_{013}}{(C_{\theta 2ac}/u_2)}\right]^{1/2}$	372.12	365.02	365.58
d_2 mm	$\dfrac{60 u_2}{\pi N}$	236.24	231.73	232.09
$Cr_2 \dfrac{m}{s}$	$\left(\dfrac{C_{\theta 2ac}}{u_2}\right)\left(\dfrac{u_2}{\tan 60°}\right)$			130.71
$C_2 \dfrac{m}{s}$	$\dfrac{Cr_2}{\cos 60°}$			261.43
α_{re12} deg	$\tan^{-1}\left[\left(1 - \dfrac{\cos\theta_{2ac}}{u_2}\right)\dfrac{u_2}{Cr_2}\right]$			46.80
$W_2 \dfrac{m}{s}$	$\dfrac{Cr_2}{\cos\alpha_{re12}}$			190.93

$$Aa_2 = \frac{1.0}{1.628 \times 130.7} = 4.699 \times 10^{-3} \text{ m}^2 = \pi d_2 b_2;$$

$$\therefore b_2 = \frac{Aa_2}{\pi 0.2321} = 6.44 \text{ mm.}$$

Radial-diffuser stability We need the Reynolds number.

$$\text{Re}_2 \equiv \frac{C_2(d_2/2)\rho_{\text{st2}}}{\mu_{\text{st2}}}.$$

From table 2 of *Gas Tables*, or appendix A1, at a mean T_{st2} of about 320 K, 575 R,

$$\mu = 1.305 \times 10^{-5} \text{ lbm/s-ft}$$

$$= 1.942 \times 10^{-5} \text{ Ns/m}^2,$$

$$\text{Re}_2 = 2.54 \times 10^6,$$

$$b_2/r_2 = 0.0555.$$

From the stability limits in Jansen's curves (figure 4.20)

b_2/r_2	$(r_3/r_2)_{\text{max}}$
0.125	4.0
0.08	2.9
0.055	2.0

80 percent of 2.0 is 1.6 for (r_3/r_2), so $d_3 = 1.6 \times 232.1 = 371.36$ mm.

Turbine rotor diameter The turbine has slightly increased mass flow, because of fuel addition, and must supply windage and bearing power in addition to compressor power;

$$\therefore \Delta h_{0e} = \frac{1.02}{1.04} 82{,}772 \text{ J/kg}$$

$$= \frac{u_5^2}{g_c} \frac{C_{\theta 5}}{u_5}.$$

According to Eck,

$$\frac{C_{\theta 5}}{C_{\theta 5\text{th}}} = \frac{1}{1 + \{(2\cos\beta_5)/Z_e[1 - (d_{s6}/d_5)]\}}.$$

Take $d_{s6}/d_5 = 0.517$ for the compressor value of d_{s1}/d_2:

$$C_{\theta 5\text{th}} = u_5,$$

$$\beta_5 = 0°,$$

$$Z_e = 13;$$

$$\therefore \frac{C_{\theta 5\,act}}{u_5} = \frac{1}{1 + [2/13(1 - 0.517)]} = 0.7584;$$

$$\therefore u_5^2 = \frac{1.02}{1.04} \times \frac{82{,}772.4}{0.7584},$$

$$u_5 = 327.17 \text{ m/s},$$

$$d_5 = 207.7 \text{ mm}.$$

Engine back pressure We need to find the turbine pressure ratio:

$$\dot{m}_e \Delta h_{0e} = 1.02\,\dot{m}_a \Delta h_{0e}.$$

We obtain the specific heat for 100 percent fuel from the gas constant and molecular weight in table 9 of *Gas Tables* (appendix A1), which is 1,172 J/kg-°K.

$$\therefore \Delta T_{0e} = \frac{1.02 \times 1.0}{1.04} \times \frac{1{,}010}{1{,}172} \times 81.953$$

$$= 69.27°\text{K},$$

$$\frac{p_{0,\text{in}}}{p'_{0,\text{out}}} = \left[\frac{T_{0,\text{in}}}{T_{0,\text{out}}}\right]^{(Cp/R)/\eta_{ts}}$$

$$= \left[\frac{800 \text{ K}}{730.73}\right]^{(Cp/R)/\eta_{ts}}$$

$$= 1.3396$$

(where, as usual, $p'_{0,\text{out}}$ and η_{ts} are defined for the same locations so that $p'_{0,\text{out}} \equiv p_{st,\text{out}}$);

$$\therefore p_{0,\text{in}} = 1.3396 \times 1.2 \times 10^5$$
$$= 1.608 \times 10^5 \text{ N/m}^2.$$

9.4 Nozzles for radial-inflow turbines

Nozzle configurations vary widely. Small turbocharger turbines often have no nozzle vanes: the acceleration and flow direction is imparted by the "snail-shell" scroll or inlet volute. We know of no proven design methods for turbine or compressor snail-shell scrolls; many have been published, and some may be effective.

Medium-size radial-inflow gas turbines normally employ fixed nozzle vanes, of the type shown in figure 9.7 in a variable-setting form. The shape of the nozzle vanes seems to be of minor importance because of the strong flow acceleration they impart. The design problem reduces to

Radial-Flow Turbomachines

choosing the solidity (a value of 1.5 based on the spacing at the trailing edge is typical) the number of vanes (10 to 14 is again typical), and predicting the flow deviation. Fairbanks (1980) reviews previous work and recommends a streamline-curvature method, so attaching more importance to nozzle-vane shape than was assumed here. For a desired flow-outlet angle of about 60° to the radial direction he found that twelve vanes appeared to give minimum loss, and the deviation with this design was 7.0°.

Large Francis (and Kaplan) water turbines always incorporate variable-setting nozzle ("wicket" gates) vanes, for which knowledge of the deviation is obviously unimportant. The hydrodynamic design of wicket gates is somewhat compromised by the requirement that, when closed, they seal the water inflow to the maximum degree possible.

9.5 Performance characteristics for radial-flow turbomachines

In chapters 7 and 8 the fundamentals of the changes in the velocity diagrams and the associated density changes which accompanied the changes in axial-flow turbines and compressors were reviewed. Examples of performance characteristics were given.

We are not doing the same for radial-flow machines because all the arguments presented for axial-flow machines apply equally well to radial-flow machines. The characteristics of the two types of machines are almost identical. If, for instance, one finds compressor characteristics for a machine of 4.0 design-point pressure ratio, there will be no strong indication that it was taken from a ten-stage axial-flow compressor or a single-stage centrifugal, except perhaps the peak efficiency should be higher for the axial machine if the mass flow is above, say, 5 kg/s for atmospheric pressure at inlet. This similarity will occur of course only if similar design limitations were used, especially the relative-Mach-number limit at the tips of the inlet blades. In turbine comparisons radial-inflow turbines with radial blades will have characteristics similar to axial-flow turbines of about 50-percent reaction having similar nozzle-outlet Mach numbers.

9.6 Conditions for flow separation in compressor rotors

It is known that the more highly angled the blades of radial-flow compressors are to the radial direction at outlet, the smaller is the extent of flow separation in the impeller. There are some who believe that the flow is so complex, and the requirements of blade loading and changes in direction so severe, that some degree of separation is inevitable. Another point of view is that, if the amount of diffusion required of the most severely loaded streamline (that at the shroud) is limited, separation can be avoided.

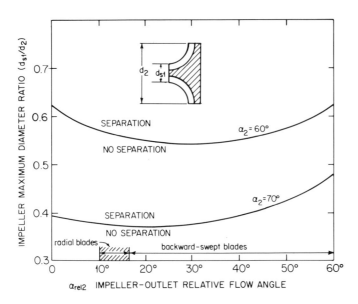

Figure 9.12 Rotor design to avoid separation.

We take this latter view. We suggest, more from intuition than from hard evidence, that a lower limit of relative outlet-to-inlet velocity ratio (W_2/W_{s1}) of 0.8 be used as a guide to the prevention of separation in rotors with subsonic inlet relative Mach numbers (W_{s1}/a_{st}) and normal Reynolds numbers. This criterion leads to the design rules shown in figure 9.12, derived as follows.

The inlet and outlet velocity diagrams, figure 9.1b and c, give the following relationships between W_2/W_{s1} and other design parameters:

$$W_{s1} = u_{s1}/\sin \alpha_{s1}, \tag{9.8}$$

$$W_2 = Cr_2/\cos \alpha_{re12}, \tag{9.9}$$

$$Cr_2 = C_{\theta 2}/\tan \alpha_2, \tag{9.10}$$

$$C_{\theta 2} = u_2 - Cr_2 \tan \alpha_{re12}; \tag{9.11}$$

$$\therefore \frac{W_2}{W_{s1}} = \frac{\sin \alpha_{s1}}{(u_{s1}/u_2)(\tan \alpha_2 + \tan \alpha_{re12}) \cos \alpha_{re12}}; \tag{9.12}$$

$$\therefore \frac{d_{s1}}{d_2} = \frac{\sin \alpha_{s1}}{(W_2/W_{s1})(\cos \alpha_{re12} \tan \alpha_2 + \sin \alpha_{re12})}. \tag{9.13}$$

Radial-Flow Turbomachines

When the inlet shroud diameter is optimized for minimum relative velocity at the shroud, W_{s1}, the relative flow angle, α_{s1}, is normally found to be close to 60°. Accordingly, figure 9.12 is plotted for $\alpha_{s1} = 60°$ and for two values of the absolute outlet flow angle, α_2, 60° and 70°. This range covers the flow angles normally used in design, as implied by the requirements for stability in the vaneless diffuser or vaneless space, figure 4.20. The abscissa is the relative outlet flow angle, α_{re12}, which is a function of, principally, the blade angle at outlet, β_2, and the number of rotor blades. Impellers that have radial blades at the periphery have relative flow angles generally between 10° and 15°, and for the higher outlet flow angle, 70°, the likelihood of separation is at a maximum. Only radial-blade compressors of very low specific speed have the diameter ratio (d_{si}/d_2) below the implied no-separation boundary of 0.38.

The implication of figure 9.12 is that there are twin benefits from decreasing the design value of absolute outlet angle α_2. First, there is a reduced likelihood of separation in the impeller. Second, a smaller flow angle into a vaneless space or vaneless diffuser gives less likelihood of flow instability in that component. A low value of outlet flow angle is given by a small value of axial blade width at the rotor periphery, b_2. Reducing b_2 incurs a disadvantage also: the clearance between blade and casing becomes relatively larger, increasing the losses associated with clearance.

References

Dean, Robert C., Jr. 1972. *The fluid dynamic design of advanced centrifugal compressors*. Report TN-153. Creare, Inc. Hanover, N.H.

Fairbanks, F. 1980. *The determination of deviation angles at exit from the nozzles of an inward-flow radial turbine*. Paper 80-GT-147. ASME, New York.

Keenan, Joseph H., and Joseph Kaye. 1948. *Gas Tables*. Wiley, New York.

Kluge, Friedrich. 1953. *Kreiselgebläse und Kreisel-verdichter radialer Bauert*. Springer-Verlag, Berlin.

Roelke, Richard J. 1973. Miscellaneous losses. In *Turbine Design and Application*. Vol. 2. Ed. by Arthur J. Glassman. Special publication SP-290. NASA, Washington, D.C., pp. 125–147.

Rohlik, Harold E. 1975. Radial-inflow turbines. In *Turbine Design and Application*. Vol. 3. Ed. by Arthur J. Glassman. Special publication SP-290. NASA, Washington, D.C., pp. 31–58.

Problems

9.1

Calculate the preliminary design dimensions—inlet hub and tip diameter, outlet diameter, and outlet blade width—of a pump suitable for a sewage-treatment plant. The specifications and design assumptions to be used follow. Sketch the meridional and transverse cross sections of the pump. Also sketch the performance characteristics, from zero flow to zero head, at constant design-point speed (for no cavitation).

Specifications

Design lift: 105 ft (flange-to-flange, total-to-static).

Flow: 10^7 gal per day (U.S.).

Design choices suggested

Specific speed, N_s: 0.10 (nondimensional).

Outlet blade angle, β_2: 60°.

Slip factor (deviation), $\tan \alpha_{re12}$: $\tan \beta_2 + (\pi \cos \beta_2 / Z \phi_{be2})$.

Number of blades, Z: 7.

Hub-tip ratio at inlet: 0.6.

Shroud diameter, d_{s1}: giving minimum relative velocity.

Inlet axial-velocity profile: $C_{xh1}/C_{xs1} = 0.75$.

Velocity at outlet flange: 10 ft/s.

Efficiency, flange-to-flange, total-to-total: 0.88.

Flow angle (absolute) at rotor exit: 65°.

9.2

Find the optimum inlet shroud diameter of a radial-flow pump having 50-degree blade angle at rotor exit, giving a relative flow angle of 55 degrees, and an absolute flow angle of 60 degrees. The nondimensional specific speed should be 0.12. The hub diameter at inlet is to be 25 percent of the rotor peripheral diameter. The axial velocity at inlet varies linearly with radius, being 50-percent greater at the shroud than at the hub. Also find the blade angle for zero incidence at the hub and shroud at inlet, and the rotor diffusion (relative-velocity ratio) along the shroud streamline.

9.3

Find the overall polytropic total-to-total efficiency, the total-to-total pressure ratio, and the ratio of the rotor-blade heights at outlet (b_2/b_1) of the two-stage centrifugal air compressor sketched. Assume that each stage has the performance characteristic shown in figure P9.3. The design point for the first stage is shown. The working point for the second stage should be chosen to give maximum efficiency for the non-dimensional speed which is calculated. The second stage will be manufactured identical to the first stage except that the blade height will be cut back further. For simplicity, assume that the fluid is a perfect gas with $\gamma = 1.4$. The inlet conditions for the first stage are 300 K, 1×10^5 N/m². In each stage, $C_{x1}/u_2 = 0.35$, $C_{\theta2}/u_2 = 0.5$, $\alpha_2 = 65°$, where 1 refers to rotor inlet and 2 to rotor outlet. Also $\eta_{p,tt,1-2}$ for each

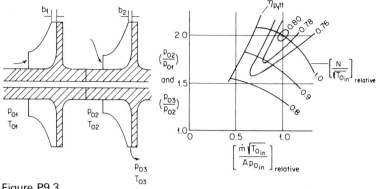

Figure P9.3

stage is 5 percentage points above the stage efficiencies given on the performance characteristic.

9.4

Find the shroud diameter at inlet for minimum relative velocity in a radial-flow water pump designed to give a static lift of 100 feet at a specific speed of 0.05 (nondimensional). Also find the zero-incidence blade angle at the shroud, the inlet hub diameter, the peripheral diameter, the rotational speed, and the rotor diffusion (W_{s1}/W_2). The following are some specifications and assumptions:

Blade angle at outlet: $60°$.

Efficiency, total-to-static, flange-to-flange: 90 percent.

Slip factor: $C_{\theta 2, ac}/C_{\theta 2, th}$ 0.88.

Hub diameter at inlet: $d_2/3$.

Inlet swirl velocity: 0.

Inlet axial velocity: proportional to (radius)$^{1/2}$.

Rotor-exit flow angle, absolute: $65°$.

Mean flow velocity at outlet flange: 15 ft/s.

Flow quantity: 500 gal/min (U.S.).

9.5

(a) Suppose that a solid particle enters the nozzle ring of a radial-inflow turbine (figure 9.6, with fixed nozzles in the position shown in broken lines in figure 9.7). The turbine is running in design conditions, with the inlet vector diagram shown at the top of figure 9.4. At nozzle exit the particle has, we suppose, been dragged along by fluid friction with the accelerating gas so that it has acquired half the gas velocity, and the same direction as the gas. Redraw the inlet velocity diagram lightly, and in heavier lines draw the absolute and relative velocities of the particle at rotor entrance.

(b) It will obviously be probable that the particle will hit—or be hit by—a rotor blade. Assume that it bounces off the rotor blade as if it were a light ray reflected off a mirror, but with only half the prebounce relative velocity. Show dotted the relative and absolute velocity vectors. What will be the likely destination of the particle?

Radial-Flow Turbomachines

9.6

The following is extracted from a typical exercise given to students, one which is recommended in general. A manufacturer's paper, report, or brochure is copied, and the task is to check the design choices from the data given. This usually includes having to make imprecise measurements from photographs or diagrams. Often there are some satisfying agreements between predictions and manufacturers' claimed performance values and some puzzling discrepancies that cannot be resolved except by assuming that a typographical or advertising error has occurred. Sometimes it becomes obvious that the machine has been sensibly optimized for part-load operation, leading to compromise with the design-point or full-load performance.

The following data were extracted from a report on an experimental automotive gas-turbine engine. The compressor data will be given here, and those for the two radial-inflow turbines in problem 9.7.

Centrifugal compressor

Number of blades: 32 (including splitter blades).

Rotational speed: 86,256 rev/min.

Total pressure ratio: 4.5 (to diffuser outlet).

Isentropic efficiency: 0.825 (to same outlet conditions).

Inlet total temperature: 288 K.

Inlet total pressure: 100 kPa (assumed).

Rotor outlet blade angle: 45° (approximate from photograph).

Vaned-diffuser inlet angle: 70° (approximate from photograph).

Hub diameter at inlet: 27.6 mm (approximate from photograph).

Shroud diameter at inlet: 61.4 mm (approximate from photograph).

Rotor outside diameter: 120.1 mm (approximate from photograph).

Rotor-outlet axial width: 6.2 mm (approximate from photograph).

Mass flow: 0.345 kg/s.

You will need to assume a polytropic efficiency for the rotor, total-to-total conditions. We found a large discrepancy in the rotor-outlet axial width and reasonable agreement elsewhere. We therefore assumed that the designer had used a very large blockage factor to allow for flow separation, blade thickness, and boundary layers. (There is a point of view which maintains that using blockage factors to increase flow areas actually promotes separation and blockage.) The inlet shroud diameter should be checked for minimum relative velocity. We assumed no inlet swirl and obtained good agreement with the manufacturer's geometry as approximately measured. The machine had inlet guide vanes which could be preset to give swirl in the inlet flow.

9.7

In table P9.7 are some data for the radial-inflow turbines driving the compressor (the "gasifier" turbine) and the output shaft (the "power" turbine) in problem 9.6. Make check calculations for the rotor outside diameters and axial widths, assuming that at design point the inlet velocity triangles are similar to the (compressor) correlations of Stodola (equation 9.4), of Eck (equation 9.8), and/or one specifically for radial-inflow turbines given by Rohlik (1975): $C_{\theta ac}/C_{\theta th} = 1 - (2/Z)$. (Note: Assume efficiencies where needed. Not all the data are required for the calculations.)

9.8

(a) Calculate the apparent polytropic turbine efficiency (flange-to-flange, total-to-total) of the hydrogen turbine driving the hydrogen booster pump of the space-shuttle main engine. Figures taken from the *Space-Shuttle Technical Manual* are given in table P9.8. The liquid hydrogen flows from the main tank to the booster pump, goes on to the main turbopump, is circulated around the rocket-nozzle coolant passages, and returns as a gas to drive first the two-stage axial turbine driving the booster pump, and then, downstream, drives the turbine for the main pump. (The turbine drives only the pump. The molecular weight of hydrogen is 2.016; the liquid H_2 specific heat is 6,240 J/kg-°K, the liquid density is 74.62 kg/m³, and the mean specific heat of the gas, Cp, is 14,822 J/kg-°K.)

(b) Also calculate the power output from the turbine blading, the ideal work of the booster pump, the pump losses, and hence the pump efficiency. Comment on the inconsistencies you find. Suggest which one property value might be in error and which, if corrected, would yield more believable results.

Table P9.7

	Gasifier turbine	Power turbine
Inlet total temperature, K	1,561	1,407
Outlet total temperature, K	1,407	1,206
Inlet total pressure, kPa	400.5	236.9
Mass flow, kg/s	0.345	0.345
Fuel-air ratio	0.124	0.124
Outlet total pressure, kPa	240	109
Power output, kW	65.77	82.5
Shaft speed, rev/min	86,256	68,156
Rotor outer diameter, mm	112.5	148.1[a]
Rotor-inlet axial width, mm	8.0	9.4[a]

a. Approximate values.

Table P9.8

	Centrifugal booster pump	Two-stage axial turbine
Mass flow, kg/s	66.77	13.789
Inlet total temperature, K	20.56	315
Inlet total pressure, kPa	206.8	29,420
Outlet total temperature, K	21.83	308.9
Outlet total pressure, kPa	1,620.96	24,318

10

Convective heat transfer in blade cooling and heat-exchanger design

A knowledge of the physics of convective heat transfer is valuable in the design of most types of turbomachinery. It is vital in the design of advanced, high-temperature gas turbines. High turbine-inlet temperatures give high thermal efficiency and high specific power, as explained in chapter 3. The turbine-inlet temperature of advanced gas turbines has been increasing at an average rate of 20°C per year, about half of the rise being through improvements in blade cooling (figure 10.1).

The strong favorable influence of high heat-exchanger thermal effectiveness on the thermal efficiency of heat-exchanger cycles was also demonstrated in chapter 3.

In each of these critical areas, and in many others that are less critical, there is more design freedom than in any other aspect of gas-turbine

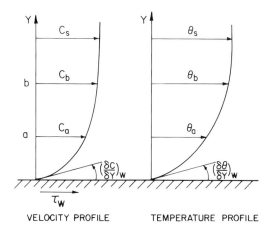

Figure 10.1 Boundary-layer velocity and temperature profiles.

design. This means that there is more scope for design skill. It also means that there are perhaps more chances of failure. For the costs of deficiencies in design in these areas are high. Inadequate blade cooling of some part of the blade profile in some conditions of operation can lead to early and disastrous failure of the entire machine. Overcooling can lead to high thermal stresses and again failure. Poor heat-exchanger design can lead also to high thermal stresses and to areas where condensation can lead to corrosion failure. The heat exchanger might also be impracticably large, or have penalizing pressure losses, or be uneconomically expensive.

In this chapter we develop some simple design methods and guidelines that can be used to arrive at a very near approach to an optimum design of critical components. Insofar as these methods are appropriate to preliminary design, they will satisfy, in particular, the requirements of heat exchangers for analysis. They will provide an understanding of the design and operation of cooled turbine blades. But the state of the art of cooled-blade design is now so sophisticated that, for purposes of preliminary design, gas temperatures would be chosen based on existing cooled-blade performances rather than being designed from first principles. As in most other areas of modern design final optimization of heat exchangers and of cooled blades is normally made using digital-computer programs. These are, however, no better than the principles on which they are based. Computer programs are all too often built upon poor foundations. Even if they are sound, they will usually arrive at a better result if the input is close to an optimum design, rather than if a somewhat random configuration were used.

We shall deal principally with problems of convection heat transfer—conduction through moving fluids. The starting point is Reynolds analogy between heat transfer and fluid friction.

10.1 Reynolds' analogy between fluid friction and heat transfer

Many of the fundamental relations correlating convective heat transfer are based on the simple statement made by Osborne Reynolds that the heat-transfer coefficient in certain classes of fluid flow is a simple multiple of the skin-friction coefficient. It can be derived by modeling the velocity and temperature profiles in an "attached" boundary layer—in other words, in an unseparated flow.

In continuum flow the molecular layer next to the wall is at rest. Outer layers either slide past in laminar flow, or exchange "packets" of high velocity fluid with low-velocity fluid displaced from the inner parts of the boundary layer. In both laminar and turbulent boundary layers, the skin-friction tangential stress at the wall, τ_w, is proportional to the angle of the

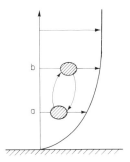

Figure 10.2 Eddy exchange in turbulent flow.

tangent to the velocity profile at the wall $(\partial C/\partial y)_w$. The viscosity, μ, is defined by this angle

$$\mu \equiv \frac{\tau_w}{(\partial C/\partial y)_w}. \tag{10.1}$$

This will be the viscosity at "static" conditions, μ_{st}, but because the Mach number in heat-transfer devices is generally small, we omit the subscripts.

In the same way all (except radiative) heat transfer between the fluid and the wall, $\dot{Q}$, must be conducted through the final layer of stationary molecules. The fluid thermal conductivity, k, is defined in terms of the temperature gradient at the wall $(\partial \theta/\partial y)_w$.

$$k \equiv \frac{\dot{Q}/A}{(\partial \theta/\partial y)_w}, \quad \text{where } \theta \equiv T - T_w. \tag{10.2}$$

There is clearly an analogy between the transfer of heat and of skin friction through the fluid layer next to the wall in laminar flow. We show in the following that in certain circumstances the analogy also holds in turbulent boundary-layer flow. (The treatment is that of Eckert 1963.) Consider a single turbulent eddy exchanging "packets" ("austaches") of fluid at a rate $\dot{m}$ between two layers a and b (figure 10.2). The momentum change will cause a tangential stress τ in the fluid:

$$\tau = \frac{\dot{m}}{g_c} \frac{(C_b - C_a)}{A}, \tag{10.3}$$

where A is the area of the wall over which the mass flow rate $\dot{m}$ occurs. The fluid packets will carry with them sensible heat. The effective heat transfer between the two layers is

$$\dot{Q} = \dot{m}Cp(\theta_b - \theta_a). \tag{10.4}$$

The ratio of heat transfer to skin friction within the turbulent boundary layer is then

$$\frac{\dot{Q}/A}{\tau} = \frac{g_c Cp(\theta_b - \theta_a)}{(C_b - C_a)}.\tag{10.5}$$

Within the laminar layer close to and at the wall this same ratio is

$$\frac{\dot{Q}/A}{\tau_w} = \frac{k}{\mu}\left(\frac{\partial \theta}{\partial C}\right)_w.\tag{10.6}$$

Obviously, these two ratios are effectively identical when $k/\mu = g_c Cp$; that is, when $g_c Cp\mu/k = 1$.

This collection of properties, $g_c Cp\mu/k$, is called the Prandtl number, Pr:

$$Pr \equiv \frac{g_c Cp\mu}{k}.$$

Gases have Prandtl numbers sufficiently close to unity for this analogy to be accepted as valid. "Close" can be 0.7 for air, for instance.

For the case of fluids with Prandtl numbers close to unity, the ratio of heat transfer to fluid tangential stress is the same throughout the laminar and the turbulent parts, if any, of an attached boundary layer. Therefore we are free to locate the planes a and b where we wish. Greatest utility is given by putting plane a at the wall interface, and plane b at the outer edge of the boundary layer. Then $C_a = \theta_a = 0$, and

$$\frac{\dot{Q}/A}{\tau_w} = \frac{g_c Cp\theta_s}{C_s},\tag{10.7}$$

where subscript s denotes the "free-stream" values. We define the heat-transfer coefficient, h_t, as

$$h_t \equiv \frac{\dot{Q}}{A\theta_s}\tag{10.8}$$

and a non-dimensional heat-transfer coefficient, Nu, as

$$Nu \equiv \frac{h_t l}{k}\tag{10.9}$$

where l is a characteristic length (for instance, the hydraulic diameter of a passage or the chord length of a turbine blade). Then

$$h_t \equiv \frac{\dot{Q}}{A\theta_s} = \frac{g_c Cp\tau_w}{C_s},$$

$$Nu \equiv \frac{h_t l}{k} = \frac{g_c Cp}{k} \frac{l\tau_w}{C_s}. \tag{10.10}$$

It is convenient to form a nondimensional skin-friction coefficient, Cf, with the free-stream incompressible dynamic head $\rho_0 C_s^2/2g_c$:

$$Cf \equiv \frac{\tau_w}{\rho_0 C_s^2/2g_c}, \tag{10.11}$$

$$Nu \equiv \frac{g_c Cp\mu}{k} \frac{lC_s\rho}{\mu} \frac{\tau_w}{\rho_0 C_s^2/2g_c} \frac{1}{2g_c} = \mathrm{Pr} \times \mathrm{Re}\frac{Cf}{2}. \tag{10.12}$$

Another nondimensional heat-transfer coefficient, the Stanton number, St, can be formed from Nu, Re, and Pr:

$$St \equiv \frac{Nu}{Re \times Pr} = \frac{h_t}{\rho Cp C_s} \tag{10.13}$$

so that

$$St = \frac{1}{2}Cf. \tag{10.14}$$

This is the formal statement of Reynolds' analogy for fluids of $\mathrm{Pr} \approx 1$. Von Karman extended the analogy to fluids with $\mathrm{Pr} = 1$ by a modification of equation (10.14); more usually, empirical correlations of Prandtl number effects are given (for example, see Eckert 1963, and Kays and London 1964).

10.2 The NTU method of heat-exchanger design

Figure 10.3 is a diagrammatic representation of the changes in temperature of the hot and cold fluids in a two-fluid heat exchanger. The diagram is drawn as if the fluids were in pure counterflow. For the purposes of the following argument, they could equally well be in crossflow or cocurrent flow, or a mixture of all three types.

The thermal performance of a heat exchanger is usually measured by its effectiveness ε_x, defined as

$$\varepsilon_x \equiv \frac{\dot{Q}}{\hat{\dot{Q}}},$$

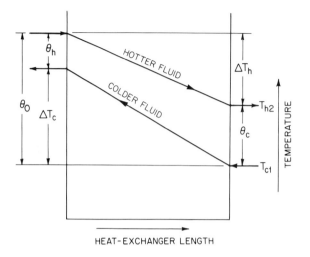

Figure 10.3 Representation of temperature distribution in a counterflow heat exchanger.

where $\dot{Q}$ is the heat actually transferred and $\hat{\dot{Q}}$ is the maximum amount of heat that could be transferred. This occurs when the outlet temperature of the fluid with the smaller heat capacity, $(\dot{m}Cp)_{\text{min}}$, reaches the inlet temperature of the other fluid:

$$\hat{\dot{Q}} = (\dot{m}Cp)_{\text{min}}\theta_0. \tag{10.16}$$

The actual heat transferred can be used to define an overall mean heat-transfer coefficient, $\bar{U}$, a mean temperature difference, θ_m, and a mean heat-transfer area, $\bar{A}_h$:

$$\dot{Q} = \overline{UA_h}\theta_m; \tag{10.17}$$

$$\therefore \varepsilon_x \equiv \frac{\dot{Q}}{\hat{\dot{Q}}} = \left[\frac{\overline{UA_h}}{(\dot{m}Cp)_{\text{min}}}\right]\left(\frac{\theta_m}{\theta_0}\right).$$

The dimensionless heat-exchanger size, $\overline{UA_h}/(\dot{m}Cp)_{\text{min}}$, is termed the "number of transfer units" (NTU):

$$\text{NTU} \equiv \frac{\overline{UA_h}}{(\dot{m}Cp)_{\text{min}}}; \tag{10.18}$$

$$\therefore \varepsilon_x = \text{NTU}\left(\frac{\theta_m}{\theta_0}\right). \tag{10.19}$$

The ratio of the mean to the overall temperature difference is a function

Convective Heat Transfer

of flow arrangement and NTU or effectiveness, and the ratio of the smaller heat-capacity rate to the larger, $C_{rat} \equiv (\dot{m}Cp)_{min}/(\dot{m}Cp)_{max}$. Kays and London, who pioneered and developed this method of heat-exchanger design for compact (gas-turbine) heat exchangers, have published charts of effectiveness versus NTU, with C_{rat} as a parameter, for each principal heat-exchanger flow arrangement (Kays and London 1964). Examples are shown in figure 10.4.

The NTU approach has resulted in an enormous simplification in heat-exchanger design over the old methods where the logarithmic mean temperature difference had to be calculated for all conditions and for variations from these conditions. Heat-exchanger design is still, however, complex. One reason that it is so is the rather strange one that the designer has many degrees of freedom. Most designers are not used to such freedom and hesitate to make what may initially be rather random choices of, for instance, flow configuration, flow surfaces, and flow velocities without reference to analytical methods. The following guidelines for heat-exchanger design give some rules for making these otherwise arbitrary choices.

10.3 Guidelines for choice of heat-exchanger passages

For low pumping-power loss in a single-phase heat exchanger transferring a given flow of heat, the fluid velocities should be low. For the overall size of the heat exchanger to be small, the hydraulic diameter of the passages should be small. The development of these guidelines is given here.

They are guidelines rather than rules because frequently other considerations, such as packaging or header (connecting duct) size, become dominant.

Analysis
The basis for the argument is Reynolds' analogy between heat transfer and skin friction (equation 10.14):

$$St = \frac{1}{2}(Cf).$$

Here $St \equiv$ Stanton number $\equiv \dfrac{Nu}{Re \times Pr} \equiv \dfrac{h_t}{\rho CCp}$,

$h_t \equiv$ heat-transfer coefficient $= \dot{Q}/A_h\theta$,

$\dot{Q} \equiv$ heat flow,

$A_h \equiv$ heat-transfer area,

$\theta \equiv$ temperature difference driving $\dot{Q}$

$\rho \equiv$ fluid density,

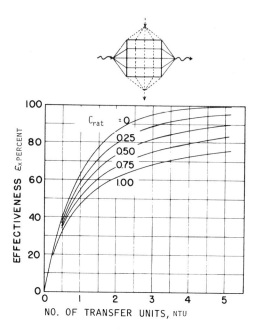

Figure 10.4 (a)

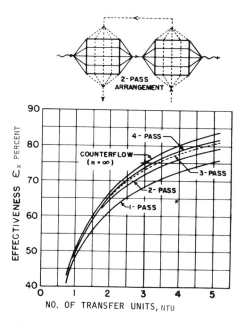

Figure 10.4 (b)

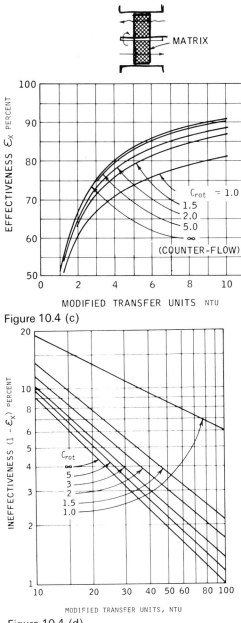

Figure 10.4 (c)

Figure 10.4 (d)

Figure 10.4 Examples of charts of effectiveness versus NTU. From Kays and London, 1964. (a) Crossflow heat exchanger with unmixed fluids; (b) multipass cross-counterflow heat exchanger; (c) periodic-flow heat exchanger, $C_{rat} = 1.0$; (d) high-effectiveness periodic-flow heat exchanger, $C_{rat} = 1.0$.

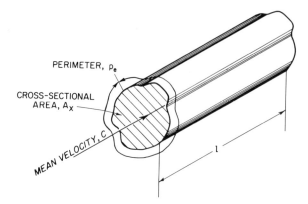

PERIMETER, p_e

CROSS-SECTIONAL AREA, A_x

MEAN VELOCITY, C

l

Figure 10.5 Heat-exchanger channel of arbitrary, but constant, cross section.

$$C \equiv \text{mean fluid velocity,}$$
$$Cp \equiv \text{specific heat at constant pressure,}$$

$$Cf \equiv \text{skin friction coefficient} \equiv \frac{\tau_w}{\rho C^2/2g_c},$$

$$\tau_w \equiv \text{wall skin friction,}$$
$$g_c \equiv \text{constant in Newton's law.}$$

Consider heat transfer from a fluid to the wall of a tube of arbitrary (constant) cross section through which it is flowing (figure 10.5):

$$\dot{Q} = A_h \theta h_t$$

$$= Z p_e l \theta h_t, \tag{10.20}$$

where $Z \equiv$ number of parallel, identical tubes,

$l \equiv$ length of each tube,

$p_e \equiv$ perimeter of tube (internal) cross section.

The pumping power $\dot{W}$ required of an ideal external pump or compressor to overcome the pressure drop is

$$\dot{W} = \dot{V} \Delta p$$

$$= Z A_x C \Delta p, \tag{10.21}$$

where $\dot{V} \equiv$ volume flow,

$\Delta p \equiv$ pressure drop across tube length,

$A_x \equiv$ tube cross-section area.

By a force balance

$$\Delta p A_x = \tau_w p_e l = \tau_w A_h;$$

$$\therefore \dot{W} = ZC\tau_w A_h;$$

$$\therefore \frac{\dot{Q}}{\dot{W}} = \frac{Zlp_e\theta h_t}{ZC\tau_w lp_e} = \left[\frac{h_t\theta}{C\tau_w}\right]. \tag{10.22}$$

Introducing Reynolds' analogy

$$\frac{h_t}{\rho C C p} = \frac{g_c \tau_w}{\rho C^2};$$

$$\therefore \frac{h_t}{\tau_w} = \frac{g_c C p}{C};$$

$$\therefore \frac{\dot{Q}}{\dot{W}} = \frac{g_c C p \theta}{C^2}. \tag{10.23}$$

Therefore for given Cp and θ and for any length, tube size, shape, or number of tubes, the mean fluid velocity should be reduced to increase the heat-transfer to pumping-power ratio.

If it is a compressible fluid, $a^2 = g_c \gamma R T$ and $M^2 = C^2/a^2$;

$$\therefore \frac{\dot{Q}}{\dot{W}} = \frac{g_c R T}{R T} \frac{C p \theta}{C^2} = \frac{(\theta/T)}{(\gamma - 1)M^2} = \frac{\theta'}{(\gamma - 1)M^2}, \tag{10.24}$$

where $\theta' \equiv \theta/T$.

For a compressible fluid therefore, the Mach number should be low to increase the ratio of heat transfer to pumping power.

Volume for heat transfer

Again consider heat exchange via a number, Z, of parallel identical tubes of arbitrary constant cross section. How should one choose the length and hydraulic diameter for minimum volume, V? This case is one for a given fluid mass flow, $\dot{m}$, temperature difference (driving $\dot{Q}$), θ, and fluid specific heat, Cp, in turbulent flow:

$$\frac{\dot{Q}}{V} = \left[\frac{h_t \theta Z p_e l}{Z A_x l}\right] = \frac{4h_t \theta}{d_h} \tag{10.25}$$

since $d_h \equiv 4A_x/p_e$. For turbulent flow in tubes the Stanton number is almost constant with Reynolds' number:

$$St \equiv \frac{h_t}{\rho CCp};$$

$$\therefore h_t = \rho CCpSt;$$

$$\therefore \left(\frac{\dot{Q}}{V}\right)_{turb} = \frac{\rho CCpSt\theta Zp_e}{ZA_x} = \frac{4\rho CCp\theta St}{d_h}.$$

The total mass flow, $\dot{m} = ZA_x\rho C$;

$$\therefore \left(\frac{\dot{Q}}{V}\right)_{turb} = \frac{\dot{m}CpSt\theta}{Z(A_x^2/p_e)}.$$

The cross-sectional area, A_x, is proportional to the square of the hydraulic diameter, d_h^2.
The perimeter p_e is proportional to the hydraulic diameter, d_h:

$$\therefore \left(\frac{\dot{Q}}{V}\right)_{turb} \propto \frac{\dot{m}StCp\theta}{Zd_h^3}. \tag{10.26}$$

The numerator is constant, or nearly so, for any case where a given tube shape is being compared at different sizes (d_h), lengths (l), and so forth. The denominator is strongly affected by hydraulic diameter, d_h, and to a lesser extent by the number of tubes, Z.

For a minimum-volume heat exchanger in turbulent flow, d_h should be minimized; Z should also be small. If a minimum value of mean through-flow velocity, C, has been chosen, ZA_x or d_h^2Z will be constant. Then the volume will be directly proportional to the hydraulic diameter, d_h, other things being equal.

In laminar flow, the Nusselt number is constant:

$$Nu \equiv \frac{h_t d_h}{k},$$

where $k \equiv$ fluid thermal conductivity. Then

$$\left(\frac{\dot{Q}}{V}\right)_{lam} \propto \frac{Nuk\theta}{d_h^2} \tag{10.27}$$

by the same procedure.

Again the hydraulic diameter, d_h, should be minimized. In laminar fow neither the number of tubes, Z, nor the throughflow velocity, C, influence the heat-exchanger volume.[1]

---

1. These guidelines are useful for unseparated flow in tubes and channels, and have somewhat lower validity for situations where there is separated flow, such as in flow across circular tubes.

Convective Heat Transfer

10.4 Guidelines for heat-exchanger design

With the NTU method of approach and data on the performance of various types of heat-exchanger surfaces, it is relatively easy to design a heat-exchanger to fulfill a required thermal performance. The checklist in this section is meant as a guide to the design method. The examples given later will illustrate some of the approaches. There are so many possible configurations of heat exchanger that it would not be practicable to illustrate the design of all of them. The adaptation of the approaches to other configurations is, however, usually obvious. Each configuration has its own set of independent and dependent variables. For instance, a counterflow heat exchanger has the same length for the hot flow and the cold flow. The flow areas must be in the ratio of the two surfaces chosen. The crossflow regenerator also has equal flow lengths, but one can choose any ratio desired for the flow areas.

To arrive at an "optimum" design (something which needs very careful definition) is not easy. There are too many nonlinearities for it to be possible to use a good analytical method. There are also many degrees of freedom available in heat-exchanger design, and often some parameter may be varied to advantage.

Optimization therefore becomes essentially a matter of designing a large number of heat exchangers and choosing what seems to be the best. Unless one has some guiding rules as to what is good, one can find oneself tackling thousands of different designs (especially with a computer program) for even a simple case. The guidelines given previously regarding the desirability of choosing low Mach numbers or velocities and small hydraulic diameters should assist in making the initial choices.

The example method given here is for a multipass crossflow configuration. Modifying it to apply to other configurations requires merely care and common sense.

Design steps for a multipass crossflow heat exchanger

Assume that the problem is to design a heat exchanger to give required outlet temperatures (obviously these must be compatible with the First Law) of each of two fluids, whose mass flows, inlet temperatures, and other properties are known. Usually there is an additional requirement: that the pressure drops shall not exceed some value. It is in general not possible to specify precisely the pressure drop required in each fluid: to do so introduces an unnatural constraint which could result in the production of sometimes strange designs. For each design, or design trial, the following process will be necessary.

1. Given the fluid temperatures, find the thermal effectiveness, ε_x (equation 10.16a, b).

Three-stage centrifugal compressor with integral intercoolers. Courtesy Sulzer Brothers Limited.

2. Specify the flow rates and properties at inlet of each fluid.

3. Calculate $(\dot{m}Cp)_{\max}$, $(\dot{m}Cp)_{\min}$, and $C_{\mathrm{rat}} \equiv [(\dot{m}Cp)_{\min}/(\dot{m}Cp)_{\max}]$.

4. Choose a configuration (figure 10.6) that seems in advance, from philosophy or experience or plain guesswork, to be favorable—such as counterflow, single-pass crossflow, or, as in this example, five-pass crossflow.

5. Choose compatible surfaces for heat transfer for each fluid—(such as 1/4-inch tubes with 0.025-inch walls, or certain configurations of finned plates); these surfaces should be chosen from among those for which heat-transfer and fluid-friction data exist (Kays and London 1964), because it is almost impossible to generate the data analytically. Choose (low) mean flow velocities and find cross-sectional flow areas.

6. Find the NTU required to give the specified effectiveness using the curves for the appropriate heat-exchanger arrangement and $(\dot{m}Cp)_{\min}/(\dot{m}Cp)_{\max}$ from curves such as figure 10.4.

7. From the NTU derive the overall transfer coefficient $\overline{UA}_{\mathrm{h}}$ $\overline{UA}_{\mathrm{h}} = \mathrm{NTU}(\overline{\dot{m}Cp})_{\min}$.

8. Choose the ratio of the conductances or resistances:

$$\frac{1}{\overline{UA}_{\mathrm{h}}} = \frac{1}{h_{\mathrm{tc}}A_{\mathrm{hc}}} + \frac{t_{\mathrm{w}}}{k_{\mathrm{w}}A_{\mathrm{hw}}} + \frac{1}{h_{\mathrm{th}}A_{\mathrm{hh}}}, \tag{10.28}$$

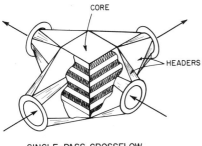

CORE

HEADERS

SINGLE-PASS CROSSFLOW
PLATE FIN

Figure 10.6 (a)

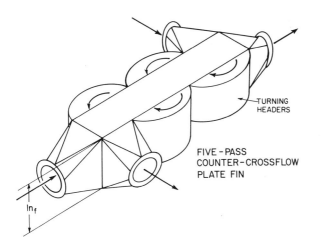

TURNING
HEADERS

FIVE-PASS
COUNTER-CROSSFLOW
PLATE FIN

l_{nf}

Figure 10.6 (b)

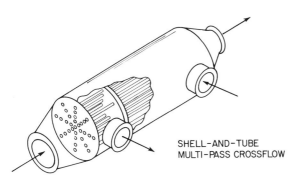

SHELL-AND-TUBE
MULTI-PASS CROSSFLOW

Figure 10.6 (c)

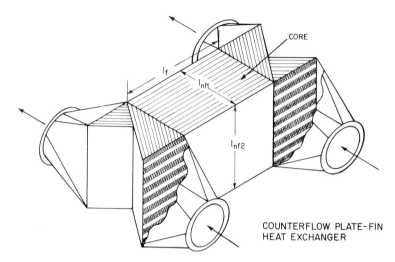

CORE

l_f

l_{nf1}

l_{nf2}

COUNTERFLOW PLATE-FIN
HEAT EXCHANGER

Figure 10.6 (d)

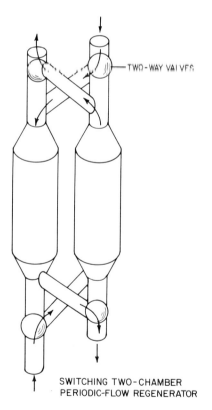

TWO-WAY VALVES

SWITCHING TWO-CHAMBER
PERIODIC-FLOW REGENERATOR

Figure 10.6 (e)

Convective Heat Transfer

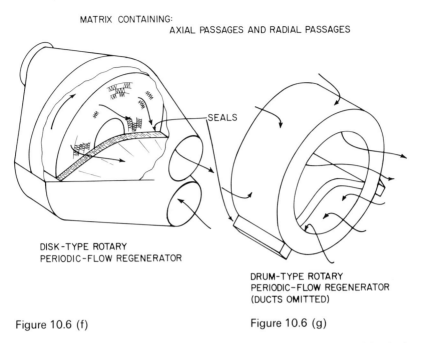

MATRIX CONTAINING:
AXIAL PASSAGES AND RADIAL PASSAGES

SEALS

DISK-TYPE ROTARY
PERIODIC-FLOW REGENERATOR

DRUM-TYPE ROTARY
PERIODIC-FLOW REGENERATOR
(DUCTS OMITTED)

Figure 10.6 (f)   Figure 10.6 (g)

Figure 10.6 Alternative two-fluid-heat-exchanger configurations: (a) single-pass crossflow plate fin; (b) five-pass counter-crossflow plate fin; (c) shell-and-tube multipass crossflow; (d) counterflow plate fin; (e) switching two-chamber periodic-flow regenerator; (f) disk-type rotary periodic-flow regenerator; (g) drum-type rotary periodic-flow regenerator (ducts omitted).

where $t_w \equiv$ wall thickness. Good single-phase heat exchangers usually have about equal film conductances, and the wall resistance is usually negligible. Therefore a good initial choice is to make $h_{tc}A_{hc} = h_{th}A_{hh}$. (For other cases, for instance, surfaces with long fins, refer to Kays and London 1964.)

9. Choose a middle-of-the-range Reynolds number for one of the two surfaces from the appropriate curves such as in figure 10.7, and find values of the Colburn modulus, $j \equiv \mathrm{St} \times \mathrm{Pr}^{2/3}$, and friction factor, Cf. For a first approximation, use the mean temperature and pressure for each fluid to find the mean velocities.

10. From the estimated Colburn modulus so found find the heat-transfer coefficient, h_t, and the associated heat-transfer area, A_h, for this surface and fluid (hot side or cold side).

11. From the geometry of the two chosen heat-transfer surfaces (A_{hc}/A_{hh}) is known, so that the two values of h_t and A_h can be found. (This restriction applies to fixed-surface but not periodic-flow heat exchangers.)

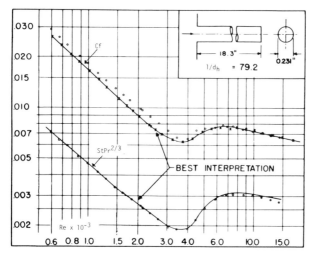

Tube inside diameter = 0.231 in.

Hydraulic diameter = 0.231 in., 0.01925 ft

Flow length/hydraulic diameter, l/d_h = 79.2

Free-flow area per tube = 0.0002908 ft^2

Figure 10.7 (a)

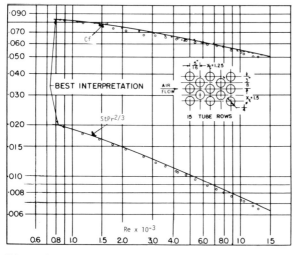

Tube outside diameter = 0.250 in.

Hydraulic diameter, d_h = 0.0166 ft

Free-flow area/frontal area, = 0.333

Heat transfer area/total volume, = 80.3 ft^2/ft^3

Note: Minimum free-flow area is in spaces transverse to flow.

Figure 10.7 (b)

Convective Heat Transfer

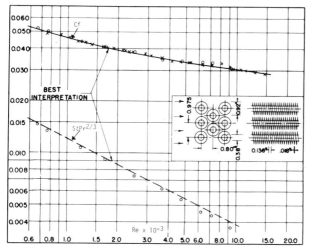

Tube outside diameter = 0.38 in.

Fin pitch = 7.34 per in.

Flow passage hydraulic diameter, d_h = 0.0154 ft

Fin thickness (average)* = 0.018 in., aluminum

Free-flow area/frontal area, = 0.538

Heat transfer area/total volume, = 140 ft^2/ft^3

Fin area/total area = 0.892

Note: Experimental uncertainty for heat transfer results possibly
somewhat greater than the nominal ±5% quoted for the other
surfaces because of the necessity of estimating a contact
resistance in the bi-metal tubes.

* Fins slightly tapered.

Figure 10.7 (c)

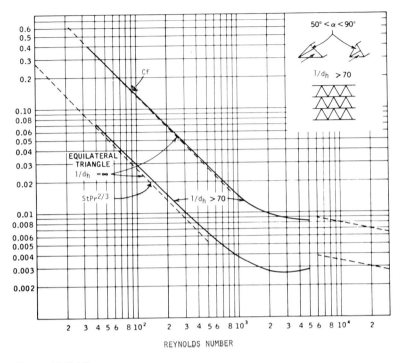

Figure 10.7 (d)

Figure 10.7 Heat-transfer and skin-friction data for surfaces. From Kays and London 1964. (a) Flow inside circular tubes; (b) flow normal to a bank of staggered tubes; (c) flow normal to a bank of staggered finned tubes; (d) flow along a matrix of triangular passages.

12. Having chosen one Reynolds number, the other can be found. Then both flow velocities can be calculated.

13. Using the calculated mean flow velocities and fluid properties, the total cross-sectional passage area, A_{ff}, and the associated "core face" areas, A_f, can be found.

14. These values and A_{hc}, A_{hh} enable the flow lengths l_{fc}, l_{fh} and the "nonflow" length, l_{nf} to be found. The configuration and dimensions of the heat exchanger are known. (In a crossflow heat exchanger the nonflow length is common to both sides.)[2]

---

2. A rectangular-solid crossflow two-fluid heat exchanger has two dimensions called "flow lengths," because they are in the directions of flow of the two fluids, and an orthogonal dimension called the "nonflow length," l_{nf}. A counterflow or cocurrent heat exchanger has one flow length common to both fluids and, for a rectangular-solid core, two nonflow lengths. See figure 10.6.

Convective Heat Transfer

15. The friction factor for each side can be found from the Reynolds number. The "core" pressure drops can be calculated. To these should be added at least the inlet and outlet losses. A simple assumption is that these are 1.5 dynamic heads.

16. The resulting heat exchanger should produce the desired thermal performance. Often, however, it is unsatisfactory for various reasons, such as the shape (it may be very long and thin, for instance). There are obviously many places where new choices can be made. With a little practice one can learn to make changes to move the results in the direction desired, although first attempts are sometimes surprising.

This simplified approach ignores many complexities which would be accounted for in a full detailed design. The large change of gas properties with temperature in a gas-turbine-engine heat exchanger makes it desirable to divide the heat exchanger into several sections rather than using overall mean properties. Axial conduction and temperature drops in fins might be important. Entrance and exit losses can be quantified, as can flow conditions in headers. All these are treated in more detail in Kays and London (1964). An alternative approach to header design is given in Wilson (1966).

10.5 Heat-exchanger design constraints for different configurations

The foregoing guidelines for heat-exchanger design have been written for a (multipass) crossflow configuration. They have also emphasized gas-to-gas or liquid-to-liquid heat exchangers where equality of conductances $(h_t A_h)$ is a reasonable first choice. The design process is different for other configurations. The design sequence and the constraints for all the principal two-fluid heat-exchanger arrangements, including, for completeness, the crossflow, are given in this section. In preliminary design the wall resistance, $(t_w/k_w A_{hw})$ in equation (10.28), is treated as being negligibly small.

Regenerator design

There are two general regenerator types: the rotating matrix (and its equivalent the rotating duct on a stationary matrix) and the two-chamber switching regenerator (hot stoves for blast furnaces, for example) figure 10.6. In the two-chamber regenerators the chambers are designed to be identical, so that not only the matrix, the hydraulic diameter, and the flow length are the same for both fluids, but also the face area, A_f, the free-flow area, A_{ff}, the heat-transfer area, A_h and the volume, V, are the same for both fluids. Given all these, one's freedom of choice is confined to the Reynolds number or velocity of one side or the other; the other Reynolds

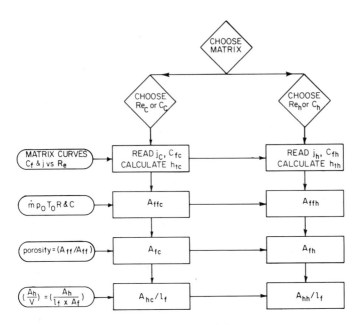

Figure 10.8 Rotating-regenerator calculation sequence.

number and velocity will be given directly from the now-known free-flow area and mass flow.

We shall write here the design progression (figure 10.8) for the more usual rotary regenerator, in which the area constraints do not apply, so that one has freedom to choose the Reynolds numbers of both sides. The flow length, l_f, is common to both sides and is unknown. From ε_x effectiveness and the rotary-regenerator NTU graph (figure 10.4c) we find

$$\overline{UA} = \text{NTU}(\dot{m}\overline{Cp})_{\text{min}}. \tag{10.29}$$

usually choosing $C_{\text{rot}}[(\dot{m}\overline{Cp})_{\text{mat}}/(\dot{m}\overline{Cp})_{\text{min}}] = 5.0$. (10.30) (This value is a compromise; choosing a higher C_{rot} increases both the effectiveness and the "carry-over" leakage.)

$$\overline{UA}^{-1} = (h_{\text{tc}}A_{\text{hc}})^{-1} + (h_{\text{th}}A_{\text{hh}})^{-1}. \tag{10.31}$$

(Alternatively, one could choose only one Re or C, calculate h_t and A_h/l as shown, set $h_{\text{tc}}A_{\text{hc}} = h_{\text{th}}A_{\text{hh}}$, and so work "backward" in the same sequence to find the other C and Re.) So we obtain l_f, A_{hc}, A_{hh}, and Δp_c and Δp_h. A_{fc} and A_{fh} determine the seal positions. The outside diameter can be chosen after independent choices are made for the hub diameter and the width of the seals.

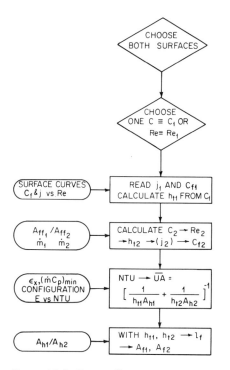

Figure 10.9 Counterflow- or cocurrent-recuperator calculation sequence.

Counterflow or cocurrent recuperators

In these the flow length l_f is the same for both fluids, as is the transverse nonflow length l_{nf1} (figure 10.6). The other nonflow length l_{nf2} consists of the stacked matrices in the case of plate-fin recuperators and of the combined matrix in the case of, for instance, shell-and-tube recuperators.

Therefore, once the matrices have been selected, the only remaining degree of freedom is the Reynolds number or velocity of one of the two fluids (figure 10.9). The ratio of the two free-flow areas will then determine the other Reynolds number and velocity. How A_{fc} and A_{fh} is configured is open for choice. They are here defined by the following:

$$(A_{fc} + A_{fh}) = l_{nf1} \times l_{nf2}. \tag{10.32}$$

(These are the two nonflow lengths.) Usually, counterflow or cocurrent heat exchangers incorporate crossflow sections at the end to bring the two fluids out, as shown in figure 10.6.

Crossflow recuperators

Figure 10.10 gives a summary of the earlier notes on heat-exchanger design.

Convective Heat Transfer

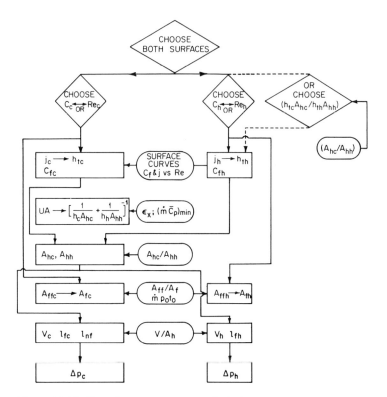

Figure 10.10 Crossflow-recuperator calculation sequence.

Example (design of rotary regenerator)

Make a preliminary design of a rotary regenerator for a solar-heated Brayton cycle. This will use pure air on both sides of the heat exchanger and will have a low pressure ratio. The cycle conditions are as follows.

$\dot{m} = 1.0$ kg/s.

$T_{01} = 300$ K (compressor-inlet total temperature).

$p_{01} = 100$ kN/m² (compressor-inlet total pressure).

$\eta_{\text{pcts}12} = 0.82$ (compressor total-to-static polytropic efficiency).

$p_{\text{st}2} = 200$ kN/m² (compressor-outlet static pressure).

$T_{04} = 1{,}400$ K (turbine-inlet temperature).

$p_{\text{st}6} = 100$ kN/m² (heat-exchanger outlet static pressure).

$\eta_{\text{pcts}45} = 0.90$ (expander total-to-static polytropic efficiency).

$\sum(\Delta p_0/p_0) = 0.08$ (sum of total-pressure drops in cycle, max).

$\varepsilon_x = 0.95$ (heat-exchanger effectiveness).

We will allow for a 2 percent leakage rate, which for simplicity we will assume

Convective Heat Transfer

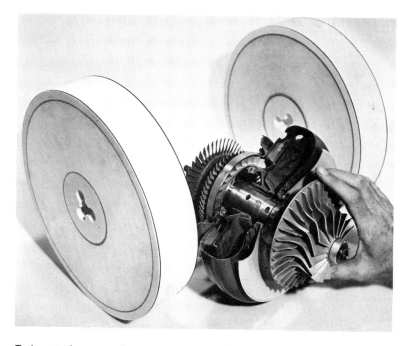

Twin rotating ceramic regenerators on "radial-axial" automotive gas-turbine engine. Courtesy Williams Research Corporation.

occurs between the compressor outlet and the heat-exchanger inlet. We will calculate the additional carry-over leakage in the heat-exchanger pores. We will calculate the pressure drops in the cores and accept the design if there is sufficient allowance within 8 percent of the total cycle pressure drops for the pressure drops in the headers and in the air heater. Other design choices follow.

Discussion and calculations

Number of transfer units. We choose the nearest appropriate curves from figure 10.4, for a periodic-flow heat exchanger of $(1 - \varepsilon_x)$ at, for this preliminary design, a heat-capacity-rate ratio C_{rat} of 1.0. We will choose a switching rate (C_{rotat}) of 5.0, which gives close to the performance of infinite switching rate with, usually, an acceptable carry-over loss. The NTU required is then approximately 22.

Heat-transfer conductances. The conductance ratio is open to our choice; we choose, initially at least, equal hot- and cold-side conductances:

$$h_{\text{tc}} A_{\text{hc}} = h_{\text{th}} A_{\text{hh}},$$

$$\frac{1}{UA_{\text{h}}} = \frac{1}{h_{\text{tc}} A_{\text{hc}}} + \frac{1}{h_{\text{th}} A_{\text{hh}}} = \frac{2}{h_{\text{tc}} A_{\text{hc}}},$$

$$\text{NTU} = \frac{UA_{\text{h}}}{(\dot{m} C_{\text{p}})_{\text{min}}} = 22 = \frac{h_{\text{tc}} A_{\text{hc}}}{2(\dot{m} C_{\text{p}})_{\text{min}}}.$$

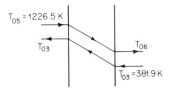

Figure 10.11 Diagram of mean temperatures through regenerator.

The mass flow allowing for leakage is 0.98 kg/s. We shall use a single mean value for the mean specific heat of the air on the cold and on the hot side. In a more detailed later calculation we could split the heat exchanger into several sections and use mean properties for each section.

Mean properties

$$\frac{T_{02}}{T_{01}} = \left[\frac{p'_{02}}{p_{01}}\right]^{[(R/\bar{C}_p)(1/\eta_{pc})]} = 2.0\left(\frac{286.96}{1{,}010 \times 0.82}\right).$$

The value of $\overline{C_p}$ can be iterated. We find $T_{02} = 381.9$ K (see figure 10.11). For the expansion,

$$\frac{T_{04}}{T_{05}} = \left[\frac{p'_{04}}{p_{05}}\right]^{[(R/\bar{C}_p)\eta_{pc}]}.$$

We include cycle pressure losses of 0.08 and again iterate on $\overline{C_p}$:

$$\frac{T_{04}}{T_{05}} = [0.92 \times 2.0]^{[(286.96/1{,}190) \times 0.90]},$$

$$T_{05} = 1{,}226.5 \text{ K}.$$

The cold side has the smaller thermal capacity: $\dot{m}_c\overline{C}_{pc} < \dot{m}_h\overline{C}_{ph}$, so that, from the definition of effectiveness, $T_{03} - T_{02} = 0.95(T_{05} - T_{02})$; therefore $T_{03} = 1{,}184.3$ K.
The mean temperatures and specific heats of each side are then

$$T_{mc} = 783.1 \text{ K}, \quad \overline{C}_{pc} = 1{,}094.3 \text{ J/kg-}^\circ\text{K},$$

$$T_{mh} = 828.8 \text{ K}, \quad \overline{C}_{ph} = 1{,}105.0 \text{ J/kg-}^\circ\text{K}.$$

The velocities in the heat exchanger will be so low that we can use total properties for the densities. We do not yet know if the pressure drops will be close to the specified maximum values for the cycle, but the following assumptions can lead to only small errors:

$$\rho_{0mc} = \frac{196{,}000}{286.96 \times 783.1} = 0.8722 \text{ kg/m}^3,$$

$$\rho_{0mh} = \frac{102{,}000}{286.96 \times 829.1} = 0.4287 \text{ kg/m}^3.$$

The heat-transfer conductances are

$$h_{tc}A_{hc} = h_{th}A_{hh} = 2NTU(\dot{m}_c\bar{C}_{pc})$$

$$= 0.44. \quad (0.98 \times 1{,}094.3)$$

$$= 47{,}186.2 \text{ W/}°\text{K.}$$

Heat-transfer coefficients. We know that the very small hydraulic diameters available for regenerator matrices will ensure laminar flows at all practicable velocities. The laminar heat-transfer relation for the matrix surface chosen is

$$St(Pr)^{2/3} = 3.0/Re \quad \text{or} \quad Nu = 3.0Pr^{1/3}.$$

The Prandtl number of air at the mean temperatures of the hot and cold sides is 0.65; $(0.65)^{2/3} = 0.75$;

$$\therefore St = \frac{h_t}{\rho C \bar{C}_p} = \frac{3.0\mu}{0.75\rho Cd_h},$$

$$h_t = \frac{4.0\mu\bar{C}_p}{d_h}.$$

Matrix surface. We choose a glass-ceramic heat-exchanger surface, Cercor L689, having the following specifications, given by London et al. (1970):

Passage packing: 1,008 passages per in^2.

Porosity: 0.708.

Hydraulic diameter, d_h: 0.001675 ft $\equiv$ 0.5105 mm.

A_h/V: 1,692 ft^2/ft^3 $\equiv$ 5,551 m^2/m^3.

Laminar-flow performance: $St(Pr)^{2/3} = 3.0/Re$, $Cf = 14.0/Re$.

Ceramic specific heat at 500 F (533 K): 0.25 Btu/lbm-°F $= 1{,}039.5$ J/kg-°K.

Ceramic specific heat at 1,500 F (1,089 K): 0.30 Btu/lbm-°F $= 1{,}247.4$ J/kg-°K.

Ceramic solid density: 141 lbm/ft^3 $= 2{,}258.8$ kg/m^3.

Heat-transfer coefficients and areas

$$h_{tc} = \frac{4.0 \times 3.616 \cdot 10^{-5} \times 1{,}094.3}{0.5105 \cdot 10^{-3}} = 310.0 \text{ W/m}^2\text{-}°\text{K},$$

$$A_{hc} = 47{,}186.2/310.0 = 152.21 \text{ m}^2,$$

$$h_{th} = \frac{4.0 \times 3.75 \cdot 10^{-5} \times 1{,}105.0}{0.5105 \cdot 10^{-3}} = 324.68 \text{ W/m}^2\text{-}°\text{K},$$

$$A_{hh} = \frac{47{,}186.2}{324.68} = 145.33 \text{ m}^2.$$

Face areas and passage velocities. In a rotary regenerator the hot and cold sides have the same flow lengths. Therefore the face areas are in the same proportion as the heat-transfer areas. We are free to choose one velocity, for instance, C_{mc}, and therefrom to calculate the face areas and pressure drops:

We choose $C_{mc} = 3.0$ m/s.

Free face area, $A_{ffc} = \dot{m}/(C_{mc}\rho_{mc}) = 0.98/(3.0 \times 0.8722) = 0.3745$ m^2.

Face area, $A_{fc} = 0.3745/0.708 = 0.529$ m^2.

Face area, $A_{fh} = A_{fc}(A_{hh}/A_{hc}) = 0.529(145.33/152.21) = 0.505$ m^2.

Face area total, $\sum A_f \equiv A_{fc} + A_{fh} = 1.034$ m^2.

We will specify that the proportion of the total "active" disk occupied by seals is 20 percent. Then the annulus area is $Aa = 1.034/0.8 = 1.2925$ m^2. We will also specify that the hub-tip ratio of the regenerator is 0.5. The outside diameter of the active disk is calculated to be

$$\frac{\pi d^2}{4}(1 - 0.5^2) = 1.2925;$$

$\therefore d = 1.481$ m.

Matrix thickness, t. Cold-side volume $= 152.21/5551 = 0.02742$ m$^3 = A_{fc} \times t$

$$t = 51.83 \text{ mm}.$$

Core pressure drops

$\Delta p_0 A_{ff} = \tau_w A_h;$

$$\therefore \Delta p_0 = \frac{A_h}{A_{ff}}\left(\frac{\tau_w}{q}\right)q,$$

where $(\tau_w/q) \equiv C_f = 14.0/\text{Re}$, and $q \equiv \rho U^2/2g_c$,

$$\Delta p_{0c} = \frac{152.21}{0.3745} \times \frac{14.0 \times 3.616.10^{-5} \times 3}{0.5105.10^{-3} \times 2} = 604.6 \text{ N/m}^2,$$

$(\Delta p_{0c}/p_{0c}) = 604.6/200,000 = 0.0030,$

$\quad\quad C_{mh} = \dot{m}/(\rho_{mh} A_{ffh}) = 0.98/(0.4287 \times 0.505 \times 0.708) = 6.39$ m/s,

$$\Delta p_{0h} = 604.6 \times \frac{3.75}{3.616} \times \frac{6.39}{3.0} = 1.336 \text{ kN/m}^2,$$

$(\Delta p_{0h}/p_{0h}) = 1.336/100 = 0.0134.$

Although pressure-drop components due to acceleration, passage entry, and passage exit must be added to those core pressure drops, the totals seem likely to remain within the specified limits.

Carry-over leakage and rotational speed

$$C_{rot} \equiv \frac{(\dot{m}\bar{C}_p)_{mat}}{(\dot{m}\bar{C}_p)_{min}} = 5.0 \quad \text{(by earlier choice, see equation 10.30).}$$

(The mean ceramic temperature for equal hot-side and cold-side conductances will be halfway between T_{mc} and T_{mh}, $= 806.1$ K.)

Mean ceramic specific heat at this temperature $= 1,149.3$ J/kg-$^\circ$K.

Ceramic mass-flow rate,

$$\dot{m}_{cer} = \frac{5.0 \times 0.98 \times 1,094.3}{1,149.3} = \frac{4.665 \text{ kg}}{s}.$$

Ceramic volume flow rate $= 4,665/2,258.8 = 0.00207$ m^3/s.

Porosity $= 0.708 = \dfrac{\text{voids}}{\text{voids} + \text{ceramic}}$.

(Therefore voids-volume flow rate $= 2.425 \times 0.00207 = 0.00501 \text{ m}^3/\text{s}$.)

Mean cold-air density $= 0.8722 \text{ kg/m}^3$.

Carry-over loss $= 0.00437 \text{ kg/s} = 0.44$ percent of compressor mass flow.

Matrix mass $= (\pi/4)(1.481)^2(1 - 0.5^2)0.05183(1 - 0.708)2{,}258.8$
$\qquad\qquad = 44.17 \text{ kg}$.

(For the mass of the full heat-exchanger, the mass of the hub, rim, seals, drive, casing and headers would be added.)

Rotation rate $= 44.17/4.665 = 9.47 \text{ s}$ (time for one rotation).

Conclusions This design seems viable as a first step. It might be desirable to recalculate with different (rather than equal) hot and cold conductance by giving the hot side a larger face area. There are many other changes that could be made in the design choices, the most significant being the flow velocity (which, in laminar flow, affects the pressure drop but not the heat transfer) and the type of matrix surface. A more accurate design could follow, in which the disk would, in concept, be divided into many thin disks in each of which uniform properties could be used.

10.6 Turbine-blade cooling

The contribution of turbine-blade cooling to the increase in cycle thermo-dynamic efficiency of gas turbines was described in chapter 1 and illus-trated in figure 1.10. In this chapter we shall limit the discussion to the fundamentals of the heat-transfer phenomena, to the design problems involved in blade cooling, and to some alternative approaches to cooled-blade design.

Heat-transfer phenomena—external
The heat transfer that occurs as a result of a hot gas stream flowing at high velocity past and through a row of cooler turbine blades is dominated and controlled by the boundary layers. In general, turbines have accelerating flow, which normally produces laminar boundary layers. However, tur-bulent boundary layers can occur even in a generally accelerating flow if, for instance, there is a local area of adverse pressure gradient producing decelerating flow, or an area of flow separation, either of which might be caused by high flow incidence angles, or by poor blade profiles as a result of manufacturing inaccuracies or of corrosion or deposition.

In addition, the high first cost of cooled blades, coupled with the deleterious effect on the cycle efficiency of supplying (usually) high-energy compressed and cooled air, gives the designer a strong incentive to use high loading coefficients to avoid the need for additional rows of cooled

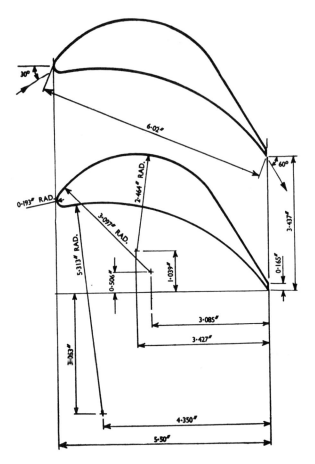

Figure 10.12 (a)

Convective Heat Transfer

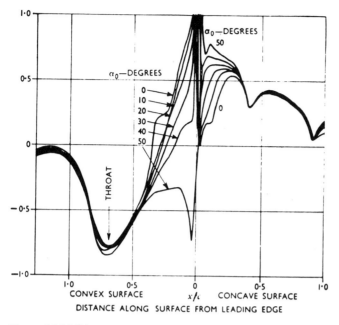

Figure 10.12(b)

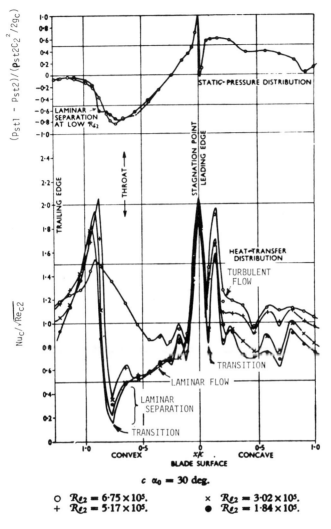

Figure 10.12(c)

Convective Heat Transfer

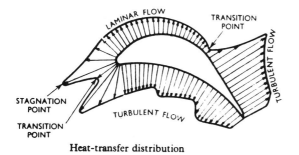

Heat-transfer distribution

Figure 10.12 (d)

Figure 10.12 Distributions of static pressure and heat transfer around a gas-turbine blade. From Wilson and Pope 1954. (a) Blade profile; (b) effect of incidence on the static-pressure distribution; (c) distributions of static pressure and heat transfer at design incidence; (d) typical distribution of heat transfer on a polar diagram.

blades. Accordingly, the reaction at the blade hub might be very small (chapter 6) which will probably lead to local adverse pressure gradients and consequent turbulence.

Turbulent boundary layers may also become established downstream of the passage throat with those blade profiles that do not result in continued acceleration of the fluid.

The distribution of heat-transfer coefficient, expressed as the Nusselt number ($h_t c/k$, where c is the blade chord) divided by the square root of the Reynolds number ($\rho W_2 c/\mu$) for such a blade profile is shown in figure 10.12. A typical heat-transfer distribution is shown on a polar diagram. The static-pressure distribution and the heat-transfer distribution as a function of Reynolds number are shown varying with x/c, where x is the distance around the blade profile along both the pressure and suction surfaces from the leading-edge stagnation point.

The stagnation-point heat transfer is very high but drops away quickly under the influence of a rapidly growing laminar boundary layer. On the convex (suction) surface the heat transfer continues to fall away until the passage throat is reached, where a strongly adverse pressure gradient is encountered. At low Reynolds numbers the laminar boundary layer leaves the wall in a separation "bubble." The boundary layer changes to turbulent and reattaches to the surface, giving a very high peak heat-transfer coefficient. The value decreases toward the trailing edge, though always above the low value of heat-transfer coefficient reached by the laminar boundary layer before transition to turbulence.

On the pressure surface of these particular blades an extremely limited but intense low-pressure region results in a steep adverse pressure gradient

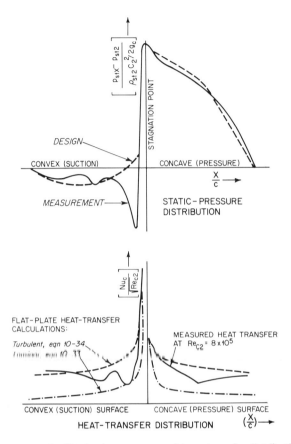

Figure 10.13 Static-pressure and heat-transfer distributions around a PVD blade.

which changes the laminar boundary layer to turbulence close to the leading edge. There is therefore a second peak of heat transfer in addition to that at the stagnation point. Subsequently, the heat-transfer coefficient falls away toward the trailing edge. There is the possibility that this turbulent boundary layer was tending toward becoming laminar again ("relaminarization") under the strongly favorable pressure gradient.

The blade profile for which these results were obtained was typical of medium-reaction medium-loading stages of up to the midsixties. Then profiles with prescribed velocity distributions (PVD profiles) began to be produced with the aid of streamline-curvature and similar computer programs, aimed at eliminating the pressure minima that convert low-loss, low-heat-transfer laminar boundary layers into high-loss, high-heat-transfer turbulent boundary layers. Polytropic turbine efficiencies increased as a result of the introduction of these PVD profiles. Pressure and

Convective Heat Transfer

heat-transfer distributions for a PVD blade from design and wind-tunnel-test values are shown in figure 10.13. The major reduction in turbulent-boundary-layer heat-transfer coefficient from those found on multiple-circular-arc blades is evident. It is not obvious, however, that the overall (mean) heat-transfer rate is reduced.

The prediction of heat-transfer coefficients by analysis with any accuracy is not, apparently, possible even in the controlled flow found in wind-tunnel cascades. Martin et al. (1978) gave two reasons for this state of affairs: uncertainty regarding the starting point and the extent of the region of boundary-layer transition from laminar to turbulent; and the peculiar form of the succeeding boundary layers, neither laminar nor fully turbulent. The strong accelerations that are induced in PVD designs appear to result in laminarization tendencies in the turbulent boundary layers. The measured heat-transfer coefficients shown in figure 10.13 lie between the predicted laminar and turbulent values for flow over a flat plate:

for laminar

$$St = 0.332 \, Re_x^{-0.5} \, Pr^{-0.67}, \tag{10.33}$$

for turbulent

$$St = 0.029 \, Re_x^{-0.2} \, Pr^{-0.67} \tag{10.34}$$

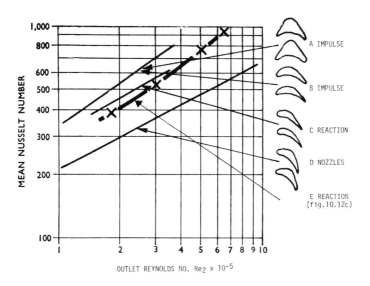

Figure 10.14 Mean heat transfer on turbine blades at design incidence.

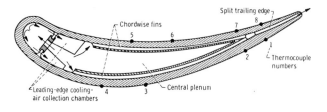

Impingement cooled leading edge; convection cooled pressure and
suction surfaces; chordwise fins in midchord region.

Figure 10.15 (a)

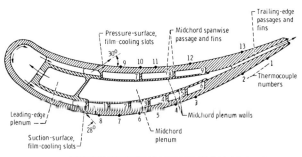

Impingement cooled leading edge; convection and film cooled pressure
and suction surfaces; spanwise fins in midchord region.

Figure 10.15 (b)

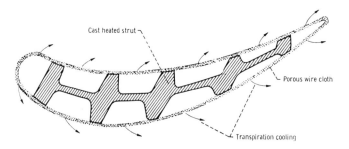

Sintered wire cloth vane.

Figure 10.15 (c)

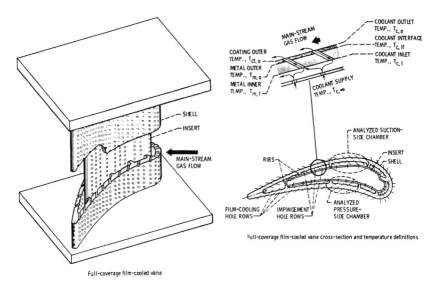

Full-coverage film-cooled vane

Full-coverage film-cooled vane cross-section and temperature definitions

Figure 10.15 (d)

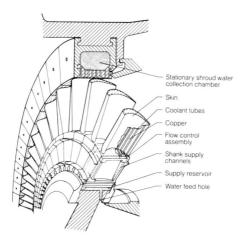

Stationary shroud water collection chamber

Skin

Coolant tubes

Copper

Flow control assembly

Shank supply channels

Supply reservoir

Water feed hole

Figure 10.15 (e)

Figure 10.15 Blade-cooling methods: (a) Impingement-cooled leading edge; convection-cooled pressure and suction surfaces; chordwise fins in midchord region. From Yeh et al. 1974. (b) Impingement-cooled leading edge; convection- and film-cooled pressure and suction surfaces; spanwise fins in mid-chord region. From Yeh et al. 1974. (c) Sintered-wire-cloth vane. From Gladden 1974. (d) Full-coverage film-cooled vane. From Meitner 1978. (e) Water-cooled blades. From Dakin et al. 1978.

where the length in the Reynolds number is the distance, x, from the stagnation point and the gas velocity and properties are the local static free-stream values. There is therefore little point in going to more accurate calculations that allow for variations in free-stream velocity. In fact Martin et al. show that a full calculation taking account of gas pressure, temperature, free-stream velocity, surface temperature, and surface radius of curvature gives considerably worse accuracy of prediction than the flat-plate equations, (10.33) and (10.34). These then are recommended, with some hesitation, for preliminary design, together with the mean-heat-transfer curves of figure 10.14.

10.7 Heat transfer with mass transfer

The considerations in section 10.6 concern heat transfer in the absence of mass transfer—that is, blades with nonporous walls. If a very large degree of cooling is required, gas or possibly liquid can be discharged through the walls, either through discrete slits or through porous materials ("film" or "transpiration" cooling). Figure 10.15 shows some alternative blade-cooling arrangements. Employing mass transfer has two large potential advantages over solid-wall cooling. First, the coolant in its intimate contact with the wall will transfer more heat from the blade. Second, the actual heat transfer coefficients around the blade can be reduced because of the modification of the temperature distribution in the "thermal" boundary layer. Therefore the blade-surface temperature can be greatly reduced for the same coolant flow, or a reduced coolant usage can be realized for the same surface temperatures as for solid-wall cooling. Experimental turbines with blades formed from rolled mesh to give a controlled porosity for cooling air have been run at design conditions with inlet gas temperatures up to 4,000 F (2,205 C).

On the other hand, mass transfer into the boundary layer adds low-momentum fluid which can make otherwise almost-laminar boundary layers fully turbulent and thereby increase flow losses and heat transfer. Film cooling applied as in some of the blades of figure 10.15 has often been found to be less effective, for these reasons, than solid-wall conduction cooling.

10.8 Internal-surface heat transfer

With air used for blade cooling, there has been a natural design progression from the crude hollow blades used in some versions of the Junkers Jumo 004 jet engine in Germany in the Second World War (figure 10.16a). These were introduced more to allow the use of non-heat-resistant metals than to attain a higher turbine-inlet temperature (see the brief history at the beginning of the book). The effectiveness of the heat exchange between

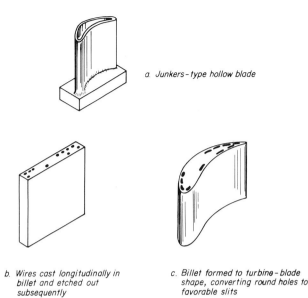

a. Junkers-type hollow blade

b. Wires cast longitudinally in
   billet and etched out
   subsequently

c. Billet formed to turbine-blade
   shape, converting round holes to
   favorable slits

Figure 10.16 Early types of air-cooled turbine blades: (a) Junkers-type hollow blade; (b) wires cast longitudinally in billet and etched out; (c) billet formed to turbine-blade shape, converting round holes to favorable slits.

the cooling air and the blade was low, and a large quantity of cold air was discharged from the blade tips, disrupting the turbine flow and requiring a large increase in compressor flow. As indicated in equations (10.26) and (10.27), the hydraulic diameter of the heat-transfer passages should be reduced when it is desired to increase the heat transfer to be accomplished in a given volume. The methods used to produce finer and more intricate flow passages have changed as new manufacturing methods have been developed.

An early method used in Britain was to cast longitudinal wires into small billets of high-temperature material, to etch out the wires, and to forge the billets into the final blade forms (figure 10.16b and c). The once-round holes would be deformed into favorably shaped slits which could be concentrated in the leading and trailing edges, regions of high external-surface heat transfer. Later, improved methods of precision casting were used to produce cooling channels that could make several passes along the blade before being discharged at the tips or along the trailing edges. These blades were often formed of an intricate internal strut and a brazed-on outer shell.

With the introduction of PVD blades the trailing-edge heat-transfer peak was reduced, but the very high heat transfer at the leading-edge stagnation point remained as a problem. The design requirement is not

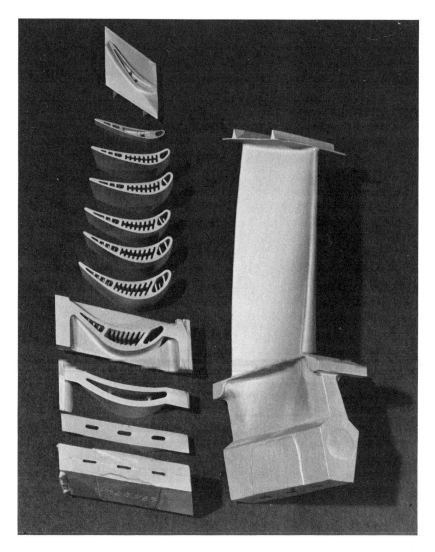

Convection-cooled turbine rotor blade with cast fins. Courtesy Detroit Diesel Allison Division of General Motors Corporation.

Convective Heat Transfer

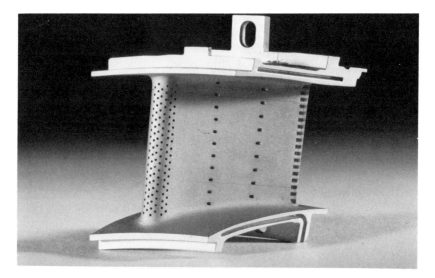

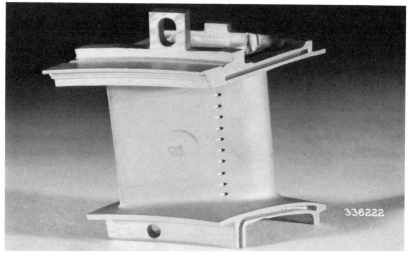

Turbine nozzle vane with impingement and film cooling. Courtesy Detroit Diesel Allison Division of General Motors Corporation.

merely to keep the blade material to a temperature below which the physical properties will be adequate but also to reduce the temperature gradients to a level below which fatigue failure will not result from thermal cycling. (This results in so-called "low-cycle" fatigue, because the stress reversals occur in general only upon start-up and shutdown or upon rapid power changes in high-performance engines, rather than at the blade natural frequencies in, for instance, bending as in "high-cycle" fatigue.) Therefore the high external heat-transfer coefficient at the leading-edge stagnation point must be matched by high heat transfer from the internal cooling passages. At the present time this is accomplished generally by producing stagnation-point heat transfer from impingement jets from the primary coolant supply (figure 10.15a).

Water can also be used for rotor-blade cooling, but the very high pressures in the centrifugal field, and the tendency for scale and other deposits to clog passages, allows less design freedom. The first known water-cooled rotor was designed by Schmidt in Germany to have a free water surface at the rotor interior (figure 10.17a). The multistage rotor and blades were machined out of a solid forging, the cooling holes drilled, and end caps were brazed on. Very high internal heat-transfer rates were assured because at design speed the water would be above critical pressure over the whole length of the blades. It was intended that the steam would be expanded in a steam turbine, condensed, and recycled for cooling. However, a wave instability of the free water surface produced dangerous rotor vibrations. In addition there were failures in the integrity of the coolant channels, and the experiments were terminated. (Unsuccessful attempts were made in Britain and the United States after 1946 to overcome these problems.)

Another form of water cooling that has been studied but not, so far as is known, built and tested is the closed thermosiphon (figure 10.17b). In one form of this system each blade carries its own heat exchanger at an inner radius, and the closed cooling passage has sealed within it a small quantity of water or other coolant. The liquid boils from the free surface near the blade tip and condenses in the hub heat exchanger. The centrifuged condensate and vapors cool the remaining parts of the passage.

A third type of water cooling is the open system, under development (1982) by General Electric, figure 10.15d. By allowing a small flow of water to pass through the blades and to be discharged at the tips, buildup of scale should be inhibited. It is intended that the unvaporized water should be collected in channels in the turbine shroud.

10.9 Concluding remarks about blade cooling

The foregoing discussion has shown that, although the use of high temperatures for future gas turbines is necessary in most cases for the attain-

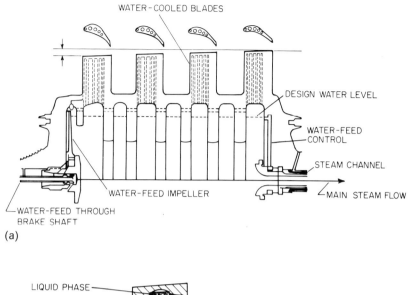

WATER-COOLED BLADES

DESIGN WATER LEVEL

WATER-FEED CONTROL

STEAM CHANNEL

MAIN STEAM FLOW

WATER-FEED IMPELLER

WATER-FEED THROUGH BRAKE SHAFT

(a)

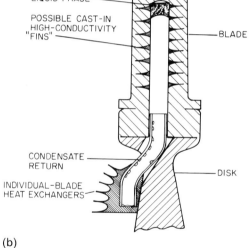

LIQUID PHASE

POSSIBLE CAST-IN HIGH-CONDUCTIVITY "FINS"

BLADE

CONDENSATE RETURN

INDIVIDUAL-BLADE HEAT EXCHANGERS

DISK

(b)

Figure 10.17 Closed-circuit water-cooled blades: (a) Schmidt turbine. From Smith and Pearson 1950. (b) Closed thermosiphon.

ment of high cycle efficiencies, no preferred method of blade cooling has emerged. Furthermore even sophisticated methods of analysis of heat transfer in the external flow have been found to be inaccurate. The same can be said in general for the analysis of the flow in the blade-cooling channels, subject to intense Coriolis accelerations, and for the analysis of transient conduction and assoicated thermal stresses in the rotating and stationary blades.

Therefore it is concluded that the preliminary design guidelines provided may be sufficient for their purposes and that an intensive development program is a requirement for any new cooled-turbine design.

References

Bayley, F. J., J. W. Cornforth, and A. B. Turner. 1973. *Experiments on a transpiration cooled combustion chamber*. Proc. Inst. Mech. Engrs., London, 187: 158–169.

Bayley, F. J., and G. S. H. Lock. 1964. Heat transfer characteristics of the closed thermosyphon. *Trans. ASME J. Heat Transfer*, paper 64-HT-6, pp. 1–10.

Bayley, F. J., and A. B. Turner. 1970. *Transpiration cooled turbines*. Proc. Inst. Mech. Engrs, London, 185: 943–951.

Brown, A., and B. W. Martin. 1974. *A review of the bases of predicting heat transfer to gas turbine rotor blades*. Gas Turbine Conference & Products Show, Zurich, Switzerland, March 30 April 4, 1974. ASME, New York.

Caruvana, A., W. H. Day, G. A. Cincotta, and R. S. Rose. 1979. *System status of the water-cooled gas turbine for the high temperature turbine technology program*. Gas Turbine Conference & Exhibit & Solar Energy Conference, San Diego, Calif., March 12–15, 1979. ASME, New York.

Dakin, J. T., M. W. Horner, A. J. Piekarski, and J. Triandafyllis. 1978. Heat transfer in the rotating blades of a water-cooled gas turbine. In *Gas Turbine Heat Transfer*. ASME, New York.

Eckert, E. R. G. 1963. *Introduction to heat and mass transfer*. McGraw-Hill, New York.

Gladden, Herbert J. 1974. *A cascade investigation of a convection- and film-cooled turbine vane made from radially stacked laminates*. Technical memorandum TM X-3122. NASA, Washington, D.C.

Horner, M. W., D. P. Smith, W. H. Day, and A. Cohn. 1978. *Development of a water-cooled gas turbine*. Gas Turbine Conference & Products Show, London, April 9–13, 1978. ASME, New York.

Kays, W. M., and A. L. London. 1964. *Compact Heat Exchangers*. McGraw-Hill, New York.

LeBrocq, P. V., B. E. Launder, and C. H. Priddin. 1973. *Experiments on transpiration cooling*. Proc. Inst. Mech. Engrs., London, 187: 149–169.

London, A. L., M. B. O. Young, and J. E. Stang. 1970. Glass-ceramic surfaces, straight triangular passages—heat-transfer and flow-friction characteristics. ASME, *J. Eng. Power*, vol. A 92.

Martin, B. W., A. Brown, and S. E. Garrett. 1978. *Heat-transfer to a PVD rotor blade at high subsonic passage-throat Mach numbers.* Proc. Inst. Mech. Engrs., London, vol. 192.

Meitner, P. L. 1978. A computer program for full-coverage film-cooled blading analysis including the effects of a thermal barrier coating. In *Gas Turbine Heat Transfer*. ASME, New York, pp. 31–38.

Smith, A. G. 1948. *Heat flow in the gas turbine.* Proc. Inst. Mech. Engrs., London, vol. 159, WEP no. 41:245–254.

Smith, A. G., and R. D. Pearson. 1950. *The cooled gas turbine.* Proc. Inst. Mech. Engrs., London, 163 (WEP no. 60):221–234.

Wilson, D. G., and J. A. Pope. 1954. *Convective heat transfer to gas-turbine blade surfaces.* Proc. Inst. Mech. Engrs., London, 168:861–876.

Wilson, D. G. 1966. *A method of design for heat-exchanger inlet headers.* Paper 66-WA/HT-41. ASME, New York.

Yeh, F. C., H. J. Gladden, J. W. Gaunter, and D. J. Gaunter. 1974. *Comparison of cooling effectiveness of turbine vanes with and without film cooling.* Technical memorandum TM X-3022. NASA, Washington, D.C.

Problems

10.1

Would Reynolds analogy be useful to you if you wanted to use the specified pressure drop in a combustion chamber to calculate the heat transfer to the walls of the chamber? Give reasons.

10.2

Assuming that it is desirable to use low fluid velocities in heat-exchanger passages if pumping-power losses are to be minimized, and that it is desirable to use small hydraulic diameters in the passages if the heat-exchanger volume is to be minimized, give some (say, four) practical limits to how small you could specify the velocity and the hydraulic diameter in the heat exchanger of a gas-turbine engine.

10.3

(a) Calculate the entropy increase in each of several categories in the gas-turbine-engine heat exchanger, whose conditions are given in figure P10.3. The effectiveness is 0.9, the mass flow 1 kg/s; the specific heat can be taken as constant at 1,010 J/kg-°K; the gas constant is 286.96 J/kg-°K, and the fluids can be treated as air and as perfect gases. Use Gibbs' equation ($Tds = du + pdv$) to find the increase of entropy at constant pressure:

from a to b,

from e to f,

from b to c,

from f to g.

(b) Add the two categories, then find the total, and comment.

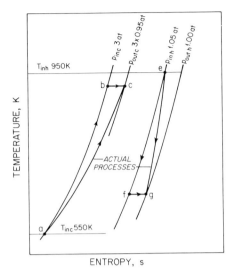

ENTROPY, s

Figure P10.3

10.4
Using figure P10.4, find the effects on cycle thermal efficiency and specific power of cooling the first-stage turbine nozzle vanes and rotor blades of a heat-exchanger cycle gas-turbine engine (CBEX) of the following specifications: $T_{01} = 540.0$ R, $\dot{m}_c = 150$ lbm/s, $r = 5.0$; $T_{02} = 876.6$ R, $T_{04} = 3,011.4$ R, $\sum \left(\dfrac{\Delta p_0}{p_0} \right) = 0.12$, $\eta_{pe} = 0.90$, and heat addition from fuel 314 Btu/lbm air, $p_{01} = 10^5 \text{N/m}^2$.

(a) The cooling air is bled from the compressor discharge. After leaving the turbine blades and vanes it neither contributes to the turbine work nor adds any losses.

(b) The vanes and blades must be maintained at a mean temperature of 1,800 F. The cooling air is fed to the blades at compressor-delivery temperature and discharges at 1,700 F.

(c) The hot gas approaches each blade row with a relative (axial) velocity of 600 ft/s and leaves with a relative velocity of 1,000 ft/s. The effective temperature of the gas is the static temperature at blade-row exit plus 0.88 of the difference between static and total temperature (use the steady-flow energy equation).

(d) Assume a mean Nusselt number ($h_c c/k$) of 900 for both vanes and blades. To save effort, find the blade chord length based just on calculating the arithmetic mean diameter from the inner and outer diameters at nozzle-vane exit (or rotor-blade entrance). Use air properties. Specify the inner-to-outer diameter ratio as 0.85, the number of blades and vanes as 48 and 49, respectively, the space-chord ratio s/c as 0.75 for both, and the blade or vane perimeter as $2.5c$.

(e) Ignore the effects of fuel addition.

(f) Use 50 percent reaction.

(g) Assume that 2/3 of the cycle pressure losses occur before the turbine.

(h) There is a 1 percent drop in total pressure from nozzle entrance to nozzle exit.

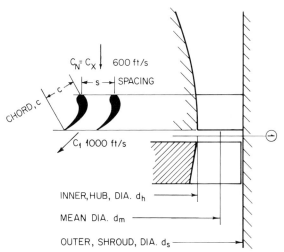

Figure P10.4

10.5

Find the number of cooling holes one millimeter in diameter, and the quantity of compressor-delivery air used, to cool the first-stage rotor blades of the solar Brayton-cycle engine given in the example in section 10.5.

(a) A four-stage turbine is used, of equal enthalpy drop per stage. The first-stage rotor has 13 blades at a mean-diameter solidity of 1.5. The mean hub-tip ratio is 0.85. The mean-diameter velocity diagram shown in figure P10.5 is that of the medium-work turbine in figure 5.7 ($R_n = 0.5$, $\psi = 1.0$, $\phi = 0.5$). This diagram may be used to find the blade speed, relative outlet velocity, Reynolds number, heat-transfer coefficient, and so forth.

(b) Use curve C, extrapolated if necessary, in figure 10.14 to obtain the mean Nusselt number of the external flow (based on chord and conductivity at local static conditions).

(c) Specify that the blade metal temperature shall be uniform at 1,200 K, for the purpose of calculating both the external and the internal heat transfer.

(d) Specify that the cooling holes make three passes in the blade. Specify also that the

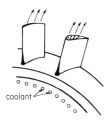

Figure P10.5

effectiveness of the crossflow heat exchanger, which the flow between the cooling air (bled from the compressor discharge) and the blade can be regarded as, shall be 85 percent. Use figure 10.4b to find the NTU required. Note that with a constant blade temperature, $C_{rat} = 0$.

(e) The flow in the cooling holes will be laminar. Assume that the mean Nusselt number, based on diameter, is 3.65. (In fact Coriolis acceleration would affect the heat-transfer coefficient on different sides of the passage.) Specify that the flow will be metered into each hole so that the desired velocities are obtained.

(f) For the external heat-transfer area, use a ratio of perimeter to chord of 2.6.

(g) Use an expander (turbine) total-to-static polytropic efficiency of 90. Ignore the effect of heat transfer on the turbine expansion for this first approximate calculation.

10.6
Find possible core dimensions for three alternative designs of ship-borne inter-coolers for a main-propulsion gas turbine. Also find the pressure drops as propor-tions of the inlet total pressure of each fluid. Common specifications for all designs are these:

Air mass flow, 30 kg/s.

Air inlet pressure, 3 at.

Air inlet temperature, 411 K.

Air outlet temperature, 300 K.

Sea-water inlet temperature, 290 K.

Sea-water outlet temperature, 310 K.

The intercooler is to be designed as a three-pass cross-counterflow unit. The three designs should use these surfaces:

(a) Compressed air inside straight tubes of 0.231 in bore (figure 10.8a). Water flows in three passes across staggered tube banks of 0.250-in outside diameter (figure 10.7b).

(b) The water flows in the tubes and the air outside the tubes of the first case.

(c) The water flows inside the tubes as in case (b), and the air flows in the same arrangement as in case (b), but the outside of the tubes are finned as in figure 10.7c.

Take all properties at the mean fluid temperatures of each side; neglect compressibil-ity of the air; neglect any tube-wall temperature difference or fin temperature drop; ignore entrance and exit effects. For the first case, choose an air-side Reynolds number at the low end of turbulent flow, and a low water velocity of a very few feet per second. Go onto a first iteration, and comment on how you would change your selections if you were to go to a second iteration. Use the results of this first case to guide you in the second and third cases.

10.7
Find the ratio (θ_h/θ_c) of the temperature differences from the hot fluid to the wall θ_h to that from the cold fluid to the wall θ_c in a parallel-plate heat exchanger with laminar flow both sides. Use the data in table P10.7, and refer to the construction details in figure P10.7:

Table P10.7

	Hot	Cold
Temperature, F	1,000	200
Pressure, at	1.2	4.0
$Nu \equiv \dfrac{h_t d_h}{k}$	6.5	3.63
$Re \equiv \dfrac{\rho C d_h}{\mu}$	1,500	900
$k \dfrac{Btu\text{-}ft}{h\ ft^2\text{-}°F}$	0.035	0.020
$\mu \times 10^5$ lbm/s-ft	2.44	1.44

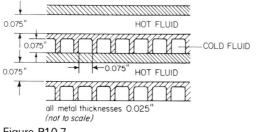

all metal thicknesses 0.025"
(not to scale)

Figure P10.7

10.8
Make a preliminary design, and one subsequent design with your own chosen conditions of flow, for a regenerative heat exchanger for a CBEX cycle with the specifications that follow. The outputs of such a preliminary design are the outside diameter of the regenerator disk, the axial thickness, the pressure losses, and the carry-over leakage (the air and gas trapped in the matrix). Cycle specifications

$\dot{m} = 1.0$ lbm/s, $r = 5.0$, $T_{01} = 540$ R, $T_{02} = 2,700$ R, $\varepsilon_x = 0.9$,

$\sum(\Delta p_0/p_0) = 0.15$, $\eta_{pc} = \eta_{pe} = 0.80$.

Some other suggested inputs are these. Use a capacity-rate ratio of 3.0 to find the NTU. Use the matrix surface the results for which are given in figure 10.8d. Take equilateral triangles for the passages with the wall thickness determined by a ratio of passage area to total matrix face area of 0.75. Use a hydraulic diameter of 0.030 in. Take the specific heat of the ceramic-core material as 0.308 Btu/lbm-°R, and the density as 141 lbm/ft³. Choose the inner (open) diameter of the matrix to be 0.4 of the outer diameter, and the seal (radial) area to increase the face area by 20 percent.

For this preliminary calculation neglect all leakage as far as determining mass flows is concerned, and find the flow conditions for each fluid at the mean of the inlet and outlet temperatures for that fluid. Also use air properties for both fluids. These choices will lead to some inconsistencies with the temperatures previously calculated: you may ignore them or remove them as you wish. The effects will be small.

Take the pressure drops as the calculated core pressure drops plus two dynamic heads, calculated again on mean conditions. All of the following choices can be used for both calculations you should make. For the first, choose the Mach number in both hot and cold passages to be 0.02. Comment on the hot-cold conductance ($h_t A_h$) ratio and the relative pressure-drop ratios ($\Delta p_0/p_0$) as well as the regenerator-disk size which are given by this and the other choices. Then, for your second calculation, make another choice of Mach numbers, Reynolds numbers, or conductances; calculate new outputs; and comment on them. (A pressure ratio of five is approximately the upper limit for rotary regenerators to have acceptable seal and carry-over leakage. A lower pressure ratio would give unacceptably high temperatures at heat-exchanger inlet.)

10.9
Find the percentage of compressor air required to give a mean blade temperature of 1,600 F in a turbine where the local effective gas temperature is 2,200 F. The coolant enters the blade root at 600 F and is discharged at 1,500 F. The blade Reynolds number of the external gas flow is 5.10^5, based on blade chord and the outlet velocity, giving a Stanton number of 0.0021.

Use the blade and passage configuration shown in figure 10.12a. Simplify the calculations with such assumptions as equalizing the compressor and turbine mass flows. Take the ratio of perimeter to chord as 2.37 and the solidity (c/s) as 1.6.

Describe the procedure you would follow, but do not make any calculations, to choose the hydraulic diameter and the configuration of the internal cooling passages.

11

Cavitation and two-phase flow in pumps

In this chapter we describe and give some guidance to predicting three separate but related phenomena. The first is cavitation damage. When a liquid enters a pump, it may produce within the blade passages tiny vapor bubbles that grow rapidly and then collapse with violent decelerations and associated pressure waves able to cause catastrophic damage to the hardest of surfaces.

The second phenomenon is that of cavitation performance loss. Vapor bubbles may grow within the blade passages to relatively large sizes and often are present in the exit stream. The pumping performance suffers even though there may be no damage to the pump.

The third phenomenon is the loss of pump performance which occurs when the incoming stream is composed of a two-phase mixture of liquid and vapor of any proportion up to an all-vapor flow.

None of these three phenomena can be predicted accurately from first principles. A combination of fundamentals with experimental data is needed. In this chapter we give the basic relationships required to use data effectively. There are too many variables involved to attempt to give more than the most approximate of experimental correlations for guidance. We refer designers to specialized texts for furture information.

11.1 Cavitation

At normal atmospheric pressure we know that we can make water boil by heating it to 100 C. If we contain the water in a kitchen pressure cooker so that the pressure is, say, 1.2 at, the water can be above 100 C and still not boil. We can make it boil by releasing the vapor to lower the pressure in the cooker. We learn to do this slowly, as the boiling can be quite violent. We know, by experience or by extrapolation, that we can similarly cause water to boil, even at temperatures close to freezing, by reducing the pressure sufficiently.

In other words, we can cause boiling, or vapor formation, in a liquid either by heating at constant pressure or by reducing the pressure at constant temperature or by a combination of both.

Cavitation is the formation of small bubbles from the reduction of stream pressure at constant temperature.

A pump can bring about a reduction in pressure in a liquid in the inlet flow in three ways.

1. The liquid may be raised through a height in a gravitational field.

2. The liquid may experience friction in flowing past solid surfaces, and in frictional dissipation by shearing forces within the liquid, and so drop in pressure.

3. The liquid may be accelerated to higher velocities, so converting static pressure to velocity pressure.

Although the situation in which cavitation bubbles can grow may be set up by gravitational and frictional losses in pressure, the immediate precursor of bubble formation is normally the drop in pressure accompanying acceleration.

We will illustrate some of the mechanisms by which pressures are reduced to below the vapor pressure of the liquid. Before we do so, we will review the physics of bubble formation, in particular so that it can be appreciated that it is frequently necessary for the pressure to be reduced to considerably below vapor pressure before bubbles are formed.

Liquid tension and bubble formation

For a bubble to form in a liquid, molecules have to be separated from one another through the pressure of a vapor. Because of the small radius of curvature a microscopically small bubble must have, the pressure in the bubble must be higher than that in the surrounding liquid by the increment $2\sigma/r$, where r is the bubble radius and σ the surface tension. Normally, this over pressure is not large because there will be scratches, fissures, and cracks in bounding surfaces that will trap gas or vapor bubbles of relatively large radius (figure 11.1). These are responsible for the streams of vapor bubbles that can be seen in water coming to the boil in a glass beaker or a metal pan, for instance. Or the liquid may have dust particules or other contaminants able to form nuclei from which bubbles can grow with only a small "superheat."

Experiments with ultrapure water heated on equally pure liquid mercury have shown that the water can exist in a metastable state without boiling even when far above the boiling point. The liquid can then be withstanding internal tensions of many atmospheres magnitude. When boiling even-

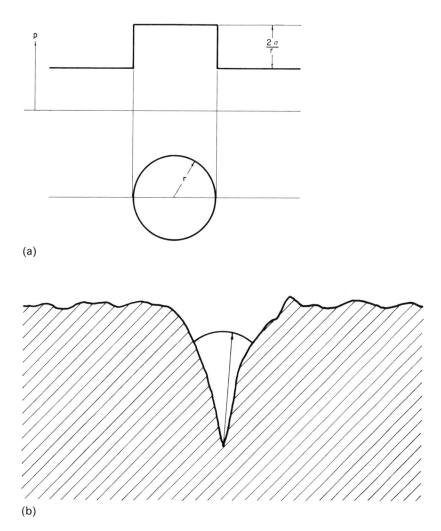

(a)

(b)

Figure 11.1 Cavitation phenomena: (a) pressure rise induced by surface tension across a bubble boundary; (b) cavitation-bubble inception—potentially large radius of surface of gas bubble trapped in fissure allows bubble growth at higher stream pressure.

tually occurs in such an experiment, it does so explosively and somewhat dangerously.

In most circumstances in which pumps are used, there will be sufficient surface scratches and nuclei in the liquid for the formation of bubbles at only a little below vapor pressure. However, if the pressure reduction is very localized, which is often the case, nuclei may not be found at the precise location needed for bubble growth, and further reductions in pressure may be possible before cavitation occurs.

There are two other reasons why cavitation bubbles may not be evident, even though there is a local reduction of pressure to below vapor pressure. These reasons are based on inertial and heat-transfer limitations. For a bubble to grow, the liquid surrounding it must acquire a radial outward velocity. In the early stages of growth the acceleration forces are small, because the incremental pressure is reduced by the surface-tension effect of $2\sigma/r$, and because the surface area is small. Bubbles may have grown to only a very small degree, therefore, when they pass out of the low-pressure region of the liquid; no further growth occurs, and they may even collapse. Bubble growth is thereby inertia limited. The collapse may occur before the bubbles are large enough to be perceptible, and/or the collapse may take place at a sufficient distance from a solid surface that the pressure waves experienced at the surface are very small.

For some liquids heat transfer from the liquid to the bubble surface may provide more of a limitation than inertia. The latent heat for vapor formation must be conducted to the surface during bubble growth and from the surface during collapse. For liquids with large evaporation enthalpies and/or small conductivities, the growth and collapse rates may be too slow for bubbles to be evident until comparatively large reductions of pressure to below the vapor pressure have taken place.

In summary, vapor bubbles do not necessarily appear immediately the local pressure is reduced to below local vapor pressure. The incremental pressure reduction necessary before bubbles not only grow but grow to a size large enough to be detected, and collapse near enough to a solid surface and with enough intensity to cause damage, will be highly dependent on the properties of the liquid and on the configuration and operating conditions of the pump.

Pressure-velocity relation in a pump

The way in which the pressure or head in a liquid changes as it approaches and passes through a pump is illustrated in figure 11.2. Both the total head and the static head are shown. It is, of course, the static head that must fall to below the local vapor pressure or head before vapor bubbles can start growing.

If the axis of the pump is horizontal, we cannot treat the flow one-dimensionally, even if it is axisymmetric with respect to velocities, because

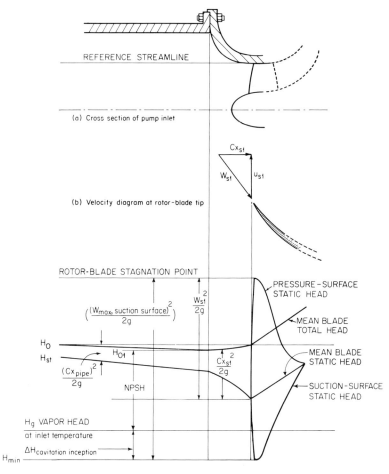

(a) Cross section of pump inlet

Cx_{st}

W_{st} u_{st}

(b) Velocity diagram at rotor-blade tip

ROTOR-BLADE STAGNATION POINT

PRESSURE–SURFACE STATIC HEAD

$\left(\dfrac{(W_{max,\ suction\ surface})^2}{2g}\right)$

$\dfrac{W_{st}^2}{2g}$

MEAN BLADE TOTAL HEAD

H_O

H_{st}

H_{O1}

MEAN BLADE STATIC HEAD

$\dfrac{(Cx_{pipe})^2}{2g}$

$\dfrac{Cx_{st}^2}{2g}$

NPSH

SUCTION–SURFACE STATIC HEAD

H_g VAPOR HEAD
at inlet temperature

$\Delta H_{cavitation\ inception}$

H_{min}

(c) Head and velocity distributions

Figure 11.2 Head-velocity changes in a pump inlet: (a) cross section of pump inlet; (b) velocity diagram at rotor-blade tip; (c) head and velocity distributions.

the pressure will be lowest in the upper part of the inlet duct and pump. The reference height for the pump is therefore defined to be the top of the inlet blading circle. The use of this location rather than the centerline of the pump can be significant for large pumps with, for instance, inlets two meters in diameter but has little significance for small pumps.

The pump in figure 11.2 can be either an axial-flow or radial-flow type, because only the inlet conditions have a deciding influence on cavitation. It is supposed that the inlet flow is led to the pump through a horizontal pipe of a diameter larger than that of the inlet blading. The head along a horizontal streamline through the top of the inlet blade circle is shown.

The total head is shown falling slowly along the inlet pipe under the effects of wall friction. The velocity will stay approximately constant until the pump inlet is reached, and the static head is shown falling at the same rate.

The flow is accelerated smoothly but sharply at the pump inlet. Friction is normally negligible under acceleration, so that the total head would remain almost constant, but the static head drops. It is possible, though not probable, that the cavitation in a pump is initiated by this acceleration-induced fall in static head.

If the pump is fitted with an initial stator-blade row, such as inlet guide vanes, there will be an additional acceleration and a lowering of the mean static head. Moreover the vanes will be loaded hydrodynamically, so that the head on the suction side of the vanes will be lower than the pressure-side head.

The diagram is drawn for the more usual situation where no inlet guide vanes are used. It would be easy, but complicating, to add the head distributions around vanes.

The flow then encounters the moving blades. The relative velocity of the blade tips at inlet, in the location of the streamline with which we are concerned, is W_{s1}. At the leading-edge stagnation point the flow is brought to rest relative to the moving blade, and the local static head at the blade surface is the stream static head plus the relative stagnation head. (At this point the relative static head is the same as the relative total head, because the relative flow velocity is zero.)

The inducer section of a radial-flow pump, or the first-stage rotor row of an axial-flow pump, is normally heavily loaded hydrodynamically. If there were zero loading, the relative flow would accelerate from the stagnation point to a value of W_{s1} on both the pressure and the suction sides of the blade. The blades would have to be infinitely thin plates aligned exactly with the flow. With loading, the relative velocity on the pressure side is less than W_{s1}, and the relative velocity on the suction side increases to a local maximum, W_{max}. (See chapter 8 and equation 8.3 for the significance of this local maximum in governing the diffusion in a pump or compressor.)

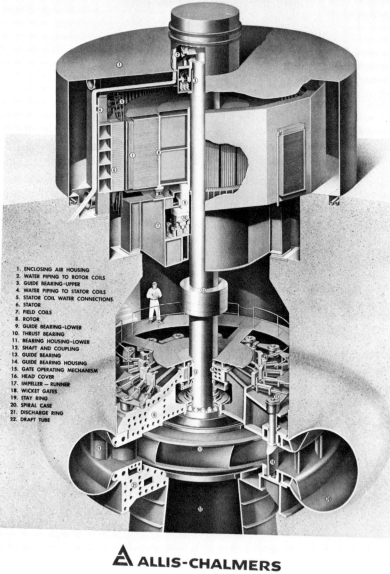

REVERSIBLE PUMP/TURBINE
and GENERATOR/MOTOR

1. ENCLOSING AIR HOUSING
2. WATER PIPING TO ROTOR COILS
3. GUIDE BEARING-UPPER
4. WATER PIPING TO STATOR COILS
5. STATOR COIL WATER CONNECTIONS
6. STATOR
7. FIELD COILS
8. ROTOR
9. GUIDE BEARING-LOWER
10. THRUST BEARING
11. BEARING HOUSING-LOWER
12. SHAFT AND COUPLING
13. GUIDE BEARING
14. GUIDE BEARING HOUSING
15. GATE OPERATING MECHANISM
16. HEAD COVER
17. IMPELLER – RUNNER
18. WICKET GATES
19. STAY RING
20. SPIRAL CASE
21. DISCHARGE RING
22. DRAFT TUBE

⊿ ALLIS-CHALMERS

Reversible radial-flow pump-turbine. Courtesy Allis-Chalmers Company.

Cavitation and Two-Phase Flow   423

The ratio of W_{max} to W_{s1} may be large for airfoils of poor configuration, or where erosion or deposits have changed the basic shape of the airfoil, or where the flow incidence angle is large. It is this local maximum-velocity minimum-pressure point that usually initiates cavitation.

Any cavitation damage will occur downstream of the minimum-pressure point. The effect of the blade hydrodynamic loading is to increase the total head and the static head (see chapter 5 and Euler's equation, 5.2). Bubbles will grow while the local static head is below the vapor pressure, and inertia will cause continued growth for a short while after the local head is above the vapor pressure although the radial acceleration will become negative. At some point downstream the bubbles will collapse, causing local pressures calculated to be hundreds of atmospheres, and a resulting spherical elastic pressure wave radiates out from the collapse point, decreasing in intensity with the cube of the distance. Whether or not surface damage is caused, therefore, depends strongly on the maximum size of the bubble, the magnitude of the pressure gradient, and the distance of the solid surface from the point of bubble collapse, as well as on the fatigue-strength and elastic properties of the surface material.

Pumps designed to handle liquids almost at their boiling points, such as the liquid-oxygen and liquid-hydrogen pumps in some rockets, have very long spiral-blade inducers with near-zero incidence angles and very low loading, so that W_{man}/W_{o1} is close to unity.

Cavitation correlations

The cavitation performance of pumps of similar design can be correlated with a Thoma cavitation number, σ, defined as

$$\sigma \equiv \frac{\text{NPSH}}{\Delta H_0},$$
(11.1)

where NPSH $\equiv$ net positive suction head (required to avoid cavitation)

$$\equiv H_{01} - H_g,$$

$\quad\quad H_g \equiv$ vapor head at inlet,

$\quad\quad \Delta H_0 \equiv$ rise in total head through the pump,

and either the pump specific speed, N_s (section 5.6), or the suction specific speed, N_{sv},

$$\text{where} \quad N_{sv} \equiv N_s \left[\frac{\Delta H_0}{\text{NPSH}} \right]^{3/4} \equiv \left[\frac{N_s}{\sigma^{3/4}} \right].$$
(11.2)

This correlation arises as follows. From figure 11.2 we see by inspection that

$$H_{\min} = H_{01} + \frac{1}{2g}[W_{s1}^2 - C_{x,s1}^2 - W_{\max}^2],$$

where $W_{\max} \equiv$ maximum relative velocity on the blade profile, which is the maximum that will be found on the suction surface. Then

$$H_{\min} = H_{01} + \frac{1}{2g}[u_{s1}^2 - W_{\max}^2]$$

$$= H_{01} + \frac{u_{s1}^2}{2g}\left[1 - \left(\frac{W_{\max}}{W_{s1}}\right)^2\left(\frac{W_{s1}}{u_{s1}}\right)^2\right].$$

In the velocity diagram, figure 11.2b, we see that

$$\left(\frac{W_{s1}}{u_{s1}}\right)^2 = \frac{C_{x,s1}^2 + u_{s1}^2}{u_{s1}^2} = (\phi_{s1}^2 + 1),$$

where $\phi_{s1} \equiv C_{x,s1}/u_{s1}$;

$$\therefore H_{01} - H_{\min} = (H_{01} - H_g) + (H_g - H_{\min})$$

$$= \frac{u_{s1}^2}{2g}\left[\left(\frac{W_{\max}}{W_{s1}}\right)^2(\phi_{s1}^2 + 1) - 1\right]$$

$$= \text{NPSH} + \Delta H_{ci}, \qquad (11.3)$$

where

$\Delta H_{ci} \equiv$ the incremental static head below the vapor pressure necessary to produce cavitation inception.

By dividing both sides of this equation by ΔH_0, the rise in total head through the pump, we will obtain Thoma's cavitation number, σ, on the left-hand side:

$$\sigma + \frac{\Delta H_{ci}}{\Delta H_0} = \frac{u_{s1}^2}{2g\Delta H_0}\left[\left(\frac{W_{\max}}{W_{s1}}\right)^2(\phi_{s1}^2 + 1) - 1\right].$$

We can substitute the head-rise coefficient ψ^x:

$$\psi^x \equiv \frac{g\Delta H_0}{u_2^2} \qquad (11.4)$$

where $u_2 \equiv$ rotor peripheral velocity. For an axial pump, $u_2 = u_{s1}$; for a radial pump, $u_2 = u_{s1}(d_2/d_{s1})$. We will write the equation for the radial-pump case, noting that for an axial pump $d_2/d_{s1} = 1.0$:

$$\therefore \sigma + H' = \frac{1}{2\psi^x} \times \left(\frac{d_{s1}}{d_2}\right)^2 \left[\left(\frac{W_{max}}{W_{s1}}\right)^2 (\phi_{s1}^2 + 1) - 1\right], \qquad (11.5)$$

where $H' \equiv (\Delta H_{ci}/\Delta H_0)$. This equation indicates that the value of Thoma's cavitation number for any pump is a function of H', which is partly dependent on fluid properties, on velocity-triangle values ψ^x and ϕ_{s1}, on the blade profile and incidence, which will govern W_{max}/W_{s1}, and, for radial-flow pumps, on the diameter ratio d_{s1}/d_2.

We know from equations (5.10) through (5.12) that specific speed is also a function of ϕ, ψ, and diameter ratio. It therefore seems reasonable to expect that for, any one fluid at a given temperature, and any one type of pump, operating at the optimum-efficiency condition, the Thoma number will be a function of the specific speed, N_s.

This is indeed found to be the case. The Thoma cavitation number for traditional types of single-suction centrifugal pumps with backward-swept blades has been found to be well correlated with

$$\sigma = 2.789 N_{s2}^{1.333}, \qquad (11.6)$$

where N_{s2} is the nondimensional specific speed for hydraulic machines:

$$N_{s2} = \frac{N\sqrt{\dot{V}}}{60(g\Delta H_0)^{3/4}} \quad \text{(equation 5.13)}.$$

Correlations of Thoma's number with specific speed for many different types of pumps can be found in specialized texts such as the *Pump Handbook* (Karassik et al. 1976). Pump efficiency or suction specific speed are sometimes used as parameters. Means for applying water-cavitation data to other liquids are also given.

This handbook (and others) also offers guidance on material selection to reduce damage from cavitation when it cannot be avoided. In general, very hard—such as Stellite—or very soft—polyurethane or neoprene—coatings are found to give the best protection.

All the previous discussion and analysis can be applied, with obvious changes, to hydraulic turbines, where cavitation tends to occur in the low-pressure, exit, part of the blading and in the downstream diffuser (draft tube).

11.2 Cavitation performance loss

When cavitation bubbles of the size that grow, collapse, and cause damage to solid surfaces are found in a pump, there is a negligible effect on pump performance. Suppose that one carries out a test in which a pump is run at constant speed and constant flow, and only the net positive suction

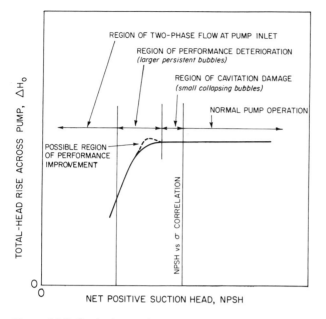

Figure 11.3 Cavitation regions.

head is slowly reduced. The head rise across the pump will remain constant up to and including the region of NPSH where cavitation bubbles first appear (figure 11.3).

When the NPSH is further reduced, more bubbles are formed, which grow larger and persist longer. Their collapse may occur downstream of the inlet blade row, or farther from the solid surface, or in a region of less-adverse pressure gradient than for the smaller bubbles, and cavitation damage lessens or disappears. The pump performance (head rise) begins to be affected, principally because of the influence of the vapor bubbles on the boundary layers. In some rare cases the initial appearance of persistent vapor cavities produces a slightly increased head. There is no agreed-upon reason for this improvement, but one possibility is that the vapor bubbles act to re-energize the boundary layer in the way accomplished by trapped toroidal vortices in ribbed diffusers (chapter 4).

Sooner or later, however, as the NPSH is reduced, the pump head falls. We define the region of cavitation performance deterioration as that in which the flow at inlet contains no vapor, although vapor forms within the pump and may persist at exit. We do not have a good correlation of cavitation performance deterioration, beyond knowing that it occurs at values of NPSH lower than that correlated by Thoma's cavitation number.

At still lower values of NPSH substantial vapor bubbles appear in the inlet flow. We refer to this condition as pump operation in two-phase flow. We are able to make approximate predictions of pump operation in certain circumstances.

11.3 Pump operation in two-phase flow

Pumps are not normally operated with substantial vapor in the inlet. The significance of being able to predict the performance in this condition of operation is that the head produced might be of critical importance in certain emergency situations. The correlations outlined here were developed to enable the pump head to be predicted after the occurrence of a hypothetical break in the high-pressure-coolant pipe of a pressurized-water nuclear reactor. In these circumstances the flow might go forward or backward through the pump, depending on whether the break is in the inlet or the outlet pipe, and the pump rotation might be reversed during the rapid outflow which would occur if the break were in the pump-inlet pipe. The combination of flow direction and direction of pump rotation is referred to as a "quadrant." Forward flow and forward rotation is the first quadrant.

The following method of correlating the performance of centrifugal pumps with two-phase inlet flow, both vapor-liquid and gas-liquid, enables the performance of certain pumps to be predicted, at least in the first quadrant (forward flow and forward rotation).

The correlating parameter is the head-loss ratio, which is the ratio of losses in two-phase flow to the losses in single-phase flow. The losses are here defined as the difference between the theoretical and the actual change of head across the pump. The limited data available show a good correlation between peak head-loss ratio and peak single-phase efficiency for the first quadrant. Therefore the present correlations may be used to make predictions of two-phase performance of centrifugal pumps for which the single-phase performance, including the peak single-phase efficiency, is known.

Outline of the correlation method
The present correlation approach is based on writing the Euler equation (equation 5.1) for a flow consisting of two parallel streams, one of liquid and one of vapor, and on a correlation due to Lottes and Flinn (Thom 1964) between losses in single-phase flow and those that will occur in a similar flow channel in two-phase flow. The consequence of this approach is to define a head-loss ratio, H^* and to show that this is a function of pump configuration, inlet void fraction, β, and flow coefficient ϕ:

$$H^* \equiv \frac{\psi'_{tp,th} - \psi'_{tp}}{\psi'_{sp,th} - \psi'_{sp}} = f \ (\beta, \phi, \text{pump configuration}),$$ \hfill (11.7)

where ψ' is the head-rise coefficient (equation 17.4), and the subscripts sp, single phase; tp, two phase; and th, theoretical. The void fraction β is the ratio of vapor or gas flow rate to total flow rate.

Correlation results

Very few two-phase data in the three quadrants of interest have been taken on reasonably large-scale models (one-fifth scale or larger) of typical circulating pumps used in pressurized-water nuclear reactors. The data taken at Babcock & Wilcox (B&W) by Winks (1977) were found to correlate well by the head-loss ratio (figure 11.4a and b). These data were for air-water mixtures. Later data from steam-water experiments on a one-fifth-scale pump at Combustion Engineering (Kennedy 1980) also correlated satisfactorily, as did experiments on much smaller pumps (Wilson et al. 1979). The relative flow coefficients ϕ', where $\phi' \equiv \phi/\phi_{\text{bep}}$, are for operation in the pump region, (figure 11.5). At higher flow coefficients a pump will operate in the "dissipation" region, in which flow must be forced through the pump (the head falls through the pump) while the shaft still requires input power. At still higher relative flow coefficients the pump works in the "turbine" region, with the shaft capable of giving an output power.

The theoretical head used for the correlation of figure 11.4 was that given by assuming that the deviation angle between the direction of the mean relative flow leaving the pump impeller at the best-efficiency point and the tangent to the blade center line at the impeller periphery is constant throughout the "pump" region of figure 11.5. This is an unsatisfactory assumption, because it is obvious that, when the pump is operating in the "turbine" region, the deviation must be reversed. The change from positive to negative deviation will also occur gradually as a function of, at least, the relative flow coefficient. Therefore, although there are data on the variation of deviation with flow coefficient for particular pumps (see Keith 1977 and Noorbakhsh 1973), we have no satisfactory correlation of this variation over a range of pumps. The best-efficiency-point correlation of deviation due to Busemann, given in Dixon (1975), has been used for figure 11.4a. Its use is somewhat too complex to give fully here. For an approximate value the correlation due to Eck (equation 9.8) may be used:

$$\frac{C_{\theta 2 \text{ac}}}{C_{\theta 2 \text{th}}} = \left\{ 1 + \frac{2 \cos \beta_2}{Z[1 - (d_{s1}/d_2)]} \right\}^{-1},$$

where Z is the number of rotor blades. The Stodola correlation (equation 9.4) may also be used as an approximation to that by Busemann.

The theoretical head coefficient, ψ'_{th}, is then given for any value of flow coefficient ϕ_2 within the pumping region by

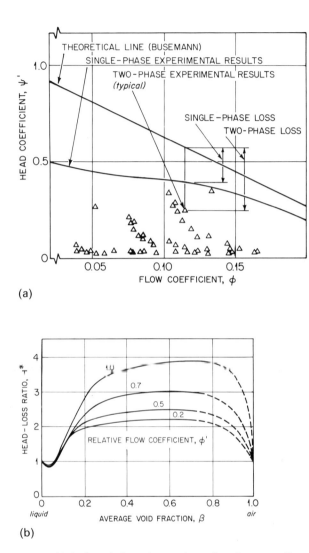

Figure 11.4 Correlation of two-phase flow in a centrifugal pump, first quadrant: (a) head coefficient versus flow coefficient; (b) head-loss ratio versus void fraction.

Cavitation and Two-Phase Flow

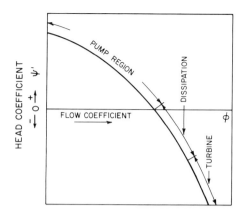

Figure 11.5 Operating regions of the first quadrant.

$$\psi'_{th} = \frac{C_{\theta 2ac}}{C_{\theta 2th}}(1 - \phi_2 \tan \beta_2). \tag{11.8}$$

The values of theoretical head coefficient obtained from equation (11.8) may be used as a close approximation for both the single- and two-phase coefficients in equation (11.7). In certain circumstances there is a significant difference between the two values, and a method of calculating them is given by Wilson et al. (1979).

Extension to prediction of new-pump characteristics

The intended method of use of the correlation method is that several centrifugal pumps covering a range of configurations (for instance, pumps with varying specific speed, blade angle, and types of diffuser) should be tested throughout the two-phase-flow range of interest, and from these tests a library of head-loss-ratio correlations would be established. Then, to predict the two-phase performance of a new pump, the single-phase performance of which would be known, one would simply select the head-loss-ratio correlations of the pump with the nearest configuration to the new pump under investigation, and apply this together with the modified Euler equation to obtain a two-phase predicted characteristic.

At present, only data from tests by Winks (1977) and Kennedy et al. (1980) can be considered to be useful guides. These can be used for first-quadrant prediction with some modifications. The magnitudes of the head-loss ratios differ for different pumps. The first-quadrant data so far analyzed from several different pump tests show that the highest levels of head-loss ratio were obtained from pumps that were relatively efficient at design point in single-phase flow, while the inefficient pumps had low maximum head-loss ratios. This seemed easily explainable because for

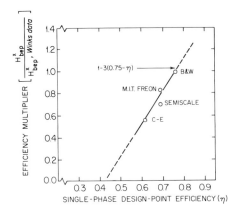

Figure 11.6 Efficiency multiplier versus single-phase design-point efficiency, first quadrant.

pumps to be efficient, they must use efficient diffusing channels in the rotor and diffuser, and diffuser performance is adversely affected by the presence of vapor bubbles.

This hypothesis is given credence by figure 11.6, in which are plotted the best efficiencies in single-phase flow as a ratio with the best efficiency of the Winks (B&W) data versus the peak head-loss ratios at best-efficiency flow coefficient for all first-quadrant test data available. These data were correlated so well by a straight line that a certain degree of good luck must be assumed.

Approximate prediction method
The full method of use of this prediction method is given in Wilson et al. (1979). The following approximate method should be sufficient for preliminary-design purposes for the first quadrant.

1. Read off the head-loss ratio from the Winks data of figure 11.4b for the desired values of void fraction and flow coefficient.

2. Multiply this head-loss ratio by the multiplier of figure 11.6 using the pump single-phase efficiency as input.

3. Obtain the single-phase value of the head-rise coefficient for the desired flow coefficient from the single-phase test curves (which have to be available).

4. Calculate the theoretical head-rise coefficient (equation 11.8) using a theoretical tangential-velocity ratio from the Busemann correlation (Dixon 1975) or from equations (9.4) or (9.8). Assume as an approximation that $\psi'_{th,tp} - \psi'_{sp,th}$.

5. Calculate the desired two-phase head-rise coefficient from

$$\psi'_{tp} = \psi'_{sp,th} - H^*(\psi'_{sp,th} - \psi'_{sp}). \tag{11.9}$$

The method may be used for other quadrants, using the head-loss correlations given in Wilson et al. (1979).

This chapter has been written descriptively rather than pedagogically, and no problems have been devised.

References

Dixon, S. L. 1975. *Fluid Mechanics, Thermodynamics of Turbomachinery.* 2d ed. Pergamon Press, Oxford, pp. 196–203.

Karassik, Igor J., William C. Krutzsch, Warren H. Fraser, and Joseph P. Messina. 1976. *Pump Handbook.* McGraw-Hill, New York.

Keith, Stephen Wayne. 1977. *Characteristics of first-quadrant one-phase flow in centrifugal pumps.* BSME thesis. MIT, Cambridge, Mass.

Kennedy, W. G., et al. 1980. *Pump two-phase performance program.* Vols. 1–7. Report NP-1556. Electric-Power Research Institute, Palo Alto, Calif.

Lakshminarayana, B., W. R. Britsch, and W. S. Gearhart (eds.). 1974. *Fluid mechanics, acoustics and design of turbomachinery.* Report SP-304. NASA, Washington, D.C.

Noorbakhsh, A. 1973. *Theoretical and real slip factor in centrifugal pumps.* Technical note 93. Von Karman Institute for Fluid Dynamics, Rhode Saint Genese, Belgium.

Thom, J. R. S. 1964. Prediction of pressure drop during forced-circulation boiling of water. *J. Heat and Mass Transfer* 7: 709–724. Pergamon Press, Oxford.

Wilson, David Gordon, Tak-Chee Chan, Juan Manzano-Ruiz. 1979. *Analytical models and experimental studies of centrifugal-pump performance in two-phase flow.* NP-677 final report (May). Electric Power Research Institute, Palo Alto, Calif.

Winks, R. W. 1977. *1/3-scale air-water pump program, test program and pump-performance results.* EPRI NP-135 (April).

12

Combustion systems
and combustion calculations

For preliminary design it is perhaps sufficient to choose from among a limited number of alternative combustion systems and to estimate the approximate volume required, and the fuel-flow rate, for a specified maximum turbine-inlet temperature. We shall briefly review the different types of combustion systems and their characteristics, including their potential for low pollutant emissions.

Precise design calculations are not yet possible for combustion systems. Design is still more of an art than a science. One can make this statement only for areas where there is considerable design freedom—and that is, in general, the case for combustors. As specifications become more stringent, especially with respect to volume and pollutant emissions, design freedoms are being narrowed as we learn more, in particular, of what not to do. It is still impossible to specify precisely what shape a combustion chamber or system must have, and combustion systems undergo in general a longer period of largely cut-and-try development after preliminary design than is required for any other component.

This is particularly true for aircraft-engine combustors, which have more stringent design requirements, than for any other gas-turbine application. The range of conditions over which they operate are

fuel-air mass ratio, 0.002 to 0.020,

inlet pressure, 0.2 to 40 at, and

inlet temperature, 245 to 945 K (440 R to 1,700 R).

These requirements seem particularly difficult when the low flame speeds that can be obtained are considered (figure 12.1).

The maximum speed normally reached in turbulent flames is shown in figure 12.1 to be little over 10 m/s. Typical axial velocities in axial-flow compressors are in the range of 125 to 200 m/s. An efficient diffuser at

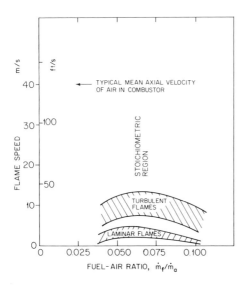

Figure 12.1 Flame speed versus fuel-air ratio.

the outlet of an axial-flow compressor cannot be designed to reduce this velocity by much more than a half (chapter 4). Therefore even in the most favorable circumstances there will be a need for the velocity to be reduced to one-sixth of diffuser-outlet velocity if a flame of even maximum speed is not to be blown downstream. We do not yet know how to reduce fluid velocities by this amount efficiently. The designer is forced to reduce the flow velocity inefficiently, by means of various forms of baffle. As a result the combustion system is usually responsible for the largest single pressure loss in a gas-turbine engine.

The flow must also be divided, because, as figure 12.1 shows, the highest flame speeds occur at near chemically correct ("stoichiometric") mixture strength. The gas turbine is the only major engine whose overall design-point fuel-air ratio is much weaker than stoichiometric. But unless catalysts are used (and attempts are being made to do so, principally, for reasons which will be obvious later, to reduce pollutant formation), combustion must be carried out in approximately stoichiometric conditions, and the resulting high-temperature gases must be subsequently diluted.

These arguments lead logically to the design of the standard form of combustion chamber, figure 12.2. A combustion system may have a single large combustion chamber, or several smaller chambers in parallel, or an annular chamber with several burners. The cross section shown in figure 12.2a applies quite closely to all these types. Combustion takes place in an enclosure within the main stream, situated in many engines at the outlet of the compressor diffuser. Holes in the upstream part of the

Combustion Systems and Calculations

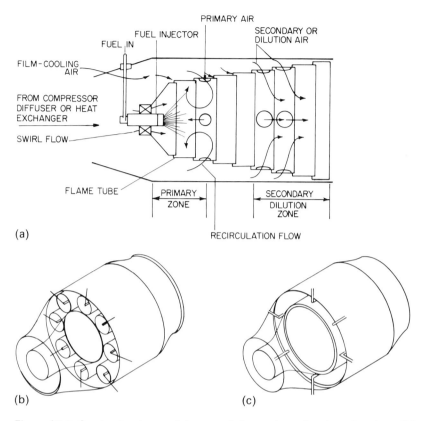

PRIMARY AIR

FUEL INJECTOR

FUEL IN

SECONDARY OR
DILUTION AIR

FILM-COOLING
AIR

FROM COMPRESSOR
DIFFUSER OR HEAT
EXCHANGER

SWIRL FLOW

FLAME TUBE

PRIMARY
ZONE

SECONDARY
DILUTION
ZONE

(a)

RECIRCULATION FLOW

(b)

(c)

Figure 12.2 Combustor types: (a) general form of combustion chamber; (b) cannular; (c) annular.

enclosure meter a proportion of the total flow to produce approximately stoichiometric conditions in this upstream region, which is called the "primary zone." Usually the enclosure (sometimes called the "flame tube") is formed from light-gauge heat-resisting metal, tubular in most early engines and annular in most current aircraft engines.

Where the holes are placed is a matter of experience and experiment. About 15 percent of the total airflow is usually passed partly through a small ring of turning vanes surrounding the fuel nozzle and partly through small holes and slits in the upstream end of the flame tube. The purpose of this wall-cooling flow is to keep the material to well below the gas temperature and sometimes also to deflect the fuel spray from the walls. (The impact of liquid, or solid, fuel on a comparatively cool wall is very likely to cause a rapid buildup of soot, or worse, hard coke, which can result in severe alterations of the air-flow pattern.) Another 15 percent of the air flow, the balance of the so-called "primary" or combustion air,

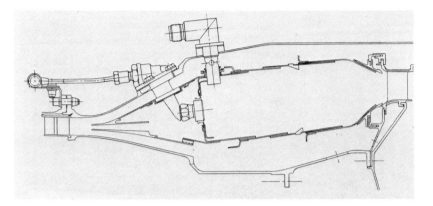

(a)

(b)

Annular combustion chamber for stationary gas turbine: (a) cross section; (b) external view. Courtesy Detroit Diesel Allison Division of General Motors Corporation.

Combustion Systems and Calculations

is led through two or more holes downstream of the fuel spray to stabilize the flame and to provide recirculation of the hot combustion products with the incoming cool air and fuel.

This recirculation comes about because the swirling air coming around the nozzle, through which the fuel emerges also in a swirling conical spray, forms a toroidal vortex, which has a low-pressure zone and upstream recirculation flow in the same manner as all intense vortices. The radial recirculation holes feed into this vortex recirculation.

The remainder of the air, called the "secondary" or dilution air, is added downstream, principally through large jets that can penetrate to the center of the hot gas stream to produce sufficient mixing and a subsequent nonpeaky temperature profile. Other secondary air is used for liner cooling.

12.1 Conservation laws for combustion

In addition to the laws of conservation of energy and of mass, a third conservation law must be observed in combustion: the conservation of atomic species.

The determination of the molal (volume) analysis of the combustion products for the burning of a hydrocarbon fuel in air is simple if these rules are followed.

1. Find the number of moles of oxygen required to burn all the hydrogen in the fuel to H_2O and all the carbon to CO_2.

2. Add nitrogen to both sides of the equation, in the amount of 3.76 moles of nitrogen for each mole of oxygen added.

3. Add excess air (oxygen and nitrogen in the above proportions) to both sides of the equation to the amount specified.

For example, to find the molal analysis of the products of combustion of benzene (C_6H_6) with 400 percent theoretical air, first we write the equation with 100 percent theoretical oxygen:

$$C_6H_6 + \text{oxygen} = 6CO_2 + 3H_2O;$$
$$\therefore C_6H_6 + 7.5O_2 = 6CO_2 + 3H_2O. \tag{12.1}$$

Then we add 100 percent theoretical nitrogen. We obtain the equation for the stoichiometric (theoretical, that is, no excess air) mixture burning in air. The principal components of air are nitrogen and oxygen, in proportion of 3.76 moles N_2 to one mole O_2:

$$C_6H_6 + 7.5O_2 + 7.5(3.76)N_2 = 6CO_2 + 3H_2O + 7.5(3.76)N_2. \tag{12.2}$$

Last, we multiply the air components by four to obtain the equation for 400 percent theoretical air:

$$C_6H_6 + 30O_2 + 30(3.76)N_2 = 6CO_2 + 3H_2O + 22.5O_2 + 112.8N_2.$$

$$(12.3)$$

12.2 Molal analysis

The number of moles of products in the foregoing example is $(6 + 3 + 22.5 + 112.8) = 144.3$. The molal proportions of the constitutents are then as follows:

Let $X \equiv$ mole fraction

$$X_{CO_2} = \frac{3}{144.3} = 0.0416,$$

$$X_{H_2O} = \frac{3}{144.3} = 0.0208,$$

$$X_{O_2} = \frac{22.5}{144.3} = 0.1559,$$

$$X_{N_2} = \frac{112.8}{144.3} = \frac{0.7817}{1.0000}.$$

$$(12.4)$$

This is the molal analysis.

12.3 Dew point of water in exhaust

If the water is all in the vapor state (and, being in a small proportion, it can be treated as a perfect gas even when about to condense), the molal analysis is also the volume analysis and the partial-pressure analysis.

We can use this partial-pressure analysis to determine the temperature at which the water would start to condense. This temperature is significant, in some circumstances, for the following reason. It is usually desirable to recover as much heat from the exhaust of industrial gas turbines as possible, first in the cycle heat exchanger if the CBEX cycle is used, and/or in a waste-heat "boiler" or process heat exchanger. However, corrosion will result if the temperature is reduced to the point where water could condense out, either in the heat exchanger or in the exhaust stack.

Suppose that the total pressure of the exhaust in the example just given is 103.4 kN/m^2 (15 lbf/in^2) absolute—a little above atmospheric pressure.

Then the partial pressure of the water vapor is $0.0208 \times 103.4 = 2.151$ kN/m^2 (0.312 lbf/in^2) for the combustion of benzene with 400 percent theoretical air.

Steam tables will show that the condensing temperature for this pressure is about 18.3 C (65 F). In engines using higher turbine-inlet temperatures the excess air will be less, and the condensation temperature will be correspondingly higher. However, gas-turbine-exhaust condensation temperatures will still be considerably below those for steam plant or for any other combustion engine using approximately stoichiometric mixtures.

This constitutes therefore an advantage of Brayton-cycle engines in that it is easier and less expensive to recover heat from their exhaust than it is for other types of engine. This is true, however, only if there is no sulfur in the fuel. Sulfur oxides in combination with excess air raise the dewpoint temperature very significantly, producing liquid sulfuric acid at well above the dewpoint temperature. Exhaust ducts and heat exchangers can be quickly corroded in this way. In a steam plant it is traditional to limit the stack-gas temperature to a minimum of about 150 C (300 F), which constitutes a significant heat loss. The same limit applies to the burning of fuels containing sulfur in gas-turbine and other engines.

12.4 Lower and higher heating values of fuels

Because in general the latent heat in the water vapor in the exhaust is not recovered, Brayton-cycle efficiencies are calculated using the so-called "lower" or "net" heating value of the fuel, rather than the "higher" or "gross" heating value. The difference is given by calculating, from the combustion equation (12.3), the kg of water per kg of fuel, and multiplying this by the latent heat, or heat of vaporization, of water (2,465 kJ/kg at 288 K).

One mole of benzene has a mass of $(6 \times 12.01 + 3 \times 2.016) = 78.108$ kg. It produces three moles of water $(3 \times 18.016) = 54.048$ kg. Therefore 1 kg fuel yields 0.692 kg water, which has a heat of vaporization of 1,706 kJ/kg.

This is then the difference between the higher and the lower heating values for the combustion of benzene. For most hydrocarbon fuels, this difference is about 4 percent.

12.5 Adiabatic flame temperature

The fundamental method of finding the flame temperature after complete combustion is to add the enthalpies of the reactants and then to add the enthalpies of the products for a range of temperatures. The actual adiabatic temperature can then be found either by iteration or by graphic interpolation. It is absolutely necessary to use thermodynamic tables with

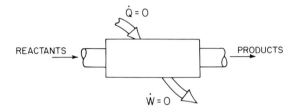

Figure 12.3 Diagrammatic representation of combustion process.

a common base temperature for the various constitutents or to make corrections for any tables with deviations. There are in fact several base temperatures in common use, and using mixed-base tables will give erroneous results.

If we use tables whose standard reference or base conditions are 25 C (77 F) and 1 at, and if we further take the example where the reactants are initially at 25 C, then the enthalpies of oxygen and nitrogen are, by definition, zero (for the reactants only) as shown in figure 12.3. We can then apply the first law for a steady-flow process:

$$\frac{Q - \dot{W}}{\dot{m}} = 0 = \Delta h_0 ;$$

$$\therefore \sum (n\hat{h})_{\text{reac}} = \sum (n\hat{h})_{\text{prod}},$$

where

$n \equiv$ number of moles $= m/MW$,

$\hat{h} \equiv$ molal enthalpy.

Let us use again the case of benzene burning with 400 percent excess air. For the reactants only the benzene has a finite enthalpy:

$n = 1$ mole,

$\hat{h} = 35,653$ Btu/lbmole   (83.76 MJ/kgmole)

(Only tables in British units were available to the author.)

$$\sum (n\hat{h})_{\text{reac}} = 35,653 \text{ Btu/lbmole}   (83.76 \text{ MJ/kgmole}) \tag{12.5}$$

For the products we can find the value of $\sum (n\hat{h})$ at various temperatures and interpolate (see table 12.1). By straight-line interpolation of the value of total modal enthalpy of the reactants given in equation (12.5) the adiabatic temperature is estimated to be 1,786.5 R (992.5 K).

Calculations of this type are tedious. Fortunately, hydrocarbon fuels

Table 12.1
Calculation of adiabatic flame temperature

Constituent	n	$\hat{h}_{1,700\,R}$	$\hat{h}_{1,800\,R}$
CO_2	6	$-158,111.1$	$-154,821.0$
H_2O	3	$-93,780.9$	$-92,803.3$
O_2	22.5	$8,930.5$	$9,760.7$
N_2	112.8	$8,449.4$	$9,226.8$
	$\sum(n\hat{h})_{\text{prod}}$	$-75,980.7$	$+53,062.9$

have rather similar enthalpies of formation for the most part, and various methods have been developed for finding the adiabatic combustion temperature for standard fuels, with average mean specific heats or enthalpies of the products and with the ability to apply corrections for variations from the average or standard conditions. One method is given in the *Gas Tables* (Keenan and Kaye 1948). Another particularly suited to the use of computers and calculators is by Chappell and Cockshutt (1974), which is summarised is appendix A1.

12.6 Combustion-system size and performance

The scaling laws for hydrocarbon-burning combustion (Clarke 1955) lead to this relationship for the volume of combustion systems.

Heat release rate per unit volume

$$\frac{\dot{Q}_b}{V_b} \propto \frac{p_{0\text{-in}}\sqrt{T_{0\text{-in}}}\sqrt{(\Delta p_0/p_{0\text{-in}})}}{(l_b/d_b)d_b}. \tag{12.6}$$

Some data collected in table 12.2 have been plotted in figure 12.4. Lines tentatively showing constant performance to agree with this scaling law have been drawn in. An important measure of performance is, however, missing: the temperature profile at combustor outlet. These data are for combustors having an approximately stoichiometric primary section, followed by a secondary dilution region in which the gas temperature is brought down to the permissible turbine-inlet temperature. This secondary section involves a large proportion of the combustor volume. Jets of dilution air must penetrate the hot stream from the primary section and mix out sufficiently well to give an acceptable temperature distribution at the turbine nozzles, figure 12.5. As the data are all for developed engine combustors, the outlet temperature profiles will have a somewhat higher temperature at the blade tips than at the blade roots where the centripetal stresses are maximum, so that an approximation to constant creep life along the length of the blade will be given (see also chapter 13 and figure

Table 12.2
Combustor data

Engine	Pressure ratio	Number of flame tubes	Flame-tube diameter, mm	$1/d$	$T_{0\,\text{in}}$, C	Combustion intensity, MW/m^3 at	$\Delta p_0/p_0$ (percent)
U.S. data							
J79–15	11.5	10	?	?	405	40.0	4.7
T56A8	9.2	6	140	4.36	318	19.5	3.6
T58	8.66	1[a]	$(d_0 = 406)$	—	304	88.0	5.5
BBC	7.2	1	1676	2.39	276	9.0	5.1
GTC–85	3.67	1	127	2.62	188	83.8	3.7
GMT–305	3.46	2	152	2.08	593	9.3	2.2
RH–TA	4.0	1	356	2.21	371	15.9	4?
GE loco	?	6	244	3.54	260	10.3	6.0
W–2000	?	12	118	5.83	246	15.5	2.7
Russian data							
Aircraft	?	?	?	2–3	?	8–19	3.5–4.5
Stationary and transport	?	?	?	2.5–3.5	?	2.2–8.3	4.5–5.5

Source: Hazard (1966) and Pchelkin (1971).
a. Annular flame tube.

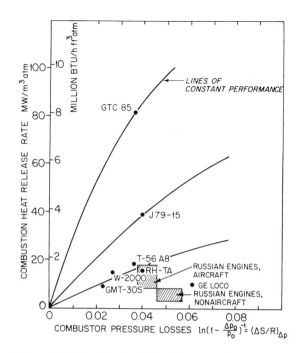

Figure 12.4 Combustion intensity versus losses.

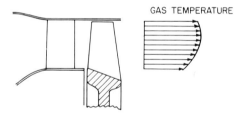

Figure 12.5 Desirable temperature distribution at combustor outlet (see also figure 13.2).

13.2). The temperature profile may be reversed, with higher temperatures at the inner diameter for blades that are strongly cooled, because the coolant temperature will be much lower at the hub where it enters the blade.

The need to have the secondary air jets penetrate the combustor gas flow to mix with it is reflected in the combustor diameter being in the denominator in equation (12.6). A combustion system comprising many small burners or flame tubes will have a smaller total intrinsic volume than a single large combustor, which in addition will have inevitably a larger volume of connecting ducting. The annular combustor (figure 12.2c) is the logical extrapolation of the concept of many small flame tubes arranged in a circle inside a single casing, and all modern aircraft jet engines employ annular combustion chambers. They still employ many individual fuel nozzles.

12.7 Pollutants and size

Gas-turbine combustors normally operate with almost 100 percent combustion efficiency. Even when aircraft were permitted to operate trailing a black plume at low altitudes, the carbon in the smoke constituted a small fraction of 1 percent of the heating value in the fuel. A modern well-developed combustor will produce negligible carbon monoxide and hydrocarbons. The sulfur emissions will depend on the sulfur content of the fuel, and this is normally low by regulation if not by market specification. The principal pollutants emitted are nitrogen oxides, NO_x. These form relatively slowly in hot gases containing oxygen and nitrogen. A virtue of small combustion systems is that the mean residence time of air and fuel molecules is small. Moreover the proximity of cooled flame-tube walls to which hot flames can radiate and thereby quickly reduce temperature also reduces the proportion of NO_x formed. An experimental automobile gas-turbine engine at General Motors was, in about 1977, the first engine type to achieve the very stringent federal limit on NO_x emission (of the equivalent of 0.25 g/km of automobile travel). To do so required that the combustor incorporated components that moved during load changes to vary air and fuel distributions. If development of catalytic combustors (figure 12.6), or of systems that involve improved mixing of, or heat exchange between, the primary and secondary air flows is successful, little or no NO_x will be produced, and there will be no need for the design inconvenience of variable configurations.

12.8 General combustor design

As intimated, combustor design involves scientific general principles, but there is a very large necessary component of experience, judgment, and

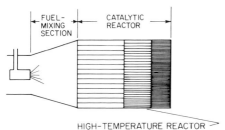

Figure 12.6 Catalytic combustor (diagrammatic).

development. Many aspects of combustor design and performance have been given little or no mention here, such as initial ignition of the fuel, and re-ignition at high altitudes in an aircraft engine; flame holding during compressor surges and other transients; flame-tube cooling, thermal stresses, and temperature transients; and many others. We cannot do justice to these topics in a book concerned with preliminary design, and we refer readers to the references for further guidance. A particularly useful general text is Lefebvre, *Gas-Turbine Combustion* (1983).

References

Borghi, Roland, Francis Hirsinger, and Helene Tichtinsky. 1978. Methodes disponibles a l'ONERA pour le calcul des chambres de combustion. In *Entropie*, 18:3–14. Paris, France.

Chappell, M. S., and E. P. Cockshutt. 1974. *Gas-turbine cycle calculations: Thermodynamic data tables for air and combustion products for three systems of units.* Report NRC 14300. National Research Council, Ottawa.

Clarke, J. S. 1956. A review of some combustion problems associated with the aero gas turbine. *J. Roy. Aero. Soc.*, London, 60: 221–240.

Deacon, W. 1969. *A survey of the current state of the art in gas-turbine combustion-chamber design.* Paper 3. Symposium on Technical Advances in Gas-Turbine Design. Inst. Mech. Eng., London.

Dooley, Philip G. 1964. *Design and development of combustion chambers for turbine engines.* Paper 64-WA/GTP-8. ASME, New York.

Graves, Charles C., and Jack S. Grobman. 1957. *Theoretical analysis of total-pressure loss and airflow distribution for tubular turbojet combustors with constant annulus and liner cross-sectional areas.* Report 1373. NACA, Washington, D.C.

Hazard, Herbert R. 1976. Combustor design. In *Sawyer's Gas-Turbine Engineering Handbook*. Vol. 1. 2nd ed. Gas-Turbine Publications, Inc., Stamford, Conn.

Keenan, Joseph H., and Joseph Kaye. 1948. *Gas Tables*. Wiley, New York.

Lefebvre, Arthur H. 1983. *Gas-Turbine Combustion*. Hemisphere, New York.

Lefebvre, A. H., and E. R. Norster. 1969. *The design of tubular gas-turbine combustion chambers for optimum mixing performance*. Paper 15. Symposium on Technical Advances in Gas-Turbine Design. Inst. Mech. Eng., London.

Pchelkin, Yu. M. 1971. *Combustion Chambers of Gas-Turbine Engines*. Transl. from Russian. N71-38542, AD 727 960 Foreign-Technology Div. Wright-Patterson AFB, Ohio.

Sullivan, D. A. 1974. *Gas-Turbine combustor analysis*. Paper 74-WA/GT-2. ASME, New York.

Tacina, Robert R., and Jack Grobman. 1969. *Analysis of total-pressure loss and airflow distribution for annular gas-turbine combustors*. Technical note TN D-5385. NASA, Cleveland, Ohio.

Szaniszlo. A. J. 1979. *The advanced low-emissions catalytic-combustor program: Phase 1—Description and status*. Paper 79-GT-192. ASME, New York.

Problems

12.1

Why is it necessary, at the present state of combustion technology, to produce a near-stoichiometric mixture in a gas-turbine combustion chamber? What new developments could relax this requirement?

12.2

Since the air velocity leaving a compressor diffuser and entering a combustion chamber is in the range of 125 to 225 m/s, why doesn't the flame blow out?

12.3

By writing down the equation for the complete combustion of ethyl alcohol (C_2H_5OH) in 250 percent theoretical air, find the percentages by volume of oxygen and nitrogen in the high-temperature combustion products. You may take air as a mixture of one mole oxygen to 3.76 moles nitrogen. Ignore dissociation effects.

12.4

If the products of combustion found in problem 12.3 were expanded through a turbine and were passed into a waste-heat boiler at 14.8 lbf/in² absolute, what would be the dew-point temperature at which the water in the mixture would start to

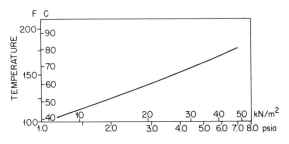

Figure P12.4

condense? Figure P12.4 shows a partial chart of the dry-saturated temperature and pressure of water vapor. (If there were sulfur in the fuel, sulfuric-acid condensation could occur at temperatures well above the water dew point.)

12.5

Describe briefly the steps you would use to find the adiabatic temperature of the combustion products of problem 12.3, given the initial temperature of the reactants and the gross or higher calorific value of the ethyl alcohol.

12.6

Calculate the final adiabatic combustion temperature, and the higher heating (calorific) value of the fuel, for the combustion of C_7H_{16} with 200 percent excess air (three times the amount of air necessary for combustion). The lower heating value of the fuel is 19,000 Btu/lbm, the initial temperature of the reactants is 100 F and the mean Cp of the products can be taken as 0.26 Btu/lbm-°F over the temperature range concerned.

12.7

(a) Write the combustion equation for the combustion of normal octane, C_8H_{18}, with 300 percent theoretical air. (b) By adding atomic weights, find the ratio (mass of products/mass of fuel). (c) Find the difference between the so-called higher and lower heating values of octane per pound of fuel if this difference is due only to the latent heat of water—1,050 Btu/lbm. (The higher heating value of a fuel is used if the water in the products is all condensed, and the lower heating value is used if the water is not condensed.) The atomic weights needed are carbon, 12.011; hydrogen, 1.008; oxygen, 16.00; nitrogen, 14.008. In air, each mole of oxygen is accompanied by 3.76 moles of nitrogen.

13

Mechanical-design considerations

In this chapter we briefly review some of the unique aspects of turbo-machinery design. We look at design criteria, material selection and forming techniques, and vibrations. It is not possible to do more than bring some of these aspects to the reader's attention for deeper analysis, guided perhaps by other treatises. Mechanical design of turbomachinery is highly specialized, and in many respects it is at the pinnacle of the engineer's art in any field. Even today, design criteria and methods are still changing, but the changes that have already occurred have greatly improved the tools available to the designer.

The state of refinement of a design method can usually be judged by the value of the safety factor that is used. Present-day safety-factor values, which are often in the range of 1.3 to 1.5, show that design methods are considerably more refined than they were two or three decades ago. This topic is discussed first.

13.1 Design criteria

In the early stage of design methods of any device whose loading is more complex than simple tension, it is common to use so-called safety factors which might have values of five or ten. These are not, of course, true factors of safety. Rather they are factors of ignorance used because the methods available for analyzing loading are highly deficient. A large safety factor might be used if a turbine disk or turbine blade were designed on the basis solely of ultimate tensile strength, for instance. We now know that turbine blades and disks can fail because one of several criteria, besides ultimate tensile strength, is exceeded. Some of the criteria that will probably be involved in the design of the disk of a high-pressure turbine rotor are indicated in figure 13.1. Criteria used for the design of high-temperature turbine blades are given in figure 13.2. Turbine disks

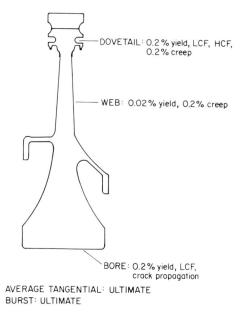

Figure 13.1 Design criteria for disks. From Anderson 1979.

und blades are the most critically stressed components of gas turbines, and this chapter will be confined to a simplified discussion of these criteria.

Low-cycle fatigue

Low-cycle fatigue is a term used for large variations in stress applied a relatively small number of times. Every time a turbomachine is started, the stresses due to rotation go from a low value to the peak value and so constitute one component of low-cycle fatigue, provided that peak value is larger than the so-called "fatigue limit." Usually, in addition there is some plastic strain in a gas-turbine disk during the first run to the over-speed proof point. The bore stretches plastically, and the rim crushes plastically around the blade dovetail roots. Then, when the machine is shut down, there is a locked-in compressive stress in the bore and a tensile stress at the rim. In subsequent starts and stops, while there should be no further plastic strain, the stress excursions may be enough to begin accumulating a fatigue history.

A more important component of low-cycle fatigue is, however, thermal stress, especially in high-temperature turbines. High-temperature materials tend to have lower thermal conductivities and higher thermal-expansion coefficients than metals used for service under 500 C (931 F), greatly intensifying the thermal stresses. Large fluctuations in thermal

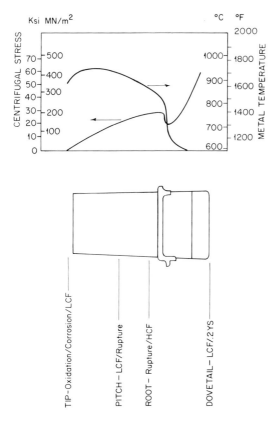

Figure 13.2 Design criteria for blades. From Anderson 1979.

stress can occur not merely upon start-up and shut-down but upon load changes. As aircraft-jet-engine performance, especially thrust-mass ratio, increases, the thermal-stress fluctuations tend to increase. This is because engines of lower performance were started and then used at near their peak-power condition throughout their missions. Higher-performance engines, especially when used in military aircraft, cruise at a lower proportion of their peak thrust and might be boosted to maximum-thrust level perhaps ten times during an average mission (figure 13.3).

The areas where low-cycle fatigue is particularly important are shown in figures 13.1 and 13.2 to be the disk bore and rim, and the high-temperature-turbine-blade dovetail, center span, and tip. Fatigue properties of typical blade materials are shown in figure 13.4.

Creep

When loaded continuously at high temperatures, metallic materials continuously deform, figure 13.5. The temperature at which long-duration

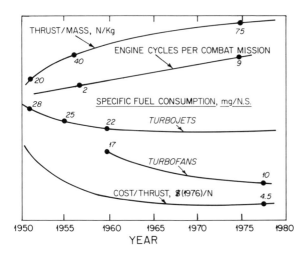

Figure 13.3 Aircraft-engine developments. From Anderson 1979 and Dixon 1979.

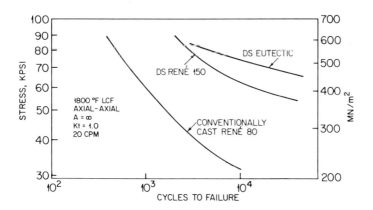

Figure 13.4 Fatigue strength of turbine-blade materials. From Anderson 1979.

Mechanical-Design Considerations

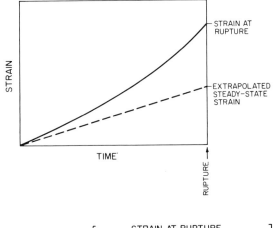

$$CREEP\ DUCTILITY \equiv \left[\frac{STRAIN\ AT\ RUPTURE}{EXTRAPOLATED\ STEADY\text{-}STATE\ STRAIN}\right]$$

Figure 13.5 Creep-strain phenomena.

deformation is significant can be far below the ultimate tensile stress. For instance, a popular turbine-blade material has, at room temperature, an ultimate tensile stress of 965 MN/m² (140,000 psi), which reduces to 453 MN/m² (65,750 psi) at 900 C (1,650 F). The stress for 0.1 percent creep strain at 816 C (1,500 F) in 1,000 hours is 69 MN/m² (10,000 psi). This stress level is close to that producing rupture in 10,000 hours at the same temperature.

A limiting value of creep often used for design is 0.2 percent strain. The stress that can be used at various combinations of temperature and life for various disk alloys is shown in figure 13.6. Large aircraft engines are currently being designed for lives of the range of 20,000 to 40,000 hours. Large industrial engines are usually designed for 100,000 hours.

Crack propagation
Crack propagation is a relatively recent design criterion used at the bore of the disks of high-pressure turbines. It recognizes that for high-performance engines, particularly those for military service, it is impracticable or uneconomic to avoid that combination of thermal low-cycle fatigue and creep which results in the formation of cracks. The materials are not notch-sensitive, and cracks continue to grow slowly. Limiting lifetimes for crack propagation are established for particular disk and blade designs, and no simple general design rules are yet available.

High-frequency fatigue
The natural frequency of vibration of blades and disks depends on the component size and shape, and the mode of the vibration, but can be of

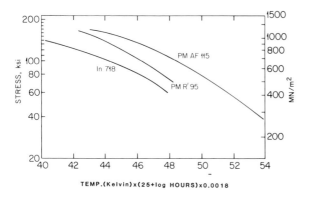

Figure 13.6 0.2-percent creep strength of disk alloys. From Anderson 1979.

the order of 10,000 Hz. The number of reversals is then of the order of 10^8 per hour, sufficient to reach the fatigue limits for many materials. Accordingly, fatigue modes that produce stresses greater than the fatigue limit must be avoided, as will be discussed more fully.

Ultimate tensile stress

The minimum cross-sectional area of a disk is set by the ultimate tensile stress at the appropriate temperature acting to carry the rotational load of the blades and rim at overspeed. The overspeed used for the burst criterion is often 20 or 25 percent over design speed. The ultimate tensile stress is appropriate because, when the disk approaches burst speed, the bore and inner parts of the disk will have yielded.

Oxidation–corrosion

Modern high-temperature turbine blades are coated to protect the base material against corrosion. The clearance between blade tip and shroud is set at the minimum practicable value because aerodynamic losses in this region are proportional to clearance. If a tip rub occurs, the protective coating is removed, and the base alloy is then subjected to a hostile corrosive environment.

The coating may also be removed through low-cycle fatigue acting on the unrestrained blade tip.

The turbine life is then set by the oxidation-corrosion rate of the base alloy without its protective coating.

13.2 Vibration characteristics

In conditions of extreme centripetal loading, it is almost impossible to incorporate mechanical damping into blade and disk construction other than that provided by the material itself. If a blade or disk receives an excitation, usually from an aerodynamic source, at a frequency close to one of the fundamental natural frequencies, large amplitudes and high vibratory stresses can result. Failure in high-frequency fatigue can then be rapid. An intense effort must be made at the design stage to avoid such excitations, because de-tuning is expensive if severe vibratory stresses are found after prototype manufacture. Nevertheless, prototype testing normally incorporates the extensive use of strain gauges, the readings from which can provide real-time and accurate information on stress fluctuations.

Avoidance of excitation of fundamental modes of compressor and turbine blades can be assisted by the construction of a Campbell diagram, figure 13.7. The principal degree of freedom in the mechanical design, once the velocity diagrams and other aspects of the aerodynamic design have been decided upon, is the number of blades in each blade row, rotor, and stator. Changing the number of blades changes the principal excitation frequency for neighboring blade rows, and also changes the natural frequency of each blade through the change in the blade chord and, hence, the aspect ratio. (The change in aspect ratio will have a minor effect on stage efficiency, figure 7.14, so long as the change in blade number is not drastic, for instance, doubling or halving the number.)

Other excitations come from asymmetry of the flow (which is certain to be present to some degree and will give at least a once-per-revolution excitation) and from upstream or downstream struts, of which there may be typically three or four.

The natural frequencies of prototype blades in various modes (fundamental bending, or "first flap"; fundamental torsion, second bending, second torsion, and so forth) can be measured in the laboratory. But the stiffening effect of rotational stress cannot be simulated accurately without rotation. Accordingly, more reliance is given nowadays to computer programs that can calculate the various frequencies of, at least, solid blades, over a full range of rotational speeds, as shown in figure 13.7.

The intersection of a fundamental mode with an exciting frequency in the usual running range of a turbomachine is normally cause for a recalculation using a different number of blades in one or more blade rows. Excitations that can be run through rapidly during accelerations up to speed can usually be tolerated, although strain-gauge readings are desirable during prototype testing to find if low-cycle or high-frequency fatigue limits could be approached during a lifetime of frequent starts and shutdowns.

Special measures must often be taken for long jet-engine fan blades, whose vibrational characteristics are often changed through the incorporation of part-span shrouds, figure 13.8.

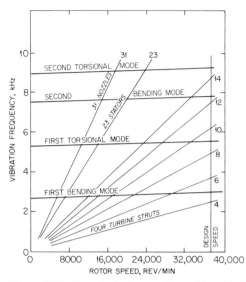

Figure 13.7 Campbell diagram for turbine-blade vibration.

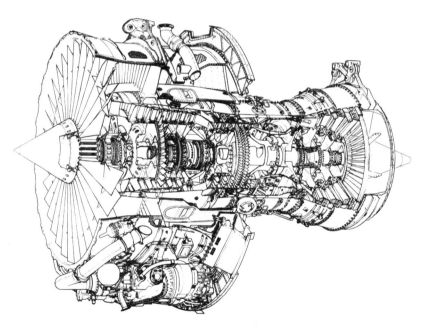

Figure 13.8 Part-span shrouds used to reduce vibration on the Rolls-Royce RB-211 fan. Courtesy of Rolls-Royce Ltd.

13.3 Material selection

The selection of materials for components like disks and turbine blades in which several design criteria must simultaneously be satisfied does not leave much freedom. Figure 13.9 shows some past and possible future choices of U.S. turbine-blade materials and some possibilities for future developments. At the time of writing (1982) there is intense interest and activity in new high-temperature materials. Virtually all candidate materials would be used with coatings, at least in turbine-blade use, for oxidation resistance in the case of materials like carbon and graphite and for foreign-object damage and mounting-stress distribution in the case of ceramics. Some of the potential materials are these:

carbon and graphite (although at room temperature the strength is low, it may actually increase with temperature),

nonoxide ceramics (silicon carbide, silicon nitride, aluminum nitride, and ternary compounds, which have high-temperature strength, low density,

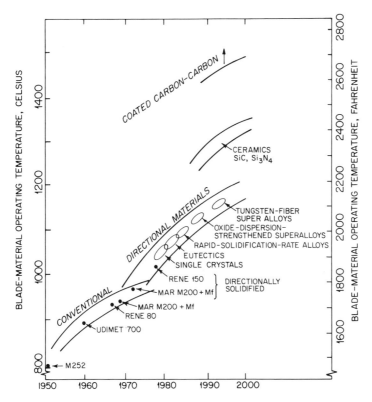

Figure 13.9 Turbine-blade materials. From Machlin 1979.

Equiaxed grain structure in cast turbine airfoils. Courtesy TRW Defense and Space Systems Group.

Mechanical-Design Considerations

Cutaway of mold for casting of turbine blades by the lost-wax method. Courtesy TRW Defense and Space Systems Group.

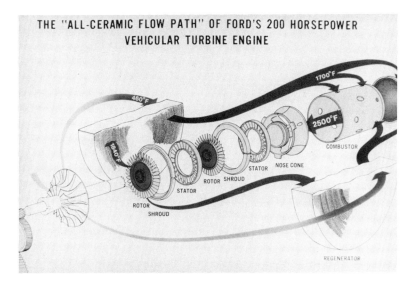

THE "ALL-CERAMIC FLOW PATH" OF FORD'S 200 HORSEPOWER VEHICULAR TURBINE ENGINE

Ceramic components under development for gas turbines. Courtesy Ford Motor Company.

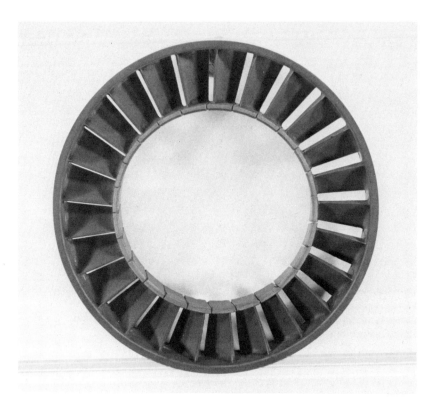

(a)

Experimental ceramic turbine stator and dual-density rotor: (a) stator; (b) rotor.
Courtesy Ford Motor Company.

and some oxidation resistance but are brittle, possibly susceptible to thermal shock, and do corrode or oxidize),

ternary oxide ceramics (of similar but somewhat lower promise than the nonoxide ceramics), and

transition metals (the refractory metals, tungsten, tantalum, molybdenum and columbium, and the dispersion-strengthened super alloys, all of which are fairly strong at high temperature and have good ductility but have high density and low oxidation and creep resistance).

This chapter has been written primarily for information rather than for purely pedagogic purposes, and no problem sets have been devised.

Mechanical-Design Considerations

(b)

References

Dixon, James T. 1979. *High-temperature combustor and turbine-design problems.* In IDA paper P-1421. Proc. Workshop on High-Temperature Materials for Advanced Military Engines. Institute for Defense Analyses, Arlington, Va.

Anderson, Richard J. 1979. *Material properties and their relationships to critical jet-engine components.* In IDA paper P-1421. Proc. Workshop on High-Temperature Materials for Advanced Military Engines. Institute for Defense Analyses, Arlington, Va.

Machlin, Irving. 1979. *High-temperature materials in turbofan engines for V/STOL aircraft.* In IDA paper P-1421. Proc. Workshop on High-Temperature Materials for Advanced Military Engines. Institute for Defense Analyses, Arlington, Va.

Freche, John C., and G. Mervin Ault. 1979. The promise of more heat-resistant turbine materials. *Product Engineering* (July).

Appendix A

Properties of air, combustion products, and water

In this appendix we give some thermodynamic data for air, combustion products and water in SI units. Comprehensive data on steam in SI as well as traditional units have been available for some years (for instance, *Steam Tables* by J. H. Keenan, F. G. Keyes, P. G. Hill, and J. G. Moore, Wiley, 1969, 1978), but although there have been several sets of short air tables published in SI units, nothing to rival the *Gas Tables* by J. H. Keenan and J. Kaye (Wiley, 1945, 1948) has appeared.

However, the National Research Council, Canada, has recently published a very useful set of data in several units: *Gas-turbine cycle calculations: thermodynamic data tables for air and combustion products for three systems of units* by M. S. Chappell and E. P. Cockshutt (NRC no. 14,300, Ottawa, 1974). The three unit systems are the British system based on the British thermal unit and degrees Rankine (its use is restricted largely to the United States); the earlier system, most widely used in places with a British tradition, based on the centrigrade heat unit (CHU) (1 CHU = 1.8 Btu) and degrees Kelvin; and the SI. We have been given generous permission to reproduce parts of the Canadian NRC report. (We have also been given permission to reproduce the *h-s* (Mollier) chart for water from W. C. Reynolds, *Thermodynamic Properties in SI*, Stanford University, 1979.)

A particularly useful feature of the report is a set of polynomial coefficients for generating thermodynamic data, and we are reproducing these as table A.1. These make it particularly easy to perform accurate cycle and process calculations on a hand calculator, particularly one with sufficient memory to store all the coefficients and the associated subroutines. The published coefficients are for the CHU system. We have added SI coefficients, kindly provided by André By and Robert Bjorge, formerly MIT graduate students.

The data and the polynomials are based on the treatment of dry air and

Table A.1
Polynomial coefficients for generating thermodynamic data in SI units

Symbol	Temperature range	
	200 K–800 K (J/kg)	800 K–2200 K (J/kg)
C_0	$+1.0189134E + 03$	$+7.9865509E + 02$
C_1	$+1.3783636E - 01$	$+5.3392159E - 01$
C_2	$+1.9843397E - 04$	$-2.2881694E - 04$
C_3	$+4.2399242E - 07$	$+3.7420857E - 08$
C_4	$-3.7632489E - 10$	$0.00E + 00$
C_5	0	0
CH^{a}	$-1.6984633E + 03$	$+4.7384653E + 04$
CF	$+3.2050096E + 00$	$7.0344726E + 00$
CP_0	$-3.5949415E + 02$	$+1.0887572E + 03$
CP_1	$+4.5163996E + 00$	$-1.4158834E - 01$
CP_2	$+2.8116360E - 03$	$+1.9160159E - 03$
CP_3	$-2.1708731E - 05$	$-1.2400934E - 06$
CP_4	$+2.8688783E - 08$	$+3.0669459E - 10$
CP_5	$-1.226336E - 11$	$-2.6117109E - 14$
H_0	$+6.2637416E + 04$	$-1.7683851E + 05$
H_1	$-5.2903044E + 02$	$+8.3690644E + 02$
H_2	$+3.2226232E - 00$	$+3.6476206E - 01$
H_3	$-2.1670252E - 03$	$+2.5155448E - 04$
H_4	$+2.4951703E - 07$	$-1.2541337E - 07$
H_5	$+3.4891819E - 10$	$+1.6406268E - 11$

Source: Chappell and Cockshutt (1974). SI coefficients supplied by André By, NREC, and Robert Bjorge, MIT.
a. For a continuous enthalpy function, use $CH = +4.7378825E + 04$.

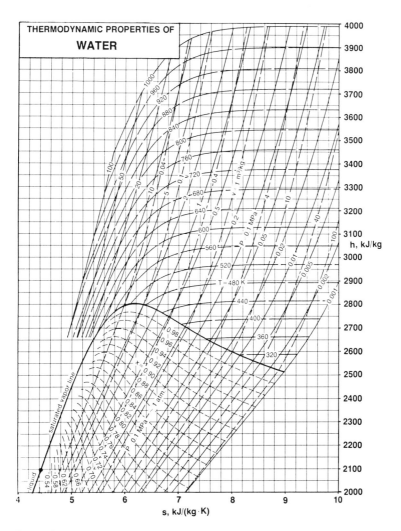

Figure A.1 Thermodynamic properties of water. © Department of Mechanical Engineering, Stanford University.

combustion products as semiperfect gases, so that the specific heat, enthalpy and entropy functions are dependent solely on temperature and are independent of pressure. The combustion products assume the use of a standard fuel of 86.08 percent carbon and 13.92 percent hydrogen by mass, which gives a molecular weight of combustion products identical to that of dry air (D. Fielding and J. E. C. Topps, *Thermodynamic Data for the Calculation of Gas-Turbine Performance*, R&M 3099, Aeronautical Research Committee, U.K., 1959).

We give here an abbreviated guide to the use of the polynomials. We omit the so-called entropy functions, because we believe that the polytropic-efficiency methods given in chapters 2 and 3 are simpler than those requiring these functions. Table A.1, giving the polynomial coefficients, and table A.2, with air data from the National Bureau of Standards, follow.

Guide to the use of polynomials

The symbols in this appendix are principally those of Chappell and Cockshutt, to be used simply for programming calculators, and are not used in the remainder of this book.

Specific heat and enthalpy of dry air

Specific heat of dry air at T, Kelvin:

$$Cp_{\mathrm{air},\,T} = C_0 + C_1 T + C_2 T^2 + C_3 T^3 \ldots$$

Enthalpy of dry air at T, Kelvin:

$$h_{\mathrm{air},\,T} = C_0 T + \frac{C_1}{2} T^2 + \frac{C_2}{3} T^3 + \frac{C_3}{4} T^4 + \cdots + CH.$$

Effective calorific value (ECV) of standard fuel
at T, Kelvin

$$ECV_T = \Delta H_R - (h_{\mathrm{air},\,T} - h_{\mathrm{air},\,288K}) - (\theta_{\mathrm{h},\,T} - \theta_{\mathrm{h},\,288K})$$

where $\Delta H_R \equiv$ reference enthalpy of reaction at 288 K
$$= 43{,}124 \text{ kJ/kg} = 10{,}300 \text{ CHU/lbm}$$

$$\theta_{\mathrm{h}T} = H_0 + H_1 T + H_2 T^2 + H_3 T^3 + \cdots.$$

Fuel/air ratio by mass, f
For a process starting with dry air at T_1 K and reaching T_2 K after combustion,

Table A.2
Thermodynamic property data for dry air in SI units

T(K)	c_p(kJ/kg · °C)	μ(kg/m · s × 10^5)	k(W/m · °C)	Pr
100	1.0266	0.6924	0.009246	0.770
150	1.0099	1.0283	0.013735	0.753
200	1.0061	1.3289	0.01809	0.739
250	1.0053	1.488	0.02227	0.722
300	1.0057	1.983	0.02624	0.708
350	1.0090	2.075	0.03003	0.697
400	1.0140	2.286	0.03365	0.689
450	1.0207	2.484	0.03707	0.683
500	1.0295	2.671	0.04038	0.680
550	1.0392	2.848	0.04360	0.680
600	1.0551	3.018	0.04659	0.680
650	1.0635	3.177	0.04953	0.682
700	1.0752	3.332	0.05230	0.684
750	1.0856	3.481	0.05509	0.686
800	1.0978	3.625	0.05779	0.689
850	1.1095	3.765	0.06028	0.692
900	1.1212	3.899	0.06279	0.696
950	1.1321	4.023	0.06525	0.699
1,000	1.1417	4.152	0.06752	0.702
1,100	1.160	4.44	0.0732	0.704
1,200	1.179	4.69	0.0782	0.707
1,300	1.197	4.93	0.0837	0.705
1,400	1.214	5.17	0.0891	0.705
1,500	1.230	5.40	0.0946	0.705
1,600	1.248	5.63	0.100	0.705
1,700	1.267	5.85	0.105	0.705
1,800	1.287	6.07	0.111	0.704
1,900	1.309	6.29	0.117	0.704
2,000	1.338	6.50	0.124	0.702
2,100	1.372	6.72	0.131	0.700
2,200	1.419	6.93	0.139	0.707
2,300	1.482	7.14	0.149	0.710
2,400	1.574	7.35	0.161	0.718
2,500	1.688	7.57	0.175	0.730

Source: U.S. National Bureau of Standards, *Circ.* 564, 1955.

$$f \equiv \frac{\dot{m}_{fuel}}{\dot{m}_{air}} = \frac{h_{air,T2} - h_{air,T1}}{ECV_{T2} + (\text{sensible heat of injected fuel})}.$$

The fuel sensible heat is small and is normally neglected. If there should be a second combustion process in a reheat combustor or afterburner, the products of the first combustion include partly vitiated air. The second fuel/air ratio, f_2, is given by

$$f_2 - f_1 = \frac{(1 + f_1)(h_{p1,T2} - h_{p1,T1})}{ECV_{T2}},$$

where $h_{p1} \equiv$ enthalpy of the products from the first combustion. The sensible heat of the fuel has been omitted.

Specific heat and enthalpy of products of combustion
at T, Kelvin

$$Cp_{prod,T} = Cp_{air,T} + \frac{f}{1+f}\theta_{Cp,T}$$

where $\theta_{Cp,T} = CP_0 + CP_1 T + CP_2 T^2 + CP_3 T^3 + \cdots,$

$$h_{prod,T} = h_{air,T} + \frac{f}{1+f}\theta_{h,T}.$$

Appendix B

Collected formulas

First law, steady flow: $\quad \dfrac{\dot{Q} - \dot{W}}{\dot{m}} = \Delta_1^2 h_0 + \Delta_1^2 (Zg/g_c)$ (equation 2.8)

Gibbs' equation for
simple substance: $\quad Tds = du + pdv$ (equation 2.28)

Equation of state for
perfect gas: $\quad pv = RT$ (equation 2.29)

Also, for perfect gas: $\quad a^2 = g_c \gamma RT$ (equation 2.31)

$$\gamma \equiv Cp/Cv$$

$$Cp = Cv + R \quad \text{(equation 2.32)}$$

$$dh = CpdT \quad \text{(equation 2.33)}$$

$$du = CvdT \quad \text{(equation 2.34)}$$

One-dimensional
isentropic flow,
perfect gas:

$$\frac{T_0}{T_{st}} = 1 + \frac{\gamma - 1}{2} M^2 \quad \text{(equation 2.39)}$$

$$\frac{p_0}{p_{st}} = \left[1 + \frac{\gamma - 1}{2} M^2 \right]^{\gamma/\gamma - 1} \quad \text{(equation 2.41)}$$

$$\frac{v_0}{v_{st}} = \frac{\rho_{st}}{\rho_0} = \left[1 + \frac{\gamma - 1}{2} M^2 \right]^{-1/\gamma - 1} \quad \text{(equation 2.42)}$$

$$\frac{\dot{m}\sqrt{T_0}}{Ap_0} = \sqrt{\frac{g_c \gamma}{R}} M_1 \left[1 + \frac{\gamma - 1}{2} M^2 \right]^{-(\gamma + 1)/2(\gamma - 1)}$$

(equation 2.45)

Polytropic relations:
$$\frac{T_{02}}{T_{01}} = \left(\frac{p'_{02}}{p_{01}}\right)^{(R/\overline{C}p_c)/\eta_{pc}} \quad \text{(compressor; equation 2.60)}$$

$$\frac{T_{02}}{T_{01}} = \left(\frac{p'_{02}}{p_{01}}\right)^{(R/\overline{C}p_e)\eta_{pe}} \quad \text{(expanders; equation 2.61)}$$

$$\eta_{sc} = \frac{(p'_{0,out}/p_{0,in})^{R/\overline{C}p_c} - 1}{(p'_{0,out}/p_{0,in})^{(R/\overline{C}p_c)/\eta_{pc}} - 1} \quad \text{(compressor; equation 2.74)}$$

$$\eta_{pc} = \frac{\ln (p'_{0,out}/p_{0,in})^{R/\overline{C}p_c}}{\ln \left[\dfrac{(p'_{0,out}/p_{0,in})^{R/\overline{C}p_c} - 1}{\eta_{sc}} + 1\right]}$$

(compressor; equation 2.75)

$$\eta_{se} = \frac{1 - (p'_{0,out}/p_{0,in})^{(R/\overline{C}p_e)\eta_{pe}}}{1 - (p'_{0,out}/p_{0,in})^{R/\overline{C}p_e}} \quad \text{(expanders; equation 2.76)}$$

$$\eta_{pe} = \frac{\ln \left[1 - \eta_{se}(1 - (p'_{0,out}/p_{0,in})^{R/\overline{C}p_e})\right]}{\ln (p'_{0,out}/p_{0,in})^{R/\overline{C}p_e}}$$

CBE, CBEX specific power:
$$\dot{W}' = \frac{\dot{W}}{\dot{m}\overline{C}p_c T_{01}} = \left(\frac{\dot{m}_e}{\dot{m}_c}\frac{\overline{C}p_e}{\overline{C}p_c}\right) E_1 T' - C_1$$

(equation 3.1)

CBE, CBEX thermal efficiency:

$$\eta_{th} = \frac{[(\dot{m}_e/\dot{m}_c)(\overline{C}p_e/\overline{C}p_c)]E_1 T' - C_1}{[(\dot{m}_b/\dot{m}_c)(\overline{C}p_b/\overline{C}p_c)][T'(1 - \varepsilon_x(1 - E_1)) - (1 + C_1)(1 - \varepsilon_x)]} \quad \text{(equation 3.2)}$$

$$E_1 \equiv \left\{1 - \left[\left(1 - \Sigma\left(\frac{\Delta p_0}{p_0}\right)\right)r\right]^{(R/\overline{C}p_c)\eta_{pc}}\right\}$$

$$C_1 \equiv \{r^{(R/\overline{C}p_c)/\eta_{pc}} - 1\}$$

Diffuser
$$Cpr = \frac{p_{st2} - p_{st1}}{p_{01} - p_{st1}} \quad \text{(equation 4.1)}$$

$$Cpr_{th} = 1 - \left(\frac{W_2}{W_1}\right)^2 \quad \text{(equation 4.7)}$$

Euler's equation:
$$g_c\frac{\dot{W}}{\dot{m}} = (u_1 C_{\theta 1} - u_2 C_{\theta 2}) \quad \text{(equation 5.1)}$$

$$g_c \Delta_1^2 h_0 = \Delta_1^2 (u C_\theta) \quad \text{(equation 5.2)}$$

Work coefficient:	$\psi \equiv \dfrac{-g_c \Delta_1^2 h_0}{u^2} = -\left[\dfrac{\Delta_1^2(uC_\theta)}{u^2}\right]_{\text{ad}}$	(section 5.2)

Flow coefficient:

$$\phi \equiv \frac{C_x}{u}$$

Reaction:

$$\begin{cases} Rn \equiv \dfrac{\Delta h_{\text{st}}}{\Delta h_0} = 1 - \dfrac{(C_2^2 - C_1^2)/2}{u_2 C_{\theta 2} - u_1 C_{\theta 1}} \quad \text{(equation 5.8)} \\[3mm] = 1 - \dfrac{C_{\theta 1} + C_{\theta 2}}{2u} \quad \text{(simple diagrams; equation 5.9)} \end{cases}$$

Specific speed:

$$Ns_1 \equiv \left(\frac{N}{60}\right)\frac{\sqrt{\dot{V}_1}}{(g_c \Delta h_0)^{3/4}} = \frac{\phi^{1/2}\sqrt{1-\lambda^2}}{\psi^{3/4} 2\sqrt{\pi}}$$

(section 5.6)

$$Ns_2 \equiv \left(\frac{N}{60}\right)\frac{\sqrt{\dot{V}}}{(g\Delta H_0)^{3/4}} \quad \text{(hydraulic machines; equation 5.13)}$$

Simple radial equilibrium:

$$\frac{1}{r^2}\frac{d}{dr}(r^2 C_\theta^2) + \frac{d}{dr}(C_x)^2 = 0 \quad \text{(equation 6.5)}$$

Slip-factor correlation:

$$\left[\frac{C_{\theta\text{ac}}}{C_{\theta\text{th}}}\right] = \left[1 + \frac{2\cos\beta}{Z(1 - d_s/d)}\right]^{-1} \quad \text{(radial-flow machines; equation 9.8)}$$

Reynolds' analogy:

$$\text{St} \equiv \left[\frac{h_t}{(\rho C p C_s)}\right] = \frac{C_f}{2} = \frac{\tau_\omega}{2(\rho C_s^2/2g_c)} \quad \text{(equation 10.14)}$$

Air mole ratio:

1 mole O_2 + 3.76 moles N_2 (equation 12.2)

Isentropic relations:

$$\left(\frac{T_2}{T_1}\right) = \left(\frac{p_2}{p_1}\right)^{(\gamma-1)/\gamma} = \left(\frac{p_2}{p_1}\right)^{R/Cp}$$

$$= \left(\frac{\rho_2}{\rho_1}\right)^{\gamma-1} = \left(\frac{\rho_2}{\rho_1}\right)^{R/C_v}$$

$$\left(\frac{1}{\gamma-1}\right) = \left[\frac{Cp}{R} - 1\right] = \frac{C_v}{R}$$

$$\left(\frac{\rho_2}{\rho_1}\right) = \left(\frac{T_2}{T_1}\right)^{(Cp/R)-1}$$

Polytropic relations (additional):

$$\left(\frac{T_2}{T_1}\right) = \left(\frac{\rho_2}{\rho_1}\right)^{n-1} = \left(\frac{p_2}{p_1}\right)^{(n-1)/n}$$

$$\frac{n-1}{n} = \frac{R}{Cp_c}\frac{1}{\eta_{pc}} \quad \text{(compression)}$$

$$= \frac{R}{Cp_e}\eta_{pe} \quad \text{(expansion)}$$

Appendix C

Conversion factors

Energy:
$$1\ \text{kWh} = 3{,}412.76\ \text{Btu}$$
$$1\ \text{Btu} = 778.161\ \text{ft-lbf}$$
$$= 1{,}054.9\ \text{J}$$
$$= 0.29302\ \text{Wh}$$
$$1\ \text{J} = 1\ \text{W s} = 0.7376\ \text{ft-lbf}$$
$$1\ \text{kW} = 0.947989\ \text{Btu/s}$$
$$1\ \text{hp} = 0.7456\ \text{kW}$$
$$1\ \text{cal} = 4.1868\ \text{J}$$
$$1\ \text{kg cal} = 3.969\ \text{Btu}$$

Specific heat:
$$1\ \text{Btu/lbm-}^\circ\text{R} = 4.187\ \text{kJ/kg-}^\circ\text{K}$$

Viscosity:
$$1\ \text{lbm/s-ft} = 1.48817\ \text{Ns/m}^2$$

Torque:
$$1\ \text{lbf-ft} = 1.3558\ \text{Nm}$$

Mass:
$$1\ \text{lbm} = 0.45359\ \text{kg}$$

Length:
$$1\ \text{in} = 25.40\ \text{mm}$$
$$1\ \text{ft} = 304.8\ \text{mm}$$
$$1\ \text{mile} = 1.609\ \text{km}$$

Area:
$$1\ \text{ft}^2 = 0.09290\ \text{m}^2$$

Volume:
$$1\ \text{ft}^3 = 0.02832\ \text{m}^3$$
$$1\ \text{liter} = 0.001\ \text{m}^3$$
$$1\ \text{gal (U.S.)} = 3.785\ \text{liter} = 0.003785\ \text{m}^3$$
$$1\ \text{gal (BR)} = 4.546\ \text{liter}$$

Force:	1 lbf	$= 4.44822$ N
Pressure:	1 lbf/in^2	$= 6.89476$ kPa ($\equiv$ kN/m^2)
Density:	1 lbm/ft^3	$= 16.0166$ kg/m^3
Thermal conductivity:	1 Btu/ft h °R	$= 1.73073$ W/m-°K

Appendix D

Some constants

Universal gas constant:	8313.219 J/kg mole °K	
	1.98592 Btu/lb mole °R	
Molecular weights:	Oxygen	$O_2 = 32.000$
	Nitrogen	$N_2 = 28.016$
	Hydrogen	$H_2 = 2.016$
	Sulfur	$S = 32.06$
	Carbon	$C = 12.01$
	Helium	$He = 4.003$
	Water	$H_2O = 18.016$
	Carbon monoxide	$CO = 28.01$
	Carbon dioxide	$CO_2 = 44.01$
	Air	28.970
Approximate constituents of air:	3.76 moles N_2 to 1 mole O_2	
Gas constant for air:	$R = 286.96$ J/kg-°K	
	$= 0.068549$ Btu/lbm-°R	

Appendix E

Miscellaneous mathematical and geometrical relationships

The symbols do not necessarily agree with those used in the nomenclature

Binomial expansion

$$(1 + x)^n = 1 + nx + \frac{n(n-1)x^2}{2!} + \frac{n(n-1)(n-2)x^3}{3!} \cdots$$

Quadratic

$$ax^2 + bx + c = 0,$$

$$x = \frac{-b \pm \sqrt{b^2 - 4ac}}{2a},$$

where a, b, c are constants.

Diameter-ratio relationships for annuli

Define

$$\lambda \equiv \frac{d_h}{d_s},$$

$$\lambda_m \equiv \frac{d_m}{d_s}.$$

If the mean diameter, d_m, is to divide the annulus into two annuli of equal areas,

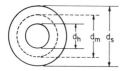

Figure E.1 Annulus dimensions.

then $\quad \lambda_m = \sqrt{\dfrac{1}{2}(1 + \lambda^2)}.$

Annulus area, $Aa = \dfrac{\pi d_s^2}{4}(1 - \lambda^2)$

$$= \frac{\pi d_m^2}{4}\left[\left(\frac{d_s}{d_m}\right)^2 - \left(\frac{d_h}{d_m}\right)^2\right]$$

$$= \frac{\pi d_m^2}{4}\left[\left(1 + \frac{1}{d_m}\right)^2 - \left(1 - \frac{1}{d_m}\right)^2\right],$$

where $1 \equiv$ blade length $= \dfrac{d_s - d_h}{2}.$

This last relationship is exact only for $d_m = (d_s + d_h)/2$ and not for the area mean diameter.

Chord of turbine blades

$$c = \frac{1}{\cos \lambda}[b + r(\cos \lambda + \sin \lambda - 1)],$$

where $c \equiv$ chord,
$\quad\quad b \equiv$ axial chord,
$\quad\quad r \equiv$ leading-edge radius,
$\quad\quad \lambda \equiv$ setting angle.

Symbols

It is impossible to avoid having one symbol serve several uses. By including the equation or figure of first use, where the symbol is usually defined, we hope to spare the reader at least some of the confusion.

a	velocity of sound (equation 2.33)
a	a constant (equation 6.7)
A	area (equation 2.44)
Aa	annulus area (equation 8.6)
b	breadth (figure 4.5)
b	a constant (equation 6.7)
b	blade-row axial chord (equation 7.2)
B	boundary-layer blockage (figure 4.9)
c	chord of blade (equation 7.6)
C	velocity (equation 2.2)
C_1	compressor temperature ratio (equation 3.1)
Cf	skin-friction coefficient (equation 10.11)
C_L	lift coefficient (equation 7.4)
Cp	specific heat at constant pressure (equation 2.30)
Cpr	pressure-rise coefficient (equation 4.1)
C_{rat}	capacity-rate ratio $[C_{rat} \equiv (\dot{m}Cp)_{min}/(\dot{m}Cp)_{max}]$ (figure 10.4)
C_{rot}	periodic-flow heat-exchanger capacity-rate ratio $C_{rot} \equiv (\dot{m}Cp)_{mat}/(\dot{m}Cp)_{min}]$ (equation 10.30)
Cv	specific heat at constant volume (equation 2.30)
d	diameter (equation 5.10)
d_h	hydraulic diameter $[d_h \equiv 4A/p_e]$ (equation 10.25)
d_h, d_s	diameter (inner and outer) (equation 5.10)
Deq	equivalent diffusion ratio (equations 8.3–8.6)

e	radius of curvature of blade convex surface at throat (equation 7.8)
E	internal energy, general (equation 2.1)
$E_{1,2,3}$	expander temperature ratios (equation 3.2)
f	power-law index for axial-velocity variation (equation 9.6)
F	force (equation 7.1)
g	gravitational acceleration (equation 2.2)
g_c	constant in Newton's law ($g_c \equiv ma/F$) (equation 2.2)
h	enthalpy per unit mass ($h \equiv pv + u$) (equation 2.3)
$\hat{h}$	enthalpy per mole ($\hat{h} \equiv h\,MW$) (equation 12.5)
h	height, blade height (equation 7.1)
h_t	heat-transfer coefficient (equation 10.8)
H	head in gravitational field (equation 2.12)
H	boundary-layer shape factor (equation 8.10)
H^*	head-loss ratio (equation 11.7)
i	incidence (angle) of flow (equation 7.7)
j	Colburn modulus: ($j \equiv St\,Pr^{2/3}$) (equation 10.7)
k	thermal conductivity (equation 10.2)
K	a constant, initially unspecified (equation 2.15)
l	a length (equation 6.2)
L	length along wall (figure 4.5)
m	mass (equation 2.2)
M	Mach number ($M \equiv C/a$) (equation 2.39)
MW	molecular weight (equation 2.48)
n	polytropic index (equation 2.60)
n	a constant (equation 6.7)
n	number of moles (equation 12.5)
N	axial length of diffuser (figure 4.9)
N	shaft rotational speed (rev/min) (equation 5.10)
NPSH	net positive suction head (NPSH $\equiv H_{0\text{-in}} - H_g$) (equation 11.1)
N_{s1}	specific speed (nondimensional) $\left[N_{s1} \equiv N\dot{V}^{1/2}/(60\,\Delta h_0^{3/4})\right]$ (equation 5.10)
N_{s3}	specific speed (dimensional) $\equiv N\,\text{rpm}\,(\dot{V}\,\text{gal/min})^{1/2}$ $(\Delta H_0\,\text{ft})^{3/4}$ (equation 5.11)
N_{sv}	suction specific speed (equation 11.2)
NTU	no. of transfer units (equation 10.18)
Nu	Nusselt number (equation 10.10)
o	opening or throat width (equation 7.8)
p	pressure (equation 2.2)
p_e	perimeter (equation 10.20)
Pr	Prandtl number (Pr $\equiv g_c C p \mu/k$) (equation 10.12)
q	quantity of heat transferred (equation 2.1)
q	dynamic head or pressure (equation 4.9)

$\dot{Q}$	rate of heat transfer ($\dot{Q} \equiv \delta q/\delta t$) (equation 2.2)
r'	relative radius ($r \equiv r/r_m$) (equation 6.8)
r	radius (equation 6.2)
r	compressor total-pressure ratio $[r \equiv (p_{02}/p_{01})]$ (equation 3.3)
R	gas constant ($pV = RT$) (equation 2.29)
$\bar{R}$	universal gas constant (equation 2.48)
Re	Reynolds number (Re $\equiv C.l.\rho/\mu$) (figure 4.10)
Rh	reheat factor (equation 2.73)
R_n	reaction ($R_n \equiv \Delta h_{st}/\Delta h_{0\ stage}$) (equation 5.6)
s	specific entropy (equation 2.23)
s	spacing or pitch (equation 7.1)
S	extensive entropy (equation 2.15)
St	Stanton number $[St \equiv h_t/(\rho Cp C_s)]$ (equation 10.14)
t	time (equation 2.2)
t	thickness (equation 8.6)
T	temperature (equation 2.24)
T'	gas-turbine-cycle temperature ratio ($T' \equiv T_{04}/T_{01}$) (equation 3.1)
Tq	torque (equation 7.22)
u	internal thermal energy per unit mass (equation 2.2)
u	peripheral speed (equation 5.1)
U	overall heat-transfer coefficient (equation 10.16)
v	volume per unit mass (equation 2.2)
V	extensive volume (equation 10.25)
w	work transfer (equation 2.1)
W	fluid velocity relative to boundaries (equation 4.8)
$\dot{W}$	rate of work transfer ($\dot{W} \equiv \delta w/\delta t$) (equation 2.2)
$\dot{W}'$	specific power $[\dot{W}' \equiv \dot{W}/(\dot{m}\overline{Cp}_c T_{01})]$ (equation 3.1)
x	distance or length, usually along axis of duct or machine (figure 2.2)
X	loss coefficient (equation 7.11)
X	mole fraction (equation 12.4)
y	normal distance (equation 10.1)
Y	diffuser width (figure 4.5)
z	compressibility factor (equation 2.70)
Z	number of blades or passages (equation 9.4)
z	height above datum in a gravitational field (equation 2.2)
α	angle of flow with axial or radial direction (equation 7.1)
β	void fraction ($\beta \equiv \dot{V}_{vapor}/\dot{V}_{total}$) (equation 11.7)
β	angle of blade or vane mean line with axial or radial direction (equation 7.7)
γ	ratio of specific heats ($\gamma \equiv Cp/Cv$) (equation 2.34)
Γ	circulation around a blade row (equations 8.6, 8.7)

δ	deviation of flow (figure 7.3)
$\delta*$	displacement thickness of boundary layer

$$\left[\delta* \equiv \frac{1}{C_{fs}} \int_0^\infty (C_{fs} - C_{bl})dy\right] \text{ (equation 8.13)}$$

Δ	change in a quantity (such as a property) between two sections ($\Delta_1^2 p \equiv p_2 - p_1$) (equation 2.4)
ε	deflection of flow (figure 7.3)
ε_i	intercooler effectiveness (equation 3.13)
ε_x	heat-exchanger effectiveness (equation 3.2)
η	efficiency (equation 2.51)
θ	angle; semidivergence angle of diffuser (equation 6.2)
θ	camber angle of blade ($\theta \equiv \beta_1 - \beta_2$) (equation 7.7)
θ	momentum thickness of boundary layer

$$\left[\theta \equiv \frac{1}{C_{fs}^2} \int_0^\infty C_{bl}(C_{fs} - C_{bl})dy\right] \text{ (equation 8.8)}$$

θ	temperature difference driving heat transfer (equation 10.2)
λ	hub-tip diameter ratio ($\lambda \equiv d_h/d_s$) (equation 5.12)
λ	blade setting or stagger angle (figure 7.3)
μ	absolute viscosity (equation 10.1)
ρ	density ($\rho \equiv v^{-1}$) (equation 2.42)
σ	solidity ($\sigma \equiv c/s$) (equation 8.4)
σ	surface tension (figure 11.1)
σ	Thoma cavitation number ($\sigma \equiv \text{NPSH}/\Delta H_0$) (equation 11.1)
τ	tangential stress (equation 10.1)
ϕ	flow coefficient ($\phi \equiv C_x/u$ or $\phi \equiv C_r/u$) (equation 5.6)
ψ	work coefficient $[\psi \equiv -\Delta_1^2(uC_\theta)/u^2]$ (equation 5.5)
ψ^x	head coefficient ($\psi^x \equiv g\Delta H_0/u^2$) (equation 5.14)
ω	rotational speed (equation 6.1)
Ω	number of alternative ways a collection of particles may be arranged (equation 2.14)

Subscripts

ac	actual (equation 4.9)
ad	adiabatic (equation 5.5)
an	annulus (equation 7.13)
be	best-efficiency point (equation 9.4)
b	burner, combustion, heat-addition (equation 3.2)
bl	boundary layer
c	compression (equation 2.60)
c	cold (equation 10.32)
ci	cavitation inception (equation 11.3)
CBE	for the simple gas-turbine cycle (equation 3.6)
cer	ceramic
diff	diffuser (equation 4.9)

dp	design point
e	expansion (equation 2.66)
f	face (equation 10.32)
ff	free-face area (figure 10.9)
fl	flow
fs	free stream (equation 8.13)
ft	flame tube
g	vapor (equation 11.1)
gl,2,3	group 1, 2, or 3 losses (equation 7.13)
h	hub or inner diameter (equation 5.10)
h	heat transfer (equation 10.17)
h	hot (equation 10.32)
i	incompressible (table 2.5)
i	intercooler (equation 3.13)
in	inlet (equation 2.74)
ind	induced (equation 7.6)
lam	laminar (equation 10.27)
le	leading edge (equation 8.12)
m	mean (equation 5.12)
mat	matrix (equation 10.30)
max	maximum
min	minimum (equation 10.29)
nf	nonflow (equation 10.32)
N	nozzle (figure P2.5)
0	total, pitot or stagnation conditions or properties (equation 2.6)
opt	optimum (equation 7.5)
out	outlet (equation 2.74)
p	polytropic (equation 2.60)
pro	profile (equation 7.11)
prod	products (equation 12.5)
q	dynamic pressure (equation 4.9)
r	radial direction (equation 6.3)
reac	reactants (equation 12.5)
rel	relative (equation 9.4)
rev	for a reversible process (equation 2.56)
ro	rotor (equation 5.6)
rot	rotation (equation 10.30)
s	isentropic (equation 2.52)
s	shroud or outer diameter (equation 5.10)
s	free-stream value (equation 10.7)
sec	secondary (equation 7.13)
sp	single phase
st	static conditions or properties (equation 2.2)

sta	stator
t	throat (equation 7.8)
te	trailing edge (equation 8.11)
th	theoretical (equation 4.7)
th	thermal (equation 3.2)
tp	two phase (equation 11.7)
ts	total-to-static (efficiency) (section 2.8)
tt	total-to-total (efficiency) (section 2.8)
turb	turbulent (equation 10.26)
T	isothermal (equation 2.56)
w	wall (equation 10.1)
x	cross-sectional (equation 10.21)
x	axial or x-direction (equation 5.5)
θ	peripheral or tangential direction (equation 5.1)

Superscripts

$\hat{}$	maximum (equation 10.16)
$-$	mean value (except $\bar{R}$, q.v.) (equation 2.62)
*	hypothetical conditions at which fluid would attain sonic velocity after isentropic process (equation 2.46)
x	special use of variable
$'$	hypothetical specified condition (equation 2.52)
$'$	nondimensional form of variable (equation 6.8)
$\cdot$	rate (equation 2.2)

Prefix multiples

tera	T	10^{12}
giga	G	10^{9}
mega	M	10^{6}
kilo	K	10^{3}
milli	m	10^{-3}
micro	μ	10^{-6}
nano	n	10^{-9}
pico	p	10^{-12}
femo	f	10^{-15}
atto	a	10^{-18}

Glossary of terms

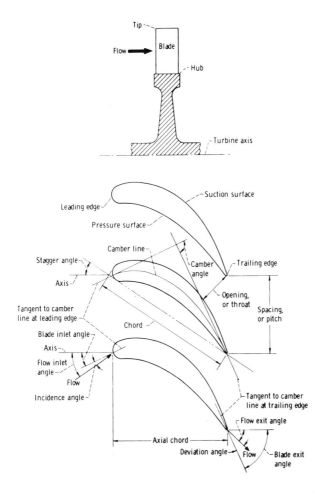

adiabatic: insulated; occurring with no external heat transfer.

aspect ratio: the ratio of the blade height to the chord.

axial chord: the length of the projection of the blade, as set in the turbine, onto a line parallel to the turbine axis. It is the axial length of the blade.

axial solidity: the ratio of the axial chord to the spacing.

blade exit angle: the angle between the tangent to the camber line at the trailing edge and the turbine axial direction.

blade height: the radius at the tip minus the radius at the hub.

blade inlet angle: the angle between the tangent to the camber line at the leading edge and the turbine axial direction.

blower: a rotary machine that produces a low-to-moderate pressure rise in a compressible fluid (usually air), usually incorporated in a duct. See "fan" and "compressor."

bucket: same as rotor blade.

camber angle: the external angle formed by the intersection of the tangents to the camber line at the leading and trailing edges. It is equal to the sum of the angles formed by the chord line and the camber-line tangents.

camber line: the mean line of the blade profile. It extends from the leading edge to the trailing edge, halfway between the pressure surface and the suction surface.

CBE: compressor-burner-expander, or the "simple" gas-turbine "cycle."

CBEX: compressor(heat exchanger)-burner-expander-heat exchanger, or the "regenerated," "recuperated," or "heat-exchanger" gas-turbine "cycle."

chord: the length of the perpendicular projection of the blade profile onto the chord line. It is approximately equal to the linear distance between the leading edge and the trailing edge.

chord line: if a two-dimensional blade section were laid convex side up on a flat surface, the chord line is the line between the points where the front and the rear of the blade section would touch the surface.

compressor: a rotary machine that produces a relatively high pressure rise (pressure ratios greater than 1.1) in a compressible fluid.

deflection: the total turning angle of the fluid. It is equal to the difference between the flow inlet angle and the flow exit angle.

deviation angle: the flow exit angle minus the blade exit angle.

diffuser: a duct or passage shaped so that a fluid flowing through it will undergo an efficient reduction in relative velocity and will therefore increase in (static) pressure.

effectiveness: a term applied here to define the heat-transfer efficiency of heat exchangers.

efficiency: performance relative to ideal performance. There are many types of efficiency requiring very precise definitions (see section 2.8).

entropy: a property of a substance defined in terms of other properties (see section 2.4). Its change during a process is of more interest than its absolute value. In an adiabatic process, the increase of entropy indicates the magnitude of losses occurring.

expander: a rotary machine that produces shaft power from a flow of compressible fluid at high pressure discharged at low pressure. In this book the only types of expander treated are turbines.

flow exit angle: the angle between the fluid flow direction at the blade exit and the machine axial direction.

flow inlet angle: the angle between the fluid flow direction at the blade inlet and the machine axial direction.

head: the height to which a fluid would rise under the action of an incremental pressure in a gravitational field.

hub: the portion of a turbomachine bounded by the inner surface of the flow annulus.

hub-tip ratio: same as hub-to-tip-radius ratio.

hub-to-tip-radius ratio: the ratio of the hub radius to the tip radius.

incidence angle: the flow inlet angle minus the blade inlet angle.

intensive property: a property that does not increase with mass; for instance, the pressure and temperature of a body of material do not double if an equal mass at the same temperature and pressure is joined to it. (The energy, on the other hand, would double.)

intercoolers: heat exchangers that cool a gas after initial compression and before subsequent compression.

isentropic: occurring at constant entropy.

isothermal: occurring at constant temperature.

leading edge: the front, or nose, of the blade.

mean section: the blade section halfway between the hub and the tip.

meridional plane: a plane cutting a turbomachine through a diametral line and the (longitudinal) axis.

nozzle blade: same as stator blade, for turbines only.

pitch: the distance in the direction of rotation between corresponding points on adjacent blades.

pressure surface: the concave surface of the blade. Along this surface, pressures are highest.

pump: a machine that increases the pressure or head of a fluid. In connection with turbomachinery it usually refers to a rotary machine operating on a liquid.

radius ratio: same as hub-to-tip-radius ratio.

recuperator: a heat exchanger, defined in this book as one with nonmoving surfaces, transferring heat from a hot fluid to a cold fluid.

regenerated cycle: see "CBEX."

regenerator: a heat exchanger, defined in this book as one having moving surfaces or valves switching the hot and cold flows.

reheat: the effect of losses in increasing the outlet enthalpy, or in decreasing the steam wetness, in a steam-turbine expansion. Also see "reheat combustor."

reheat combustor: a combustor fitted between two turbines to bring the gas temperature at inlet to the second turbine to approach the temperature at inlet to the first.

root: the compressor or turbine-blade section attaching it to its mounting platform. Rotor-blade root sections are normally at the hub, and stator-blade roots at the shroud.

rotor: the rotating part of a machine, usually the disk or drum plus the rotor blades.

rotor blade: a rotating blade.

separation: when a fluid flowing along a surface ceases to go parallel to the surface but flows over a near-stagnant bubble, or an eddy, or over another stream of fluid.

shroud: the surface defining the outer diameter of a turbomachine flow annulus.

solidity: the ratio of the chord to the spacing.

spacing: same as pitch.

specific: per unit mass (when applied to an extensive property, such as enthalpy, energy, entropy).

specific fuel consumption: mass of fuel consumed per unit work output.

specific power: the power output of an engine divided by the mass-flow rate of the working fluid.

stagger angle: the angle between the chord line and the turbine axial direction (also known as the setting angle).

stall: the condition of operation (usually defined by the incidence) of an airfoil, or row of airfoils, at which the fluid deflection begins to fall rapidly and/or the fluid losses increase rapidly.

static (conditions): conditions or properties of fluids as they would be measured by instruments moving with the flow.

stator: the stationary part of a machine, normally that part defining the flow path.

stator blade: a stationary blade.

suction surface: the convex surface of the blade. Along this surface, pressures are lowest.

surge: the unstable operation of a high-pressure-ratio compressor whose stalls propagate upstream from the high-pressure stages or components allowing reverse flow and the discharge of the reservoir of high-pressure fluid, followed by re-establishment of forward flow and a repetition of the sequence.

tip: the outermost section of the blade or "vane."

total (conditions): conditions or properties of fluids as they would be measured by stationary instruments that bring the fluid isentropically to rest.

trailing edge: the rear, or tail, of the blade.

transverse plane: the plane normal to the axis of a turbomachine.

turbine: a rotary machine that produces shaft power by extracting energy from a stream of fluid passing through it, using only fluid-dynamic forces (as distinct from "positive-displacement" or piston-and-cylinder-like machines).

turbomachine: as for "turbine," except that the shaft power may be produced or absorbed, and the energy may be extracted from or added to a stream of fluid.

working fluid: is the fluid that undergoes compression, expansion, heating, cooling, and other processes in a heat-engine cycle. In an open-cycle gas turbine the working fluid is air.

Index

ERRATA

THE DESIGN OF
HIGH-EFFICIENCY TURBOMACHINERY
AND GAS TURBINES

David Gordon Wilson

ERRATA*

p. 58, line 2:   Eliminate "$\dfrac{1}{2g_c}$". Should read "$dmC^2/2g_c$",

p. 110, line 5:   Insert negative sign. Should read

$$= 1 - \{[(1 - \Sigma\,(\Delta_{p_0}/p_0))r]^{-(R/\overline{C}_{p_e})\eta p_e}\},$$

p. 142, line 33:   Add the following: Sum of total-pressure losses, $\Sigma\,(\Delta_{p_0}/p_0)$, = 0.12

(after Regenerator thermal effectiveness)

p. 161, last line: Change $\therefore P_{o2} = 492.68$ kN/m^2

<u>to</u> $\therefore P_{o3} = 492.68$ kN/m^2

p. 162, line 1:   Remove this line from the top and insert   after $(P_{st3} - P_{st2}) = 32.389$ kN/m^2

p. 186, line 1:   Should read kN/m^2, 44.5 C

p. 197, line 7:   Should read $\Delta h_o \equiv h_{o2} - h_{o1} = (u_2\,C_{\theta 2} - u_1\,C_{\theta 1})/g_c$

p. 212, last line: Change last part of the equation. Should read

$$\left[1 - \left(\frac{d_h}{d_s}\right)^2\right]^{1/2} = \frac{\sqrt{\pi}}{30}\frac{|\psi|^{3/4}}{\phi^{1/2}}\left[\frac{N\sqrt{\dot{V}}}{(g_c\Delta h_o)^{3/4}}\right] = \frac{|\psi|^{3/4}}{(\pi\phi)^{1/2}}N_{s1}$$

p. 214, line 4:   Addition to equation 5.13. Should read

$$N_{s2} \equiv \frac{2\pi\,N\,\sqrt{\dot{V}}}{60(g\Delta H_o)^{3/4}}\,.$$

p. 215, line 5:   Equation 5.14 should read

$$N_{s2} = \frac{|\phi|^{1/2}(b/d)^{1/2}\,2(\pi)^{1/2}}{|\psi^x|^{3/4}}$$

p. 215, line 9:   Equation 5.15 should read

$$N_{s3} = 2732\,N_{s2}$$

*(Counting typewritten lines only.)

p. 222, line 11:   Add the following paragraph after 0.75:

For cases a, b, and c, the drop in total pressure in the diffuser is 12.5% of the difference between total and static pressures at inlet.

p. 222, line 27:   Add "loss". Should read

There is a frictional head loss of 0.25 m in the pipe additional to this lift and a kinetic-head <u>loss</u> equal to that leaving the pump diffuser.

p. 256, line 7:   Change the word "incompressible" to "<u>compressible</u>". Should read as follows:

as a ratio with the outlet (theoretical compressible) dynamic head, the isentropic efficiency can be found as follows.

p. 256, line 9:   Change the word "incompressible" to "<u>compressible</u>". Add equations. Should read as follows:

The relative outlet theoretical compressible dynamic pressure, q_2 from the rotor is

$$q_2 = P_{o2} \left[1 - \left(1 + \frac{M_2^2}{2\left(\frac{\overline{C}_p}{R} - 1\right)} \right)^{-\frac{\overline{C}_p}{R}} \right]$$

where $M_2 \equiv W_2/a_{st2}$, and

$$a_{st2} = [(g_c T_{o2} - C_2^2/2\,\overline{C}_p)/(1/R - 1/\overline{C}_p)]^{1/2},$$

where W_2 is the relative rotor outlet velocity taken from the velocity diagram.

p. 257, top: Add equations. Should read

$$\text{where} \quad q_{out,sta} = P_{o1} \left[1 - \left(1 + \frac{M_1^2}{2\left(\frac{\overline{C}_p}{R} - 1\right)} \right)^{-\frac{\overline{C}_p}{R}} \right]$$

$$M_1 \equiv C_1/a_{st1}$$

$$a_{st1} = [(g_c T_{o1} - C_1{}^2/2\overline{C}_p)/(1/R - 1/C_p)]^{1/2}$$

the subscript 1 refers here to stator-outlet conditions.

p. 269, line 2: Add equations after =0.0720. Should read

$M_1 = 0.6732$ $\qquad$ $M_2 = 0.692$

$\dfrac{q_{out,sta}}{P_{o1}} = 0.2565$ $\qquad$ $\dfrac{q_{out,ro}}{P_{o2}} = 0.2643$

p. 269, line 5: Change equation. Should read

$$\Sigma(\Delta_{P_o}) = \Sigma(X)_{ro}(q_{out,ro}) + \Sigma(X)_{sta}(q_{out,sta})$$

p. 269, line 6: Change equation. Should read

$$\therefore \ \Sigma(\frac{\Delta_{P_o}}{P_o}) = 0.0501$$

p. 269, line 8: Change numbers following the equation to 0.8723 (rather than 0.9037).

p. 269, line 11: Change numbers. Should read

Isentropic efficiency, $n_{se} = 0.8723 \times 0.984 = 0.8583$

p. 269, line 12: Eliminate one bracket in the equation and add another. Should read

$$\text{Polytropic efficiency, } n_{pe} \ \frac{\ln\left[1 - n_{se}\left(1 - \left(\frac{P'_{o5}}{P_{o4}}\right)^{R/\overline{C}_{pe}}\right)\right]}{\ln\left(\frac{P'_{o5}}{P_{o4}}\right)^{R/\overline{C}_{pe}}}$$

(equation 2.76) = 0.847

p. 269, line 13:   Change "close" to "far from" and "no" to "another".
                     Should read

                     This is sufficiently far from to the initial estimate of
                     0.90 for another iteration to be necessary (in preliminary
                     design).

p. 270, line 18:   Add brackets plus the index 1/2 in equation 7.26.
                     Should read

$$\frac{\dot{m}}{A} = [2g_c P(P_o - P_{st})]^{1/2}$$

p. 278, last line   Remove the words "the mean". Should read

                     Find the number of stages required, the mean-diameter
                     blade speed,

p. 279, line 1:   Eliminate   the first half of the sentence. Should
                     read

                     , and the optimum solidity for

p. 279, line 2:   Eliminate "10 k W (power from blading)." Should
                     read

                     the first stage of steam turbine for a deep-submergence

p. 279, line 12:   Add a few lines after 0.1. Should read

                     Zweifel's criterion. Use $t_{te}/s = 0.05$ and $r_{le}/s = 0.1$,
                     and $(s/e) = 0.4$.

p. 279, line 13:   Add one line (before 7.7). Should read

                     Use the low-Mach-number correlation for outlet angle.

p. 379, line 23:   Add one sentence after "whole annulus." Should read

                     Make your own design choices.

p. 362, line 12:   Change 0.10 to 0.628. Should read

                     Specific speed, N_s: 0.628 (nondimensional)

p. 362, line 25:   Change 0.12 to 0.754. Should read

                     The nondimensional specific speed should be 0.754.

p. 362, line 30:   Add one line (before 9.3). Should read

$$\dot{V} = 4.73 \ m^3/s, \ \Delta_{Po} = 1724 \ k \ N/m^2; \ and \ \eta_{hydraulic} = 0.85$$

ERRATA (Con't.) - Page 5

p. 363, Fig. P9.3:   Add the following lines:

                        (across the stage)

                        Design point

                        (see example below)

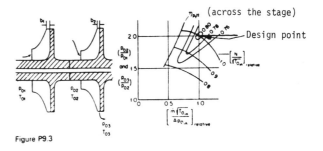

Figure P9.3

p. 363, line 3:   Add one sentence. Should read

                     (This is the total-to-total efficiency across the rotor.)

p. 363, line 6:   Substitute 0.05 with (N_{s2}) 0.314. Should read

                     pump designed to give a static lift of 100 feet at a
                     specific speed   (N_{s2}) of 0.314

p. 364, last line: Add one sentence after "calculations". Should read

                     Both turbines have twelve blades.

p. 415, line 23:   Change figure 10.8d to figure 10.7d. Should read

                     Use the matrix surface the results for which are given
                     in figure 10.7d.

p. 482, line 31:   Change equation. Should read

                     $$[N_{s1} \equiv \omega \dot{V}^{1/2}/(g_c \, \Delta h_o)^{3/4}]$$